Rome at War

STUDIES IN THE HISTORY OF GREECE AND ROME
Robin Osborne, P. J. Rhodes, and Richard J. A. Talbert, EDITORS

Rome at War

*Farms, Families, and Death
in the Middle Republic*

by NATHAN ROSENSTEIN

THE UNIVERSITY OF NORTH CAROLINA PRESS
CHAPEL HILL AND LONDON

© 2004 The University of North Carolina Press
All rights reserved

Designed by Per Jegebäck
Set in New Baskerville
by Tseng Information Systems, Inc.

Publication of this work was aided by a generous grant from the College of Humanities of The Ohio State University.

The paper in this book meets the guidelines for permanence and durability of the Committee on Production Guidelines for Book Longevity of the Council on Library Resources.

Library of Congress Cataloging-in-Publication Data
Rosenstein, Nathan Stewart.
Rome at war : farms, families, and death in the Middle Republic / by Nathan Rosenstein.
 p. cm. — (Studies in the history of Greece and Rome)
Includes bibliographical references and index.
ISBN 978-0-8078-2839-7 (cloth : alk. paper)
ISBN 978-1-4696-1107-5 (pbk.: alk. paper)
 1. Agriculture—Rome—History. 2. Agriculture—Economic aspects—Rome—History. 3. Farms, Small—Rome—History. 4. War and society—Rome—History. 5. Rome—History—Republic, 510–30 B.C. I. Title. II. Series.
S431.R67 2004
630′.937′6—dc21   2003008542

For Anne and Zoë

Contents

Acknowledgments		ix
Chapter 1.	Introduction: Agriculture in Italy from Hannibal to Tiberius Gracchus	3
Chapter 2.	War and Agriculture: A Critique of the Conventional View	26
Chapter 3.	War and the Life Cycles of Families: Three Models	63
Chapter 4.	Mortality in War	107
Chapter 5.	Military Mortality and Agrarian Crisis	141
Appendix 1.	The Number of Roman Slaves in 168 B.C.	171
Appendix 2.	The Accuracy of the Roman Calendar before 218 B.C.	174
Appendix 3.	Tenancy	181
Appendix 4.	The Minimum Age for Military Service	183
Appendix 5.	The Proportion of *Assidui* in the Roman Population	185
Appendix 6.	The Duration of Military Service in the Second Century B.C.	189
Appendix 7.	The Number of Citizen Deaths as a Result of Military Service between 203 and 168 B.C.	191
Notes		193
Bibliography		289
Index		321

Tables and Figures

TABLES

1. Available vs. Required Labor in a Hypothetical Family of Smallholders — 71
2. Combat Deaths, 200–168 B.C., as Reported by the Ancient Sources — 110
3. Combat Outcomes, 200–168 B.C., by Year and Theater — 120

FIGURES

1. Frequency of Triumphs by Month, 298–222 B.C. — 34
2. Household Labor Potential vs. Household Subsistence Demand for Labor in a Hypothetical Family of Smallholders — 72

Acknowledgments

Learning new tricks is never quick or painless for an old dog, and although this is a short book, the knowledge of unfamiliar fields and the application of unaccustomed methods that its composition required has made the writing long and arduous. Bringing it to completion has been due in no small measure to the help I have received along the way, and I take great pleasure in being able at last to acknowledge it publicly. Early drafts of Chapters 2 and 3 benefited greatly from the acute criticisms and sage advice of my friends Robert Morstein-Marx and Kurt Raaflaub as well as of my two highly numerate colleagues here in the Department of History, Randy Roth and Kenneth Andrien. In addition, I received much helpful guidance from Norm Rask of Ohio State's Department of Agricultural Economics. At a later stage, when matters turned demographic, I was fortunate indeed to meet Elio Lo Cascio and Walter Scheidel, whose extraordinary generosity in sharing their reactions to my ideas and copies of their own work in advance of publication stand as models of scholarly collegiality. Walter Scheidel also kindly agreed to read the whole of the manuscript under what were certainly very trying circumstances, and his advice and criticisms proved invaluable in strengthening the discussions of agriculture and demography. In the pursuit of comparisons between the effects of Rome's massive mobilization during the late third and second centuries on Italy's women and families and the consequences of a similar rate of conscription by the Confederacy during the American Civil War, I have been lucky to be able to draw on the wealth of knowledge of two outstanding historians of the latter era, my colleagues Joan Cashin and Mark Grimsley. I am also grateful to my colleagues Steven Conn for suggesting the comparison in the first place and Carla Pestana, who guided me through earlier periods of American history. Russell (Darby) Scott generously offered advice on the mysteries of field surveys and their meaning. Sincere thanks, too, to Andrea Gnirs for help with mummies, Myles McDonnell for sharing his paper on Roman family structure in advance of publication, James Quillin for careful and perceptive comments on Chap-

Acknowledgments

ter 3, Brendon Reay for all the far-ranging and enlightening discussions, Robert Rush for help with conditions of military service in early modern Prussia, Richard Saller for advice on men's age at marriage in the middle republic, and Brent Shaw for an offhand remark during a difficult interview that led to Chapter 4. Steve Siebert of Nota Bene repeatedly provided prompt and user-friendly technical help. And I am deeply grateful to John Rich who gave the manuscript in its penultimate stage a shrewd and thorough reading and offered a wealth of acute and salutary criticisms that have enormously improved the final product. Doubtless none of these scholars will agree with all of the conclusions I have reached; and, needless to say, for any errors and fallacies that persist despite their efforts to eradicate them, I alone am responsible.

The origins of this book go back to a warm, autumn afternoon in 1995 when I was walking past the Campanile at the University of California at Berkeley and realized that the dates preserved in the *fasti triumphales* might allow the reconstruction of the seasonal rhythms of republican warfare in the third century. That eventful year in Berkeley was made possible by a generous sabbatical leave granted by The Ohio State University's College of Humanities, for which I am very grateful. I thank Donald Mastronarde, then chair of the Classics Department, and the others on the Berkeley faculty whose warmth and hospitality made the homecoming of two alumni so welcoming, most especially Erich Gruen, friend, mentor, and inspiration. Deepest thanks, too, are owed to the American Council of Learned Societies for the award of a fellowship in 1999–2000 that provided a year free from the pleasant burdens of teaching and greatly advanced the composition of the book, and to the College of Humanities and the Department of History at The Ohio State University for generous financial assistance toward its publication.

Finally, my greatest debts are to my wife, Anne Jewel, for all her affectionate forbearance and loving support over the course of my work on this book, and my daughter, Zoë, who has taught me just how much families mean.

Columbus, Ohio
March 4, 2003

Rome at War

οὐδεὶς γὰρ οὕτω ἀνόητός ἐστι ὅστις
πόλεμον πρὸ εἰρήνης αἱρέεται. ἐν
μὲν γὰρ τῇ οἱ παῖδες τοὺς πατέρας
θάπτουσι, ἐν δὲ τῷ οἱ πατέρες τοὺς
παῖδας.

For no one is so insane as to prefer war to
peace. For in the one sons bury fathers,
but in the other fathers bury sons.
— *Herodotus 1.87.4*

Chapter 1

Introduction: Agriculture in Italy from Hannibal to Tiberius Gracchus

Limits on aristocratic competition for honor, glory, wealth, and power protected the corporate interests of Rome's governing class as well as the well-being of the people it ruled during most of the middle and late republic.[1] What was remarkable about the republican system was the fact that the elite had to impose these controls upon itself, unlike monarchies in which the interests of a ruler always set firm boundaries to his or her subjects' self-aggrandizement. By and large, the aristocracy's efforts were successful. Limits allowed aristocratic rivalry to help Rome win an empire and yet enjoy stable government until quite late in the game. But in one respect this process might appear to have fallen seriously short—indeed, no attempt to insist on a limit seems evident at all—and that was in the republic's propensity to go to war. Warfare and conquest constituted the paramount arena for the display of aristocratic *virtus* and the acquisition of prestige as well as the more tangible benefit of great wealth. Generals and others who served the republic by defending its interests and enlarging its *imperium* garnered *laus* and *fama* and laid the basis for a lasting *auctoritas* and often higher office. The aristocracy had an interest, therefore, in going to war often in order to provide its members with opportunities to advance themselves in the contention for eminence.[2] But in allowing these competitive drives to be played out year after year in increasingly distant theaters of war, the aristocracy gradually undermined first the social and economic, then the military, and finally the civic foundations of the republic. Or so many historians aver. For nearly every scholar who has sought to explain the social and political turmoil of the Roman Republic's last hundred years has traced its origins to the impact of the city's second-century wars on Italy's small farmers[3]—the men who manned the legions and furnished the army's allied contingents—when the city's demands for soldiers began to conflict fundamentally with the needs of husbandry.[4]

Prior to 200 B.C. (or perhaps the Hannibalic War—opinions differ), conventional wisdom holds that war and agriculture blended together seamlessly. Campaigns were short, conducted close to home, and fought mainly in the summers when the crop cycle left farmers with little to do in their fields. The arrival of autumn brought an end to the fighting. Soldiers were mustered out of their legions and returned home to plant and cultivate the next year's crops until the following spring when military duty would again call them from their plows.[5] All that changed with the wars of the late third and second centuries, however. Armies fighting abroad could not be discharged in the fall and then reconstituted at winter's end. Logistical and strategic imperatives dictated keeping them overseas year-round. Smallholders therefore lacked regular opportunities to return to their land before their terms of service expired or the war ended, and as a result their farms lacked the labor necessary to work them. Starvation threatened the families left behind; debts accumulated; and when (or, indeed, if) the men returned, they often could not pay them off owing to the years of fallow and neglect that had rendered their fields incapable of being easily returned to productivity. Families therefore sold or abandoned their lands or had them foreclosed. P. A. Brunt in his seminal article, "The Army and the Land in the Roman Revolution," starkly illustrated their degradation:

> The pathetic story Valerius Maximus (iv.4.6) tells of the consul Regulus will be remembered. During his absence for a year in Africa the steward of his farm of seven *iugera* had died; his hired man had run away with the farm stock, and his wife and children were in danger of starvation. Such must have been the fate, not of a consul and a noble in the third century, but of many a peasant in the second and first centuries. Thus even when the legionary was a man of some property, army service would soon reduce him to the same economic level as his proletarian comrades.[6]

At the same time, dramatic changes elsewhere in the agricultural economy were completing the ruin of Italy's smallholders, developments that, ironically, the victories these same men were winning overseas had set in motion.[7] Many members of Rome's and Italy's upper classes had grown rich from the spoils of war and the profits made in the course of the republic's conquests during the first half of the second century, particularly in the Hellenistic East. Lacking other outlets for their newly acquired capital, they began to invest it in the land that military service was forcing small farmers to relinquish. But, instead of establishing these men as tenants on their estates, wealthy proprietors preferred to work them with servile labor, of which not coincidentally the captives that Rome's armies had taken were furnishing an abundant supply for Italy's slave markets. And from the same conquests came the wealth that enabled potential investors to buy them. In addition, the kinds of estates being created in this way constituted a new and very different sort of agricultural enterprise in Italy. Termed "plantation agriculture" or the "slave mode of production," farms of this type were much larger, run almost entirely with slave labor, and geared primarily toward producing cash crops like wine, oil, grain, and livestock for Italy's burgeoning urban markets and the republic's armies.

Consequently, pressures built on small farms from two directions following the Hannibalic War. The burdens of conquering an empire began to cause many of them to become no longer economically viable, while those that held out faced increasing challenges in the form of competition from the new, slave-staffed estates. In some cases, the availability of a purchaser induced smallholders in difficulties to sell out. Elsewhere, large landowners drove out their weaker neighbors and occupied their holdings or else simply absorbed whatever land became vacant when smallholders departed or died. Even when small farmers fought to remain on their land, their inability to hold their own in the marketplace against the greater efficiencies of large-scale production, slave labor, and cheap grain imported from abroad led many to abandon an unequal struggle.[8] Some may have drifted into the cities where they became the consumers for whom the slaves now toiling on their land grew food, but most remained in the countryside as a desperately poor, landless proletariat. To these men (and women) reformers, beginning with Tiberius Gracchus, appealed for support to pass land reform and other measures while the Roman senate generally fought to block any remediation of their ills. Ultimately, however, the *patres*' intransigence or impotence in finding a solution to the problem of landlessness was requited by the overthrow of their rule. The poor, disappointed in their hopes of legislative relief and, after Marius's reforms, enrolled in the legions, came to constitute the armies of Rome. With these men behind them, the great generals of the late republic, first Marius and Sulla, then Pompey and Caesar, and finally Antony and Octavian, were able to challenge the collective control of their peers and in the end erect a monarchy on the ruins of the republic. Seen from this perspective, then, the aristocracy's refusal to impose limits on the competition that animated its relentless drive to conquer during the second century exposed the citizens and allies who fought these wars to their corrosive effects at home. That failure, in turn, furnished the instruments that eroded and finally destroyed the ruling class's ability in the late republic to contain the ambitions of its most powerful members and so guard its

Introduction

most vital interest, its own continued supremacy at Rome. Again, Brunt strikingly summarizes the *communis opinio*:

> The fundamental cause of regression [in the size of Italy's free population] was in my view the impoverishment of the mass of Italians by continuous wars. It is hard to overestimate the fearful burden that conscription imposed on the Italian people with little remission for 200 years, the loss of lives, the disruption of families, the abandonment of lands; in the end Italy suffered as much or more than the provinces which her soldiers and officials without mercy pillaged. But the upper classes profited, and used their profits to import hordes of slaves. The competition of slave labour completed the economic ruin of the majority of Italians, and made them politically the pliant instruments of unscrupulous leaders whose rivalries were to subject all the "rerum dominos" to one man.[9]

Although this reconstruction is internally consistent, supported by ancient literary evidence, and explanatory of much that caused the fall of the Roman Republic, doubters have increasingly questioned whether the growth of vast, slave-run estates in fact led to a crisis among smallholders during the early and middle decades of the second century.[10] As early as 1970 Frederiksen placed the problem on an entirely new footing when he observed that although the archaeological record for the Italian countryside in the second and first centuries B.C. ought to reflect some trace of this massive decline in the number of small farms and their replacement by large estates worked by slaves, surveys of the remains of rural habitations in this period have strikingly failed to detect evidence that would confirm this hypothesis.[11] Instead, the surveys have uncovered a complex situation that resists blanket characterization and cautions against monocausal explanations for declines where these occurred. Although evidence for small farmsteads is scarce in some areas, it abounds in others and may therefore indicate that independent farmers continued to work these holdings.[12] On the other hand, few villas of the type associated with the new plantation agriculture appear in the literary or archaeological record before the mid-second century at the earliest. Evidence for their existence only becomes widespread more than a half century subsequently, in the age of Sulla.[13]

Testimony on viticulture and the large-scale commercial production of wine fully accords with this sequence of development. These activities represent the slave mode of production par excellence as it existed in ancient Italy, but their great efflorescence now seems clearly to begin no earlier than the last third or so of the second century. The Dressel Type 1 am-

Introduction

phoras that carried the bulk of this wine to market only appear after this date, and the enormous numbers of these vessels in the archaeological record clearly reflect a massive increase in the production of Italian wine that began only in the final decades of the second century.[14] The commercial manufacture of wine in Italy was certainly not unknown earlier, but this dates back well into the third century. The post-Hannibalic era marks no watershed in this regard. Equally important, very little evidence suggests that Rome's political elite took much interest in such activities at this time. Testimony for senatorial involvement in wine production does not become common before the Julio-Claudian period. In other words, those with the greatest access to the wealth the republic derived from its second-century conquests showed little inclination to put it to work to exploit the new economic opportunities that this type of agricultural enterprise would have afforded. Instead, most wine making and marketing seem to have remained, as they long had, in the hands of small and medium-sized producers, the local elites of the Roman *municipia* and allied towns.[15] Archaeological investigations have similarly shown that the Second Punic War did not represent a turning point in the agricultural history of southern Etruria and Latium. Here in the *suburbium* of Rome itself, a variety of physical evidence appears to indicate that the countryside had lost nearly the whole of its population of smallholders by the early second century, a process that began well before Hannibal's invasion. The result was not their replacement by estates staffed by gangs of slaves, however, but rather desolation.[16] The absence of any sign that plantation agriculture was taking root in the outskirts of Rome at this date is particularly striking because this region, if any, ought to have seen its rapid development in view of the growing market for its products that the metropolis represented.[17]

The continuities that the archaeological record suggests characterized Italy's agrarian economy in the period between the outbreak of the Hannibalic War and Tiberius Gracchus's tribunate find confirmation in other sorts of evidence as well. The *lex Claudia* of circa 218 clearly indicates that the republic's upper classes were involved in the production of crops for the market prior to that date, and on many if not most of their farms they undoubtedly employed slave labor.[18] The Elder Cato, we are told, as a young man worked alongside his slaves in the fields.[19] By the time of the Second Punic War, the republic's servile population was numerous enough to permit the senate to raise two emergency legions of slave volunteers following Cannae, press 20,000 to 30,000 more into service to row its warships, and yet still leave enough in the fields to contribute substantially to raising the food the city's war effort required.[20] But a substantial slave population was no recent development. The best explanation for the abolition of debt bondage at the close of the fourth century that ended the Struggle of the Orders is the hypothesis that chattel slavery came to substitute for a dependent labor force made up of Roman citizens on the estates of the rich.[21] The mass enslavements that occurred in the wake of the republic's victories in the early second century therefore did not create a "slave society" at Rome; one already existed.[22] Even Cato's *De Agricultura*, which is sometimes claimed as evidence for the introduction of a new type of slave-based agriculture in the second century, may better be understood as an ideological rather than an economic signpost. Its publication probably represented more an effort by its author to position himself within contemporary cultural debates than an attempt to teach his contemporaries about new ways of making money.[23]

Several other factors further limited the pace at which the number of large villas worked primarily by slaves increased during the early and middle years of the second century. The rate at which Italy's commercial agriculture could expand was largely a function of the pace at which markets for its products grew, and in the second century Rome and the republic's military needs largely set this.[24] Rapid population increases in the other cities and towns of Italy, which would dramatically expand the demand for the products of plantation agriculture and so fuel its growth, took place mainly in the first century.[25] The great increase in Italian wine production began only a generation or two earlier, a development that followed closely the opening up of the Gallic market in the later second century with the advent of a more or less permanent Roman presence in southern Gaul.[26] Certainly, some large landowners may have been eager to buy slaves and invest in land to avail themselves of what opportunities for profit did exist at this time. But in practice their ability to do so depended upon access to capital, which remained in short supply immediately upon the close of the Hannibalic War.[27] In this light, the fact that livestock grazing emerged as the most attractive form of slave-based agriculture for prospective investors following the war takes on considerable significance.[28] The Elder Cato's firm conviction in the first part of the second century that such investments represented the road to riches surely resulted in part at least from the fact that it was among the least labor-intensive methods of exploiting large tracts of land.[29] The high returns that could be expected, in other words, may have derived in no small measure from the low initial investment for the slaves and livestock required compared with that for grapes or arboriculture at a time when money was scarce.[30]

In the years that followed the war, prosperous farmers in the republic's municipalities and allied towns—those who, as already noted, had long been involved in the production of wine for the market—are not likely to have shared much in the "profits of empire." Most of these men and their sons usually served in the infantry or, for the richest, as cavalry troopers, not as generals or officers. The gains they stood to make out of booty and donatives scarcely placed in their hands great sums for investment.[31] Even among the republic's political elite, the generals who led these men may have had far less leeway in helping themselves to the spoils from their victories than is usually supposed.[32] But even if their prerogatives in this regard were in theory unfettered, political competition had by the 180s made controlling the extent to which victorious commanders could enrich themselves a major issue. Complaints on this score led to damaging attacks on M'. Acilius Glabrio and Cn. Manlius Vulso and ultimately the trial of the Scipios.[33] And these object lessons seem to have had the desired effect: when Aemilius Paullus conquered the kingdom of Macedon, he showed himself a model of restraint.[34] Certainly the great run-up in the personal fortunes of the *principes civitatis* seems to have occurred mainly in the first century, not during the early and middle decades of the second, as one would expect if aristocrats in that era had been freely fattening their purses with loot.[35] Although some Roman *equites* and wealthy Italians may have participated in the companies of *publicani* that handled the disposal of war captives and booty for the republic's armies, to assume that the money they made from these activities brought about the growth of plantation agriculture is little more than a *petitio principi* without independent corroborating evidence of that growth.[36]

The notion that the number of estates engaging in the slave mode of production was growing rapidly during the first two-thirds of the second century also depends on the belief that a massive influx of slaves to labor on them was taking place at the same time, but the basis for claims to that effect is really quite weak. Overall, the size of Italy's slave population certainly declined during the conflict with Hannibal, both because Rome conscripted a substantial number of slaves to serve as soldiers or rowers and because, strapped for cash by taxes to fund the war itself, slave owners often will have been hard pressed to purchase replacements for those who had died or became enfeebled with age.[37] A not insignificant portion of

Introduction

new slaves obtained after 200 therefore probably just made good the losses that had occurred during the course of the preceding eighteen years of struggle. Even the great slave hauls of Ti. Gracchus in Sardinia and Aemilius Paullus in Epirus may have boosted the overall slave population far less than generally believed in light of the severe toll the epidemic of 175–174 had taken on rural agricultural workers.[38]

Recent studies of Italian demography have further increased doubts about a rapid expansion of the peninsula's servile population in this era. No direct evidence exists for the number of slaves in Italy at any time.[39] Brunt has little trouble showing that Beloch's estimate of 2 million during the reign of Augustus is without foundation.[40] Brunt himself suggests that there were about 3 million slaves out of a total population in Italy of about 7.5 million at this date, but he readily concedes that this is no more than a guess.[41] As Lo Cascio has cogently noted, that guess in effect is a product of Brunt's low estimate of the free population in Italy in A.D. 14.[42] That is, Brunt must assume that the slave population had come to comprise nearly 40 percent of the population of Italy by the time of Augustus because he believes that the nonservile population of Italy had only managed to stay even between 225 B.C. and A.D. 14.[43] At the same time, however, the number of residents of cities and towns throughout the peninsula and especially of Rome itself was skyrocketing. Consequently, without the supposition that slaves made up a very high percentage of the total population, not enough people would have been left in the countryside to produce the food needed to feed those in the towns. The basis for the supposition that slaves in Italy numbered as many as 3 million by the reign of Augustus in other words really consists of nothing more than a kind of elaborate circular argument in which the low free population "explains" the high number of slaves, which in turn "explains" how there could be so few free men and women in Italy.[44]

Brunt also advances the claim that the Romans owned about 500,000 slaves circa 212, which suggests that, in his opinion, the slave population of Italy might have seen an average annual net gain between then and the end of the first century B.C. of perhaps 12,500 individuals. But the starting point for postulating such a rapid rate of increase is also based on a similar piece of guesswork. After noting that, by his reckoning, the Romans had mobilized about 11 percent of their free population in that year and mentioning the comparisons that other scholars had made to the 10 percent of their populations that some Balkan states in 1913 and Germany in

1914 had mobilized, Brunt continues, "We have only to suppose that the Romans owned not far short of half a million slaves to reduce the proportion of men in the armies and fleets far below 10 percent, even after allowing that 20,000–30,000 slaves may have been used after 214 as rowers."[45] In other words, a slave population of 500,000 is necessary to bring the ratio of men under arms to the civilian population down into a range that Brunt finds acceptable. He makes no attempt to discover what might constitute a maximum rate of mobilization for a society such as Rome's in this period except to state that productivity per person was lower than in Germany and the war lasted longer than the modern conflicts. Of course, the cost of equipping and maintaining an army was much lower for the Romans as were the economic requirements of the civilian population. And one might suppose that more men could be spared from a simple agrarian economy like ancient Rome's than from a complex industrial one like early twentieth-century Germany's.

Consequently, Brunt's figures offer no basis for assuming that a dramatic rise in the number of Roman slaves—and hence in the number of the plantations that employed them—was getting under way during the early second century. To be sure, Livy records a depressing litany of enslavements by Roman armies in the course of their conquests in this period.[46] But it does not necessarily follow that these would have helped bring about the sixfold increase in the Roman slave population by the reign of Augustus that Brunt postulates. Given the usual assumption in modern scholarship that male slaves significantly outnumbered females, the slave population would have been incapable of reproducing itself at full replacement level.[47] As a result, the Romans regularly had to import substantial numbers of new slaves just to keep the slave population from shrinking. Scheidel has shown that on the assumption that slaves in 225 numbered 500,000 and were declining by only 1 or 2 percent per year, far more new slaves would have been required simply to replace current slaves who died than to generate a net increase of 2.5 million in the total slave population by 25 B.C.[48] As large as the enslavements of this period were, therefore, they cannot in and of themselves demonstrate a rapid rise in Italy's slave population along the lines Brunt supposes. It is also worth bearing in mind that not all of those whom Rome's armies captured will have wound up in Italy, for this by no means constituted the only market for slaves in the early second century. Agriculture and manufacture in Carthage, Sicily, and elsewhere in the Hellenistic world made extensive

Introduction

use of slave labor, and the same factors of imbalanced sex-ratios and low birthrates that created a very high demand for replacement slaves in Italy may well have been operating in these areas also.[49]

However, one piece of negative evidence, to which Scheidel has also drawn attention, provides an intriguing hint that conventional estimates of slaves making up as much as 40 percent of Italy's population by the late first century B.C. may be far too high.[50] An analysis of the genetic makeup of Italy's modern population argues that the various distinctive genetic combinations currently found in different regions within the peninsula by and large track the linguistic distribution that resulted from the migrations of the Iron Age.[51] No data indicate the subsequent large-scale infusion of new genetic material into the populations of these regions except in the case of southern Italy and eastern Sicily, which is explained by the well-documented Greek migrations there. If this finding is correct, then the slave population of Italy even at its greatest extent must have been far smaller than Brunt imagined, perhaps no more than a million. Otherwise, one must suppose that a very large number of slaves existed but made no contribution to the peninsula's genetic composition because they simply failed to reproduce themselves. Yet a very large number of slaves, on the order of 3 million, presupposes that this population was fairly successful at reproducing itself because it could never have reached that size in the first place and then maintained those numbers for centuries through imports alone. As already noted, the majority of new slaves brought into a servile population that was not reproducing itself completely would only have replaced old slaves who had died. But if a population of 3 million slaves, representing as much as 40 percent of Italy's inhabitants in the first century B.C., was successfully reproducing itself, it would surely have left its mark on the genetic makeup of contemporary Italians. That it did not argues strongly for a very low rate of natural reproduction among Italy's slaves, which in turn is difficult to reconcile with the hypothesis that the number of slaves ever grew large enough to comprise 40 percent of the Italian population.

If a dramatic rise in Italy's servile population during the second and first centuries is beginning to appear increasingly questionable, the decline in the numbers of free men and women that is supposed to have been its corollary is also being viewed with a growing skepticism. The census returns of 70 and 28 B.C. represent the linchpin for this pessimistic assessment of the condition of Italy's smallholders. For many years Brunt's powerful defense in *Italian Manpower* of Beloch's view that these

Introduction

totals demonstrate a drop in the free population of Italy remained unchallenged, even though the numbers themselves, around 900,000 in 70 and over 4 million in 28, would seem to reflect precisely the reverse. But Beloch and Brunt argue that the latter figure represents free men, women, and children, whereas the censors in 70 had counted only adult male citizens. When the totals are adjusted and allowances made for enfranchisements between 70 and 28 and citizens overseas, the result is a net decline in the free population.[52] When these figures are in turn compared with the census returns of 225, the general regression in Italy's free population becomes patent, a regression that Brunt traces to the damage that Rome's wars and the importation of slaves inflicted on Italy's farmers.[53]

In a provocative article, Lo Cascio has asked how it is possible to make demographic sense out of the Beloch-Brunt thesis.[54] The argument they advance must assume that the population between 70 and 28 was declining annually by .5 percent, and the implications of such a decline, Lo Cascio believes, are unacceptable. Beyond question, the urban population of Italy increased dramatically during the middle of the first century, and any rise in urban numbers, with the possible exception of Rome itself, had to come from the rural population. In the preindustrial world, however, an urban population does not grow without a sustained growth in the rural free population whose economic products support it. Thus Lo Cascio argues that unless we are prepared to suppose that the ratio of urban to rural dwellers in Italy between 70 and 28 was far in excess of preindustrial norms—and there is no good reason to do so—the Beloch-Brunt interpretation of the census total for 28 cannot be made plausible. For it must assume that a dramatic and unparalleled *drop* in Italy's nonurban population was occurring at a time of unprecedented urban growth. Consequently, the figure of 4 million must represent only adult, male citizens just as had been the case in earlier republican censuses. If that is so, then as Tenny Frank long ago argued, the free population of Italy must have been growing vigorously during the second and first centuries.[55]

Lo Cascio's article certainly will not be the final word on the controversy surrounding Beloch's and Brunt's thesis, but the mere fact that this critical prop is now being challenged renders claims about a crisis among Italy's small farmers due to war and the introduction of plantation agriculture all the more open to question.[56] From a different perspective, Morley, too, has raised additional doubts about the conventional view. He notes that the populations of early modern cities generally could not reproduce themselves; they depended instead upon a large, steady influx of immigrants

Introduction

from the countryside to reach and then to maintain their size. Rome, he believes, would have been no different. Therefore the swelling of the city's inhabitants to nearly a million over the course of the second and first centuries B.C. and the stability of their numbers at roughly that level over the ensuing centuries cannot be attributed to a single, discrete event like the displacement of smallholders after the Hannibalic War. Such an episode would create a temporary increase, but then the process would slow, perhaps even reverse course, and the city would shrink as its population gradually died off.[57]

Critics have also undermined another of the pillars that have long sustained belief in the ruin of Italy's small farmers. Gabba and Brunt erected this when each argued that the progressive lowering of the minimum census necessary to qualify as an *assiduus* (that is, a Roman citizen liable for legionary service) over the course of the late third and second centuries reveals a precipitous decline in the number of prosperous small farmers who met this criterion.[58] Gabba and Brunt claimed that the senate made these reductions because the gradual impoverishment of smallholders due to the hardships of military service and competition from slave labor caused a corresponding shrinkage in the number of potential conscripts for Rome's armies as more and more men became too poor to qualify as legionaries. Concurrently, the increased military burdens on the remaining *assidui* provoked bitter resistance to the draft. The senate's lowering of the property qualification thus attempted to obviate both problems by increasing the pool of recruits. Yet Rich in an incisive study has shown that little evidence points unequivocally to a shortage of military manpower in the second century. While complaints about the levy arose from time to time and draftees occasionally resisted conscription, the difficulties and unprofitability of the wars then in prospect seem best to account for such episodes rather than the undue strain that the levy placed on a shrinking number of smallholders. When victory and booty were in the offing, plenty of volunteers came forward to serve.[59] On the other hand, the evidence adduced to demonstrate the progressive lowering of the census qualification poses many problems. Three ancient authors record three different figures for the minimum value of property that a citizen needed to be registered in the fifth census class and so qualify as an *assiduus*: 11,000, 4,000, and 1,500 *asses*. But nothing warrants the presumption that these figures are to be arranged in a descending sequence and so used to confirm a general picture of economic decline and social dislocation among smallholders.[60]

Doubts about the incompatibility of small-scale farming with plantation agriculture further weaken the case for seeing the latter as the cause of a crisis among the former. Conflict between the two was anything but inescapable. Rathbone has acutely pointed out the need of large estates for the surplus labor represented by neighboring smallholders during critical points in the crop cycle, particularly the harvest, and his point is now generally accepted.[61] Equally important, capitalist agribusinessmen had to get their crops to market. Water transport was the preferred means of taking these to their destinations, but moving bulky commodities to the docks required short-haul conveyance overland, and this was again provided mainly by small farmers at points in their own agricultural year when they had time on their hands.[62] Nor should one underestimate the sociopolitical value of preserving a network of nearby smallholders in the eyes of their wealthy neighbors. In a world where the ability to exercise patronage was one of the prime measures of a man's worth, small farmers constituted the body of its potential recipients, out of whose gratitude locally prominent figures constructed their claims to prestige, power, and status. Those with the economic capital to create slave-run estates, in other words, may often have hesitated to act in ways that would displace those on whose acknowledgment their accumulation of symbolic capital depended.[63] Finally, because land transportation costs were high, the estates that typified the slave mode of production tended to occupy a particular spatial niche in Italy. To flourish, proximity to cheap water transport, especially access to the seacoast, was highly desirable; failing that, location near a major road was essential.[64] This constraint imposed significant geographical limits on plantations of this kind. They developed primarily in areas within twenty or thirty miles of the coasts, near major rivers, and along the great trunk roads the republic constructed into the interior of the peninsula.[65] Consequently, the many parts of Italy that did not meet these criteria along with the small farms located in them remained for the most part unaffected by competition from big, slave-staffed enterprises.

Yet even where such competition occurred, pessimistic assumptions about small farmers' inability to hold their own are unwarranted.[66] Most small farms could never become completely self-sufficient. They had to raise cash for taxes and the purchase of those items the household could not produce for itself or obtain by exchange. At times therefore they may have had to sell their products in markets along side those offered by large, slave-run estates. But in such situations smallholders would not inevitably have found themselves at a disadvantage. Certainly plantations could produce larger surpluses than smallholders, they might have more efficient

Introduction

labor configurations, and often they occupied more advantageous locations for transporting their crops to market. But these factors scarcely doomed small farmers to economic defeat.[67] What principally mattered for a family of smallholders was not to minimize its costs per unit but only whether the income that the household could derive from its surplus crops and labor was enough to meet its modest needs for cash.[68] The family's foremost aim was simply to ensure its own survival. As long as it could produce enough to achieve that end, it made little difference if its profit per unit of goods or services was smaller than what a slave-owning neighbor took in on the same items. And although large producers could reduce their costs through the economics of scale or more efficient transport, subsistence farms enjoyed a significant advantage in pricing their goods because families did not count their labor as a cost. It made no difference to smallholders if it took them longer to produce their goods or bring them to market as long as in the end they acquired whatever money they needed.

Coupled with this cost advantage was the very high productive potential that households on small farms enjoyed. The response of a smallholding family to adversity is to work harder and intensify production. Certainly, there are limits to how far it can go in this direction, but the alternative of starvation is a powerful stimulus to self-exploitation. Entrepreneurs who invested in agriculture on the other hand expected to reap a profit from the crops they grew, and the slaves and equipment purchased to work their land represented a major charge against their bottom lines. And slaves died or grew too old to work or were freed. If the number of their progeny did not fully replace them, as is likely often to have been the case, then their owners will have had to buy new ones, adding further costs.[69] Compulsion could make slaves work quite hard, and so they were more profitable on large estates than free laborers.[70] But peasants, too, can outproduce wage laborers, rendering it questionable whether slaves under the lash would be more productive than the members of a smallholding household under the goad of hunger.[71]

Finally, claims that large estates, unlike small farms, possessed storage facilities that enabled their owners to hold their products back until prices were high and so to enjoy a significant competitive advantage are unpersuasive. Smallholders certainly did not lack storage facilities; how else would they have kept grains, fruits, and vegetables for their own consumption throughout the year? Grain was placed in storage jars or underground pits while fruits and vegetables were dried or otherwise preserved.[72] Would a large plantation owner have done anything different with his surpluses if he wanted to store them until market conditions were ripe?

Continuity rather than change therefore more appropriately characterizes Italian agriculture in the years between the close of the Hannibalic War and the tribunate of Tiberius Gracchus. Little commends the notion that the number of large estates employing slave labor and producing crops for market was increasing rapidly in these decades. The heyday of villa agriculture and the slave mode of production it defined would begin only in the age of Sulla. The sorts of more modestly sized, slave-staffed farms that Cato the Elder described in his *De Agricultura* had long existed in Italy; more important, nothing suggests that such establishments would have necessarily been inimical to the interests or threatened the survival of neighboring small farms. Rather, the free population of Italy may well have been growing robustly during the republic's last two centuries; certainly claims that Rome's assiduate class was shrinking must now be discarded. And the rapid growth of Italy's servile population in this era seems anything but certain. The conclusion to which all these considerations point, however, is simply that the rise of plantation agriculture did not bring about the crisis that Tiberius Gracchus addressed, not that a crisis did not exist. It is important to emphasize this point because of the danger of assuming the reverse: that because the results of archaeological and other studies over the past thirty years have ruled out some of the causes to which Toynbee and others ascribed what he termed the "deracination of the Italian peasantry," the condition of smallholders throughout the peninsula generally must have been flourishing. But the events that Gracchus unleashed become unintelligible if the *lex agraria* did not address a serious grievance among the agrarian population. To doubt this is impossible.

Certainly, the results of field surveys have revealed areas where the number of small farmsteads remained constant or even increased in the second century. Yet the continuous occupation of small sites alone cannot demonstrate that no mass of impoverished men and women existed in the countryside to whom Gracchus appealed. Archaeological evidence of this sort is ambiguous in the absence of an overall interpretative framework. Pottery scatters and other artifacts in and of themselves tell us little about the status of those who left them behind.[73] They may have been prosperous, independent small farmers or slaves working lands left vacant when its free occupants were forced off. Or they may have been freeborn Romans or Italians who had lost their land and been reduced to the status of tenants or seasonal wage laborers on neighboring farms. Ironically, by ruling out the rise of plantation agriculture as the cause of the rural crisis that Tiberius Gracchus sought to address, work done over the past thirty years on the economic and demographic developments of the second century has only made understanding its origins more difficult than ever.

Overpopulation may be the answer, although this solution poses difficulties of its own. Fertility regimes in preindustrial populations tend toward stability over the long run, balancing births against deaths in such a way as to keep the size of a population roughly within the capacity of its environment to support it.[74] A sudden, dramatic increase in the birthrate raises the obvious questions of why and how it occurred. But before such problems can be explored and explanations along these lines pursued, the conflict between Rome's demands for military manpower and small farms' need for labor needs to be faced squarely. Earlier generations of scholars placed this at the root of the problem, and that judgment remains unquestioned even among some who otherwise raise doubts about the traditional view.[75] It has all but unanimously been assumed that warfare after Hannibal's invasion imposed a very different kind of burden on smallholders than what they had shouldered prior to 201 or 218 and that as a result farms failed for lack of the men needed to work them.[76] The consensus among scholars that the rise of plantation agriculture did not cause the suffering that Gracchus sought to alleviate might therefore make the impact of military changes in the second century seem all the more likely to have been its source.

Surprisingly, the nature of that impact has only been asserted, never demonstrated. Even Brunt's magisterial *Italian Manpower*, the strongest and most sustained argument for the decline of Italian small farmers due to the burdens that Rome's conquests imposed upon them, pays scant attention to the actual business of small-scale farming and how the character of warfare in the second century might have affected it.[77] But the anecdote concerning Regulus, quoted earlier, that Brunt employs to illustrate what he believes were the deleterious consequences of overseas duty might, on the face of it, suggest that small farmers drafted for service abroad could have had recourse to hired help to fill whatever gaps in the farm's work force their absences might create. On the other hand, studies of peasant agriculture elsewhere in the world have identified underemployment among the farming population, not a shortage of labor, as among its most salient characteristics, and Roman historians have usually presumed that the situation was the same in the middle republic.[78] One might therefore

Introduction

imagine that Roman imperialism, far from creating a deficit in the agricultural labor force, on the contrary temporarily drew off an unneeded surplus of rural manpower.[79]

Still, the strongest evidence for the disruptive impact of Rome's second-century wars has always been the very dislocation, impoverishment, and decline that they are invoked to explain. The risks of circularity inherent in arguing from presumed effects back to undemonstrated causes are patent, and in fact, as this study's second chapter will demonstrate, a conflict between war and agriculture was nothing new during and after the Second Punic War. Roman warfare had become incompatible with the requirements of small-scale husbandry long before Hannibal set foot in Italy. Evidence from the annalistic tradition indicates that from at least the late fourth century onward Roman armies regularly campaigned well into the autumn and winter months—long past the planting time for the fall-sown cereal crops that formed the mainstays of the Roman and Italian diet. The *fasti triumphales* offer important confirmation for this pattern, as do accounts of fighting in Italy during the war against Hannibal and those against the Gauls and Ligurians that followed. Strategic considerations, too, furnish further arguments for a long-standing pattern of all-season warfare.

The conflict, therefore, that military service after 200 or 218 posed for citizens and allies with small farms to run differed little if at all from what had been the case beginning around 300 B.C. or earlier. Yet no evidence indicates that Roman demands for men to fight its wars were undermining the viability of smallholders during this period. Because the traditional view is based largely on chronological juxtaposition—that is, the post hoc propter hoc claim that the crisis among Italy's small farmers developed following a major shift in the temporal rhythms of Roman war making—the demonstration that no such alteration in fact occurred deprives this argument of much of its force. But this conclusion only raises two much more important and much less discussed questions. The first is simply how Rome was able to conscript men without leaving their farms short of the labor necessary to work them. For herein, clearly, lay one of the keys to the military success that made possible the growth of Rome's *imperium*. Second, what was this system's maximum potential: how far could Rome go in taking men off their farms for its armies before causing the smallholding sector of the economy to collapse? For it might still be possible to salvage much of the traditional explanation for the Gracchan crisis by supposing that the critical change lay in the quantity rather than the seasonal pat-

19

Introduction

terns of the military service that smallholders performed. In other words, even though Rome had developed a way to obviate the conflict between the needs of warfare and agriculture toward the end of the fourth century, the republic's limited demands for soldiers during the hundred or so years that followed did not unduly tax this system. But perhaps the massive increase in the city's need for recruits after 218 overwhelmed its capacity to supply men without damage to their farms and so brought about the widespread distress that formed the focus of Tiberius Gracchus's tribunate.

In order to answer these questions, it is necessary to ascertain the labor requirements of small farms and how the removal of one or more men for long periods of military service might affect them. However, the absence of quantitative data, the near total lack of evidence from literary sources, and the great variety of geographic and climatic conditions under which small farms operated rule out a direct, empirical approach to this problem. In order to get around these limitations, Chapter 3 employs the somewhat unusual strategy of attempting to construct models of hypothetical family farms in order to test various assumptions about the balance between labor supplies and demands in the event of conscription. The food requirements of these hypothetical household configurations can be estimated with a reasonable degree of plausibility based on modern studies of nutritional requirements. And ancient agricultural writers such as Varro and Columella provide considerable data on the amount of labor needed for the different operations involved in planting, cultivating, and harvesting wheat and other crops. The time and manpower figures these authors preserve, it is true, are for slaves laboring on large estates, but, as noted, there is no reason to believe that small farmers would not, if necessary, have worked as hard or harder in order to feed themselves. On the basis of this information, therefore, it is possible to estimate in some detail how many days of work a family would need to perform in order to produce the food it required to sustain itself. These models reveal quite clearly that families would often have disposed of more than enough labor to meet their needs, even under the most pessimistic assumptions about yield rates and the labor inputs necessary to bring in a crop. These surpluses of labor in turn were what enabled them to send one or more sons to war for extended periods of time without imperiling their survival.

The critical variable for smallholder households, rather, was not when in the crop cycle their participation in warfare occurred but where in their family life cycles. Mature families with a surviving father and adult sons had workers to spare, but in young families the father represented most of the available adult labor force. Consequently, the latter disposed of little or no surplus. Patterns of family formation, therefore, prove to be much more critical in understanding Rome's manpower potential in the middle republic than the rhythms of the agricultural year. The age at which men typically married in this period was fairly late, around thirty years old. Men became liable to conscription much younger, however, at the age of seventeen. The practice of deferring marriage well beyond the age of eligibility for the draft thus created a lengthy period in young men's lives when their labor was often superfluous on their natal families' farms but not yet required to support families of their own. This pattern of family formation consequently left Rome free to conscript young men for lengthy terms of service without causing undue hardship for the families that had raised them. And by the time they were getting ready to take on the obligations that would come with marriage, a young family, and an independent smallholding, those they owed to the republic's military were coming to an end. Marriage practices, in other words, provided the key to resolving the conflict between participation in warfare and agriculture in these cases.

However, this same late age of male first marriage along with the low life expectancy prevalent among Roman and Italian smallholders also produced a high number of households in which the father was deceased by the time his sons reached adulthood. Consequently, not every seventeen-year-old conscript would have left an older man behind to carry on the farmwork. On many smallholdings, men between the ages of seventeen and thirty composed the entire adult male labor force. In such cases, what would have happened to the women and younger siblings when a son went off to war? One might imagine the worst: the family members left behind, lacking the ability to carry on the work of running the farm by themselves, would have faced starvation as in the example of Regulus's family, cited earlier. But while hardship cannot be doubted, analyses of women's and children's food requirements and their labor potential, comparative evidence for women undertaking fieldwork when necessity demanded it, and the existence of other sources of male labor combine to argue strongly that claims of destitution and helplessness among the women and children left behind on farms are overdrawn.

But how frequently Rome would have had to levy young men from families in which they constituted the sole adult male worker is itself worth asking, for it raises the second, and equally critical, problem that Chapter 3 considers. If the means by which the republic resolved the conflict between its military needs and those of the small farms from which it drew

Introduction

its manpower lay in taking men from them at the points in their families' life cycles where their labor could best be spared, where did the limits of this system lie? The effort to develop an answer begins by using model life tables to establish the approximate age structure of Rome's adult male citizen population. In conjunction with the census figures from the late third and second centuries, these tables permit a rough estimate of the number of men in the age-cohorts that usually performed military service. Computer simulations developed by Richard Saller model the effects of late male marriage and low life expectancy on the kin structure of the total male population and, in particular, of those in the age range from which Rome drew its soldiers. These models allow estimates of the proportions of men in these age-groups who would not have had a living father, other older male relative, or one or more male siblings to whom the running of a farm might have been entrusted in their absence. Against these figures can be set what is known of Rome's demands for citizen conscripts. Estimates of these can be developed based on the number of legions fielded at various dates and generally accepted estimates of the numbers of citizens each contained along with the age structure of the legionaries themselves, for the soldiers who composed a legion fell into several distinct categories and age ranges. Taken together, these figures show that the system would have worked during most of the late third and second centuries. Save for a brief period during the worst of the Hannibalic War, Rome could remove nearly all the citizens it needed for its armies from the ranks of precisely those *assidui* whose farms could most readily get along without them. Although the same modeling cannot be attempted for Rome's Italian allies, it must be assumed that similar practices could have worked for them, since nothing suggests their family structures or agricultural practices differed radically from those of Roman smallholding households.

The conclusions reached in Chapter 3 demonstrate that even with the heavy demands for manpower that Rome imposed in the late third and first two-thirds of the second centuries, lack of labor is not likely to have caused small farms to go under in large numbers. But such a finding fails to address the question of what effects Roman warfare in this period did have upon the smallholder class. For these cannot have been have inconsequential. In the course of its struggle against Hannibal and in the decades that followed, the republic mobilized a far higher proportion of Italy's population for war than any other Western power would recruit from its citizenry down to the American Civil War and for far longer periods of time. It is difficult to believe that a military effort on this scale played no role in creating the problem that Tiberius Gracchus sought to solve. Chapter 4 explores the possibility that the most immediate impact of Roman warfare following the republic's victory over Hannibal was demographic rather than economic by attempting to establish how many of the men Rome sent to war between 200 and 133 never returned. The chapter collects all figures preserved in the ancient sources for Roman and Italian deaths in battle and argues that these offer a defensible starting point from which to reconstruct military mortality levels. Obviously, any numbers preserved in the ancient literary sources are highly problematic and need to be approached with great circumspection. Therefore, the chapter begins with an attempt to understand how casualty figures might have been generated and recorded in the first place. Here a good case can be made that the logistical and tactical practicalities of warfare necessitated keeping track of the dead. We also know that controversy sometimes arose in the senate when commanders' claims to triumphs were met with complaints about excessive Roman deaths in a victory. General interest could lead to enumerations, too, as when Scipio Nasica noted the eighty soldiers who died at Pydna in 168 in a letter recounting the battle. A more serious problem is to understand how such figures found their way into the historical record and how Roman annalists treated them. The latter enjoy a very poor reputation as recorders of casualties, but this is due principally to inflated claims for the numbers of enemies killed in Roman victories. The latter are undeniable, but a variety of clues suggests that the annalists may not have been quite so cavalier in recounting Roman dead.

However, casualties are recorded only for a small number of the engagements the Romans fought in this period, and one must establish whether these constitute a representative sample. For if they do, their simple average might be treated as constituting a "typical" casualty rate that could then be generalized to the great majority of battles for which no figures survive. Unfortunately, the distribution of the preserved casualty figures in the sources indicates that the answer is probably no. The mean of all known deaths in victorious battles between 200 and 167 is significantly higher than the median. The figures fall broadly into two clusters—those that are very high and many, like the eighty who died at Pydna, that are strikingly low. This distribution strongly suggests that Roman historians tended to memorialize particularly heavy losses and to use low numbers to underscore that some victories were comparatively bloodless (at least for their side). The mortality rate for a typical victory therefore probably would have been a percentage lying roughly between the median and the

Introduction

mean of all recorded numbers of deaths in battle. Defeats, on the other hand, seem to pose fewer problems. Interestingly, Livy always cites figures for the numbers of Romans killed when he recounts defeats in the second century, and the proportions are fairly close to those preserved in Greek sources for battles in the classical period. Coupled with a tabulation of the frequency with which Roman armies fought battles in the years 200–167, these estimates allow overall losses due to death in combat to be quantified, at least in approximate terms.

Still, immediate death in combat represents only one element of military mortality. Complications arising from wounds and death from disease also contributed significantly to the overall death rate. Unfortunately, evidence for these is even scarcer than for deaths in battle, but comparative data can help to establish parameters. In the American Civil War, for example (the first conflict from which comprehensive statistics survive), complications arising from infections carried off 17 percent of wounded Union soldiers. Yet against this fact must be set the substantial differences between combat in the gunpowder age and in the Roman Republic, and Chapter 4 argues that mortality from wounds was probably far below the 17 percent figure that Union casualties might suggest. The effects of disease on early modern and nineteenth-century armies could also be grave: more than twice as many Union soldiers perished from sickness than enemy action during the Civil War, about the same ratio as during the Boer War. But once again, significant differences in the epidemiological context need to be kept in mind. Remarkably few references to outbreaks of disease of any type among Roman armies appear in the ancient sources. The good order and discipline of Roman camp life deeply impressed contemporary observers, and a high degree of mobility characterized Roman warfare in this period. Consequently, Roman military camps may have been much healthier places than many of their more recent counterparts.

Still, even on the most cautious estimates of mortality rates due to disease and wounds as well as death in combat, the massive numbers of men Rome sent to war year after year multiplied their impact enormously. If the estimates established on the basis of analysis offered in Chapter 4 are any guide, the rate of death among conscripts is likely to have been staggeringly high and to require scholars to alter profoundly their understanding of the era's demographic trends, particularly when coupled with the great numbers who perished while serving with the legions during the Second Punic War. But what would the consequences have been for the class of smallholders from whose ranks they came? Common sense might suggest that the high rate of military mortality Rome and Italy sustained during the first two-thirds of the second century renders overpopulation highly implausible as the cause of the Gracchan crisis. On the contrary, large numbers of deaths might seem to support the conventional belief that the size of the assiduate class was shrinking in the years leading up to Gracchus's tribunate. But population growth is a complex process, and a variety of factors exert an influence on it—women's age at first marriage, the birth intervals of their children, prospects for remarriage, infant exposure, nutrition, and the prevalence of epidemic diseases, among many others. Chapter 5 considers the possibility that, paradoxically, an ongoing mortality crisis such as Roman warfare between 218 and 133 represented could have significantly improved the near-term economic prospects of those who survived and so increased their prosperity, opening the way for a greatly accelerated birthrate. But even if claims along these lines are ultimately unpersuasive, military service also held the potential to diminish greatly the wealth of smallholders, although principally in the long rather than the short term. By removing the labor of sons for extended periods of military service, Roman warfare altered the normal course of cyclical mobility within the small farming class and so impeded families' ability to reproduce themselves. Conquest in this way brought poverty to a significant proportion of the men who won Rome its empire in the years following the Hannibalic War; in combination with a growing population, this process would have provided more than enough tinder to ignite the political conflagration of Tiberius Gracchus's tribunate.

Chapter 2

War and Agriculture:
A Critique of the Conventional View

A scholarly consensus has long held that before the Hannibalic War the complementary rhythms of warfare and agriculture enabled farmers to meet the demands of both. Once the Roman Republic's armies began fighting in Spain and the Greek East, however, the two came into conflict and helped bring about the agrarian crisis that Tiberius Gracchus and others sought to solve. This chapter examines that hypothesis in detail and offers a sustained critique. It argues first that a conflict between the demands of military service and the needs of small-scale husbandry was nothing new in the second century. The evidence for the seasonal patterns of Roman warfare from the late fourth century down to the eve of the war against Hannibal plainly refutes the view that fighting in that era usually ceased when the time came to plow and sow. Nor is there any reason to believe that soldiers received furloughs in the autumn to attend to their farms with orders to report back for duty in the spring. The evidence fails to bear out such a contention, and strategic considerations also militate strongly against it. Dating the origins of the antagonism between war and small farming at least a century before Hannibal entered Italy in turn poses a serious challenge to claims that their opposition played a major role in bringing about the rural distress evident in the later second century. For if that opposition were harming smallholders then, it ought to have been producing similar effects earlier. Yet, to the best of our knowledge, it did not.

Still, this finding does not rule out the possibility that some other difference in conditions of military service between these two periods explains the absence of discontent among smallholders before the Hannibalic War despite the damage conscription was causing. Perhaps Rome's regular dispatch of colonies throughout the third century palliated the suffering enough to prevent a crisis, but once colonization ceased after 181, pressures mounted. Or the key may be that the republic's greatly increased demands for conscripts after 218 had forced it to begin recruiting men significantly poorer than in earlier times. This decision, it is sometimes claimed, devastated smaller farms because they could ill-afford the loss of a man to the legions that wealthier estates could more easily weather. Or possibly the twin burdens of military service and taxation, far greater in the wake of Hannibal's invasion than before, combined to overwhelm smallholding households. Explanations along these lines in effect attempt to uphold in different terms the basic premise of the traditional view that some crucial change—if not in the character of warfare, then elsewhere—marked a watershed in Italy's agricultural history and sealed the fate of small farmers. As the second part of the chapter shows, however, they fail to withstand careful scrutiny. To suppose that colonies recompensed smallholders for the farms they lost while on military service creates more problems than it solves. Nothing indicates that the men who fought for Rome following the outbreak of the Hannibalic War were significantly poorer than recruits earlier. And the financial strains on small farmers became, if anything, progressively lighter, not heavier, as the second century wore on. These conclusions render attempts to salvage the conventional view of war's impact on small-scale agriculture during the great period of Roman expansion futile. What is needed instead is an alternative model that explains not how they came into conflict but how they were able to coexist so successfully in these centuries.

The Mediterranean agricultural cycle is well known: farmers sow their principal field crops in the fall and harvest them the next summer. Plowing in Italy usually begins in early September.[1] Sowing, according to the Roman agricultural writers, might take place in some areas as early as the autumn equinox (September 22) and continue until the winter solstice (December 22), with "most people" apparently sowing in October.[2] Winter is a time of reduced agricultural work, although late-winter- and spring-sown field crops were not unknown in Roman Italy, often constituting "insurance" or emergency measures because of concerns about the success of the main, fall-sown crops.[3] Italy's harvest gets under way in early summer at dates ranging from early June in the warmer areas to as late as early August at northern latitudes and at higher elevations.[4] During the long, dry summer, little agricultural work is done until it is time to begin preparing for the autumn's crop again.[5]

How a farmer could integrate military service into this agricultural cycle without harm to his husbandry is also fairly well understood from the pattern of warfare in archaic and classical Greece down to the Peloponnesian War.[6] Arable crops—along with vines, orchards, farm buildings, and

27

equipment—formed the primary objects of an aggressor's attack, particularly the wheat and barley that were the mainstays of the ancient Mediterranean diet. Attacking an enemy's grain fields aimed at two complementary objectives: compelling the defenders to leave the safety of their city walls and fight for control of the open country in order to protect their most important source of food for the coming year; and sustaining the attacking army during the invasion as the invaders could harvest the enemy's grain to feed themselves (attackers rarely brought with them enough food for more than a few weeks, and under conditions of ancient transport sending grain overland any distance was difficult and expensive).[7] Yet this strategy imposed a significant limitation on the timing of attacks: these could only be launched during the fairly short period when grain was ripe enough for human consumption but had not yet been harvested and stored out of reach, thus in the late spring and early summer.[8] Additionally, standing grain at this point had begun to dry out enough to burn, the most efficient way of destroying what the attackers would not eat themselves.[9] Campaigns were typically brief because even with the provisions gained by ravaging their enemy's fields invaders ran out of supplies after four or five weeks. Moreover, the fact that the attackers, too, were farmers also restricted the duration of warfare for their own crops might be ready to harvest at the same time the enemy's were ripening.[10] Finally, defenders could usually be counted on to march out to confront the invaders sooner rather than later. The threats to their food supply, political tensions within the city arising because ravaging injured some segments of the citizenry more than others, and an agonal mentality coupled with simple revulsion at the sight of strangers trampling on their fields all combined to impel citizens to meet their attackers in the open field and resolve the issue in a single, decisive battle pitting one hoplite phalanx against another.[11] This type of combat was itself well suited to the needs of farming. Hoplites in a phalanx fought in a dense formation that certainly required great courage and stamina but little in the way of drill or specialized weapons training that would in other seasons require cultivators to be away from their fields for any length of time.[12]

Short campaigns waged by amateurs that would quickly provoke a decisive battle did not much inhibit the normal course of agriculture. Invaders inflicted minimal damage to the defenders' crops, apart from whatever they had destroyed in the initial stages of the attack, and once the battle was over both sides returned to the business of agriculture and gathered in their harvests. Certainly fighting could and did take place during the summer slack season and might continue into the fall, but in such cases the lack of resources again precluded campaigning beyond a few weeks. This inability to stay long in the field minimized the damage that could be inflicted on an enemy's crops or agricultural infrastructure. No wheat or barley remained in the fields, and the destruction of vines, olive trees, or farm buildings required far more time to accomplish than an army that could sustain itself on campaign for only a few weeks could devote to the enterprise, particularly without the food that the enemy's standing grain provided in the early summer. Again, relatively brief incursions, even in the fall, did not disrupt too much the defenders' agricultural operations. On the other hand, invaders needed eventually to plow and plant their own fields in the fall, and this task imposed a natural terminus on operations for the year. Thereafter, although troops can usually be kept in the field until about January in a Mediterranean climate, campaigning generally ceased once winter rains rendered roads impassable and perhaps because transporting the additional gear necessary for winter encampments proved irksome.[13]

Although similar evidence is lacking for Rome during the fifth and most of the fourth centuries, the same pattern undoubtedly held. Roman soldiers in this period, like Greek hoplites, were small farmers who furnished their own equipment and paid their own expenses in the field.[14] Their agricultural practices and crops were largely those of farmers in classical Greek poleis. The climate of Italy is for the most part similar enough to that of Greece to impose the same agricultural rhythms on its inhabitants. Likewise, although Italy's topography is generally more favorable for arable culture than Greece's, the differences between the two are not great enough to alter the basic similarity in the farming practices that characterized each area. Hence we ought to expect that during the early republic, as in classical Greece, war and agriculture coexisted without fundamental conflict between the requirements of each. That had changed, however, by the later fourth century. At that point, Rome's enlarged diplomatic ambitions, its provision of pay and food for its troops, and the introduction of the manipular system of combat converged to bring about a quite different and far less compatible relationship between war and agriculture.

Intervention in Campania in 343 marked the first time the republic carried war well beyond the confines of Latium, and over the next century the city regularly sent its armies into increasingly distant theaters. Combat against enemies in Samnium, Apulia, Lucania, northern Etruria, Umbria, southern Italy, and ultimately Sicily and North Africa brought with it a host of changes in the way Rome conducted its military operations, but none was so essential as the dramatic increase in the amount of time that soldiers were required to spend at war. It took armies much longer to reach enemies in these regions, particularly those whose strongholds lay in the Appennine highlands where access was difficult and easily obstructed.[15] These opponents, moreover, proved to be among the toughest the republic ever encountered. Defeating them was never a matter of winning a single victory, accepting the submission of the vanquished, and then returning home in triumph. Indeed, victories all too frequently proved elusive in these struggles. Campaigns now began to require armies to spend months or even years in the field rather than the days or weeks that had been common when the republic fought its wars close to home. To enable them to do so, the Romans now began to subsidize their soldiers, if they had not done so before, by paying a wage to citizens so that they could purchase grain and by furnishing rations gratis to allied contingents.[16] A secure commissariat freed Roman armies from the operational constraints imposed by the ripening and harvest dates of the enemy's crops. However, it also required the development of a sophisticated logistical system in order to ensure that armies would be regularly supplied with food, particularly when they operated in regions and at times of the year when depending on local sources or living off the land would have been risky strategies at best.[17] Roman forces thereby acquired a staying power comparable with that of the mercenary armies of fourth-century Greece and the Hellenistic world. Now, the legions' ability to campaign year-round enabled them to attack effectively an enemy's economic infrastructure, including its vineyards, orchards, agricultural buildings, and similar targets that required a great deal of time for soldiers using only hand tools to destroy.[18] Lengthy sieges also became more feasible, as did preventing an enemy from sowing crops in the fall.

The later fourth century also saw the development of the system of combat based on maniples that was to secure for the legions a decisive advantage in open-field warfare. Manipular legions, unlike a hoplite phalanx, fought according to a complex tactical system that emphasized both individual weapons-handling ability and unit maneuverability.[19] A Roman army's success in battle now depended on each soldier holding his own in man-to-man combat, on maintaining his place within his maniple under the stress of battle, and on the maniples' discipline and cohesion as their members moved forward individually to take the place of weary fighters in the front ranks and as the maniples themselves, arrayed in three successive lines, moved up to or retired from the zone of direct contact with the enemy. Legionaries thus had to exhibit a far higher degree of discipline and swordsmanship under the pressures of combat than fighters in a phalanx, and these skills did not come easily. Mastering them demanded a considerable investment of time, and the longer soldiers were kept under arms, the more they could be drilled and trained to make this system work.[20] But for the Romans to realize this potential as well as regularly to conduct wars well beyond Latium required the city frequently to keep its citizen-soldiers with the standards for long periods of time, well past the late summer when the fighting should have ended to allow the soldiers to return to their fields to begin work on the following year's crop.

The annalistic tradition clearly reflects the greatly increased length of time armies began to spend in the field as a consequence of these changes. Livy reports that the consuls of 335, 334, and 333 took over the armies of their predecessors, and these notices imply that the legions involved had spent the winter under arms.[21] The opening of the Second Samnite War found one of the consuls of 327 wintering with his army among the Volsci, and once again Livy notes that at the beginning of the consular years 316, 315, 314, 313, and 310 the new commanders received armies from the previous year's magistrates.[22] Winter encampments are also on record in the Third Samnite War for 297/6, 296/5, and 294/3.[23] In the next year, Livy indicates that campaigning again extended well into the winter months: the soldiers were suffering from the cold and snow that covered the ground before the legions withdrew from Samnium. The senate then sent two legions into Etruria to stifle unrest there while the others remained in Campania through the winter.[24] Although the loss of Livy's second decade deprives us of similar testimonia for the rest of the Third Samnite War, a Roman army probably also remained in the field over the winter of 292/1 as well. In 292 the Samnites defeated the consul Q. Fabius Maximus Gurges, who only avoided being removed from his command when his father, Rullianus, offered to serve as his legate. Subsequently, so the story goes, the two men won a victory after which Gurges's command was prorogued. In the next year they captured the stronghold of Cominium Orcitum for which Gurges triumphed on July 1, 291. The proconsulship would imply that Gurges retained command of his legions over the winter.[25]

To be sure, no scholar would take accounts of military events in the fourth or early third centuries at face value. But hyperskepticism toward this material is equally misplaced. Accounts of events in this period preserved in Livy or other historians who drew on the annalistic tradition need to be evaluated on a case-by-case basis, and given what we know of the context in which the armies that fought these wars were often operating, reports of armies kept in being over the winter are entirely plausible. Levying and equipping a new army could easily consume more than a month before it was prepared to march.[26] Travel to the enemy's homeland could take additional weeks of walking before the legions faced a tough, stubborn foe ensconced in rugged country far from easy access to supplies. Under these conditions, it made little sense to start from scratch every year. Keeping the legions in barracks instead of sending them home brought strategic advantages as well. Enemies pacified could be kept in submission, new allies protected, and hard-won gains from previous years' fighting preserved. Legions passing the winter under arms functioned in this respect like garrisons, of which the annalistic tradition also preserves numerous reports from this same period. And there is little reason to imagine that the latter—critical instruments in securing Rome's grip on its *imperium*—were sent home every fall so that the soldiers could see to their planting.[27] No a priori reason exists, therefore, to reject this testimony out of hand.

With Pyrrhus's invasion our evidence becomes a bit better, and here, too, military necessity often compelled the senate to keep the republic's armies together well into the winter months and beyond. L. Aemilius Barbula, consul in 281, spent the winter of that year at Venusia watching developments at Tarentum before returning to Rome to triumph on July 10, 280. Presumably, his legions remained at Venusia with him.[28] In that same year, the consul P. Valerius Laevinus lost heavily to Pyrrhus; the senate punished his troops by ordering them to spend the winter in the field under canvas.[29] It also dispatched Valerius's colleague, Ti. Coruncanius, from Etruria to assist him in the South where he and his army apparently remained until well into the winter: Coruncanius triumphed for his earlier victories in Etruria only on February 1.[30] The *fasti* indicate that the consuls of 266, D. Iunius Pera and N. Fabius Pictor, campaigned first in Umbria, celebrated triumphs for their victories there on September 26 and October 5, and then marched south to fight in Calabria whence they triumphed again on February 1 and 5.[31] If these campaigns are genuine, it would be highly surprising for the consuls to have dismissed their legions after the first triumphs and then levied new ones for the campaign in the South.[32] Rather, they would have led their armies in their triumphs and then led them on to their next assignment. Yet even if these four triumphs be rejected, the alternative is to suppose that one consul fought in the North and the other in Calabria. The latter's triumph certainly fell on one of the two February dates recorded in the *fasti*, meaning that his army did not return to Rome until midwinter.

Rome's long conflict with Carthage in Sicily, as one might expect, also regularly necessitated keeping the legions on the island through the planting season. Thus the siege of Agrigentum, begun in June 262, continued through the end of that year and concluded in January 261, at which point the consuls withdrew to Messene for the remainder of the winter.[33] The legions of P. Cornelius Scipio Asina, consul in 260, also seem to have stayed on the island over the winter.[34] The consul C. Aquilius Florus, too, remained with his troops in Sicily over the winter of 259/8 where the consuls of the following year found them when they arrived to take up their commands after May 1.[35] Likewise the consuls of 256 arrived in Sicily with a fleet to find the land forces encamped at Ecnomus, having apparently wintered somewhere in the vicinity.[36] The following winter, the consul Regulus remained in Africa with his army, some 15,000 troops, and again in 254 one consul stayed in Sicily with one or both consular armies until spring.[37] Three years later when one of the consuls of 251 returned to Rome, his colleague L. Caecilius Metellus stayed behind in Sicily and won a great victory at Panormus in June of the following year. Certainly, his legions spent the winter of 251/0 on the island with him; possibly his colleague's army did so as well.[38] Finally, in that same year the Romans began the siege of Lilybaeum, which in turn soon spawned a long positional struggle around Mount Eryx. Both contests were still going on when Catulus defeated the Punic fleet in 241 to end the war, and it is difficult to believe that each did not require the year-round presence of Roman troops to prevent a breakout by the enemy during the intervening nine years.[39]

The literary evidence for the late fourth and third centuries, therefore, while not abundant and of uneven quality, presents a picture of republican warfare completely at odds with the needs of farmer-soldiers for release from military duties in the late summer or early fall. Confirmation comes from the *fasti triumphales*. The *fasti* regularly record the dates on which Roman generals celebrated their triumphs, and the bulk of these entries survive for the third century. Scholars generally accept their reliability for this period.[40] Figure 1 graphs all datable triumphs between 298, when the Third Samnite War began, and the triumph of M. Claudius Marcellus in 222, after which the *fasti* break off. The results are striking:

War and Agriculture

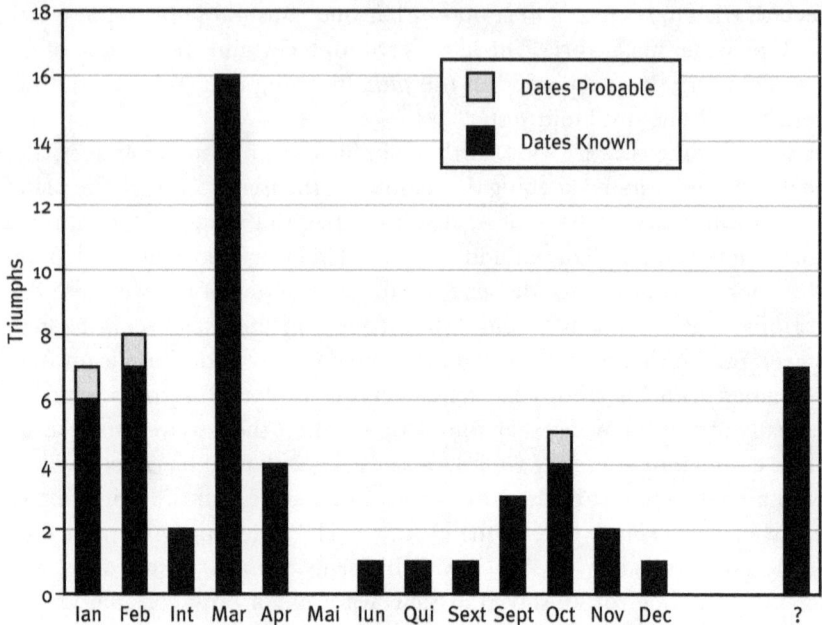

FIGURE 1. *Frequency of Triumphs by Month, 298–222 B.C.*

nearly half of them, twenty-three or twenty-four out of between fifty and fifty-two, fall into just two months, February and March.[41] If one includes the six or seven triumphs celebrated in January, the total rises to between 56 and 62 percent, while the two held during intercalary months increase the percentage to as high as two-thirds.[42] The total reaches 70 percent if those in April are brought into the reckoning. On the face of it, therefore, these winter and early spring triumphs indicate that Roman soldiers regularly remained under arms well past the fall planting season in the century preceding the Hannibalic War.[43]

One might, however, object that the dating of these triumphs cannot be trusted because of the notorious imprecision of the Roman calendar in the second century when it was as much as four months in advance of the seasons.[44] If similar displacement was also occurring in the third century, then calendar dates in the winter would in fact correspond to seasonal dates in the fall and these triumphs would offer no evidence of armies wintering in the field. Chronology is extremely difficult for the third century, lacking as we do Livy's text for most of it. However, what evidence can be gleaned offers little support for the notion that the calendar diverged seriously from seasonal time during this period. It was running approximately true in 218–217, in the first years of the Hannibalic War.[45] Likewise, the chronological indications we have do not suggest any serious problems with the calendar during the First Punic War.[46] Livy's mention of winter's onset before the consuls of 293 withdrew from the highlands of Samnium and celebrated their triumphs in January and February similarly suggests that the calendar was running close to the seasons at the beginning of the century.[47] Thus, while certainty is impossible, nothing in the evidence would suggest that the calendar in the third century had gotten as badly off the mark as happened in the second.[48]

The *fasti* therefore bear out the impression obtained from the literary evidence of a basic disjunction between the rhythms of republican warfare in the late fourth and third centuries and the requirements of subsistence cultivators for time to plow and sow their fields in the fall. Campaigns lasting well into the winter will have raised grave difficulties for farmers. Yet more than one scholar has claimed that soldiers serving in Italy during the Second Punic War regularly obtained furloughs during the fall and winter months that enabled them to make at least some contribution toward keeping their farms running.[49] If this practice was common in the struggle against Hannibal, despite the crisis Rome faced, then it is easy to assume its existence during the century preceding the war, when conditions were far less urgent. And that presumption, in turn, opens the way to a very different interpretation of the evidence. The *fasti*'s preponderance of triumphs dating to the winter months is perfectly comprehensible if soldiers regularly obtained leave at the end of summer with orders to report back to the standards once their planting was done.[50] Only after their seed was in the ground, around the end of December, were they free to come to Rome to celebrate a triumph.[51] Other soldiers would simply have returned to their barracks and remained there until the new magistrates assumed command and renewed the fighting. This would account for notices in the sources of winter billeting and of consuls turning their armies over to successors at the ends of their terms of office. If this explanation can be accepted, then the emphasis that the traditional view places on the shift to overseas theaters of combat like Spain and Greece seems entirely justified. While furloughs in Italy obviated the conflict between war and agriculture, campaigns in far-off regions made these impractical and so caused farms in Italy to suffer. The question of furloughs, therefore, is of critical importance to any understanding of the relations between warfare and farming during the third century. The evidence requires a detailed review, and will show not only that the belief in regular fall and winter furloughs for soldiers during the Hannibalic War is completely unjustified but also that nothing in the record for the second century would suggest that these were a common feature in that era either. In light of this analysis, no basis exists for presuming that leaves of absence in the fall resolved the conflicting demands that warfare and farming placed on soldiers.

De Sanctis has argued that generals regularly dismissed their troops during the war against Hannibal on the basis of a single incident: late in 215 the consul of that year, Q. Fabius Maximus, ordered the proconsul Marcellus to garrison Nola and then dismiss the remainder of his legions.[52] Yet closer examination occasions doubts. Livy remarks explicitly that Fabius acted "to avoid having the soldiers be a burden to the allies and an expense to the republic." He says nothing to suggest that Fabius intended to enable Marcellus's troops to attend to their fall planting.[53] The timing of Fabius's order makes it highly unlikely that the soldiers could have done so even if that had been the consul's purpose. Livy's report of the dismissal is embedded in a brief description of Fabius's activities in the latter half of 215: after indecisive skirmishing in Campania, the consul withdrew temporarily in order to allow the Campanians to do their sowing. When the grain was tall enough to furnish fodder, he returned and, gathering what he needed of this, transported it to the *castra Claudiana* above Suessula where he constructed his winter quarters. Fabius's order to Marcellus to dismiss his troops, already noted, follows.[54] Obviously, this order occurred in the context of Fabius's preparations for the winter, which Livy has just recounted. If so, then Marcellus's troops will have received their furloughs *after* the Campanians had plowed and sown their crops for the following year and *after* these had grown tall enough to be harvested for fodder. If we assume, as seems likely, that the Campanians will have done their sowing, at the earliest, in early October (although later dates down to the winter solstice are certainly possible, particularly in warmer areas), then it is extremely difficult to see how wheat could have grown tall enough to harvest for fodder much earlier than late winter. Wheat requires both light and heat to grow, and although winter temperatures in Campania usually do not fall below a daily mean of 40 degrees Fahrenheit (about 4.5 degrees Celsius), below which wheat becomes dormant, the cold weather there from November through February will certainly have slowed the plants' development, as would the long stretch of days with fewer than eleven hours of sunlight during that same period, which likewise inhibits the growth of wheat.⁵⁵ Hence Fabius's orders to gather fodder for the winter encampment and his instructions to Marcellus will have come, at the very earliest, in January or February. Thus, while the Campanians were preparing their fields and planting, the Roman forces opposing them were still under arms. By the time Marcellus's men were on their way home, it was far too late in the year for them to sow their own autumn crops.

Far from reflecting a practice developed to meet the needs of farmers to return to their fields to plow, Fabius's orders to Marcellus to send the bulk of his troops home for the winter arose out of the republic's straitened circumstances at that moment. Rome was desperately short of funds following Cannae: there was no money to pay or clothe the armies serving in Sardinia or Sicily in 216, none for the forces in Spain in 215.⁵⁶ The shortage of cash forced the senate to turn to private contributions to man a fleet at the beginning of 214, and the coinage was becoming increasingly debased.⁵⁷ Fabius furloughed the major part of Marcellus's army primarily to save the cost of paying the Romans and feeding the allies that winter.⁵⁸ At the same time, the situation at Nola was delicate. The city was a critical ally in the struggle for Campania, and Fabius could keep down the price of food there by reducing the demand the presence of Marcellus's troops represented. Eliminating this potential source of friction as well as the others that a large garrison of soldiers represented would help prevent the open revolt of a Nolan population already sympathetic to Hannibal.⁵⁹

However, De Sanctis, following Kromayer, claims further support for a furlough for nearly all of Rome's troops in Campania over this winter in the supposed absence of any significant Roman forces in the vicinity when Hannibal threatened Nola in the spring of 214. On this view, not only had Fabius dismissed Marcellus's legions for the winter but his own, too.⁶⁰ But this is illusory. The consular *comitia* for 214 returned Marcellus to that office *in absentia* "because he was with the army."⁶¹ Because Fabius had ordered Marcellus to dismiss his own forces, this army can only have been Fabius's, which he had encamped at the *castra Claudiana* and probably placed Marcellus in command of when he returned to Rome at the end of the year to conduct the elections.⁶² After his election, Marcellus returned to Rome to enter his consulate, and while both he and Fabius were away Hannibal entered Campania.⁶³ His arrival caught the new consuls by surprise, and Marcellus, who had left Fabius in Rome to see to the election of censors, rushed from Cales to Suessula in a single day and thence to Nola with 6,000 men.⁶⁴ Kromayer presumes that Marcellus had brought these 6,000 men with him from Cales, where news of Hannibal's arrival had reached him. Because no Roman force had opposed Hannibal prior to Marcellus's arrival, there cannot in Kromayer's view have been any Roman troops nearby. However, an army was in Campania with Marcellus at the end of the previous consular year when he was elected consul, and it is scarcely credible that these men were dismissed soon thereafter only to be reassembled a few weeks later when Marcellus took office. It is far more plausible that the army remained in its encampment and did not take the field against Hannibal because the subordinates Marcellus had left in charge were under orders not to engage the enemy until the consuls returned. Yet even without explicit orders, what legate or military tribune left in command in the consuls' absence, even with two full legions and allied forces, would have risked coming down into the open country to oppose Hannibal, particularly with the example of Minucius's near catastrophe in 217 before his eyes? Only with Marcellus's return will the troops have moved down into the plain to face the enemy.

Some have also claimed to detect in the events surrounding Hannibal's march on Rome in spring 211 another instance of legions furloughed for the winter. Providentially, Polybius notes, when Hannibal arrived before the city, "Gnaeus Fulvius and Publius Sulpicius [the consuls of 211] had completed the enrollment of one στρατόπεδον, and had engaged the soldiers on their oath to present themselves in arms at Rome exactly on this day, and they were now engaged in enrolling and testing the men for a second στρατόπεδον; and the consequence was that a large number of men were spontaneously collected in Rome just when they were needed."[65] The case for a furlough turns on the translation of στρατόπεδον. Walbank, following De Sanctis, renders it "legion" and understands the passage as referring to the levying of the two urban legions that the senate had authorized for that year: the enrollment of one had been completed while that of the other was still going on.[66] Brunt however takes the word to mean "army" and, following Gelzer, states, "The army already enrolled was in my judgment formed of the 2 *legiones urbanae* raised in 212; we can presume that they had just returned from winter furlough. The army now being raised must have been composed of 2 new *legiones urbanae*."[67] Yet, as Brunt is forced to acknowledge, Polybius's language is plainly against such an interpretation, for he says that the consuls of 211 raised the first στρατόπεδον, and, unless Polybius (or his source) is in error, this must mean that it can only have come into existence after March 15 of that year. That Polybius is accurate here gains support from the rest of the passage, where the historian notes that the soldiers of the first στρατόπεδον had sworn an oath to be in Rome on that day with their weapons.

A Roman levy, as Polybius describes it in book 6, took place in three stages. First, potential recruits were arranged in groups of four. These groups were then called forward one at a time by the military tribunes of that year. The tribunes of each of the four legions selected one man from each group of four recruits in succession to serve in their legion, rotating the order of choosing among the four legions. When each legion had its quota, its tribunes administered an oath to their men to be present at a particular place and time *without arms*.[68] Next, at the appointed place and time the tribunes assigned their men to the various categories of soldiers that composed a legion, the *velites, hastati, principes,* and *triarii*, as well as the cavalry. They also appointed centurions and other officers and then, after ordering them to arm themselves appropriately, dismissed the recruits to their homes.[69] Finally, the legionaries assembled *with their arms* at the time and place fixed by the consuls, having previously sworn an oath promising to do so.[70] There seems little doubt that the situation Polybius describes in 211 corresponds exactly to this third stage of the levy, for the soldiers have come to Rome at the time appointed by Fulvius and Sulpicius, they are there in consequence of an oath sworn to the consuls, and they now have their weapons. Clearly, therefore, these soldiers had been levied by the consuls of 211 and, hence, after the latter had entered office on the Ides of March. Consequently, this episode offers no support for the notion that winter furloughs were a regular feature of warfare in this period. If στρατόπεδον in this passage is to be understood as meaning "army" rather than legion (and this seems the best translation in the context), the situation is certainly anomalous, for of course the urban legions of 212 ought to have been levied in that year rather than in the following, as the senate had in fact directed.[71] Marchetti's suggestion therefore is very attractive, that the *inopia iuniorum* the previous year's consuls had encountered in their recruiting efforts had caused the levying of that year's two urban legions to be deferred until 211, by which time the two commissions the senate sent out in the preceding year to scour the countryside for additional men of military age had completed their tasks.[72]

Neither of these passages therefore demonstrates that furloughing troops in Italy during the Hannibalic War to tend their farms was usual, and little support for the notion exists elsewhere in the sources. Nothing in the evidence for the war's early years so much as hints at such a practice. Rome's initial strategy in 218 would have ruled out any return to their farms for the legionaries in that year. One army was sent to fight in Spain; the other went to Sicily to prepare for an invasion of Africa.[73] Hannibal's entry into Italy, of course, forced the abandonment of those plans and led to the battle at the Trebia River sometime in December or January. Afterward, the legions spent the remainder of the winter of 218/17 in Placentia and Cremona.[74] Knowledge that Hannibal was on his way to Italy and his arrival in October or November will have made dismissal unthinkable for Roman forces that fall.[75] During the autumn of the following year, as Hannibal made his preparations for the winter, the legions watched and attacked whenever an opportunity arose. Once the Carthaginians were encamped for the winter, Rome's forces remained under arms, skirmishing with the enemy until spring 216.[76] After Cannae (August 2), Hannibal entered Campania. He incited Capua to revolt, launched attacks on Naples and Nola, stormed Nuceria and Acerrae, and placed Casilinum under siege before retiring into winter quarters. It is inconceivable that in the midst of this crisis Rome's commanders sent their troops home early in the fall to attend to their planting. Instead, the dictator Pera's army seems to have remained near Casilinum during the winter attempting to resupply the garrison in that city, while Marcellus's forces established themselves at Suessula in the *castra Claudiana* to protect Nola.[77] The winter of 215/14, discussed earlier, also saw regular skirmishing between Hannibal's forces and those of the consul Sempronius in Apulia operating out of their winter encampments.[78] Nothing can be said of the following year.

In late April or May 212 according to the Roman calendar a delegation of Beneventines bearing news of Hanno's effort to send grain into Capua found both consuls encamped at Bovianum in Samnium. To intercept the wagon train carrying the grain the consuls arranged that one of them should lead his army to Beneventum and surprise the Carthaginians.[79] Because the consuls had taken over two armies of the previous year, one of which had been at Arpi, the other at the *castra Claudiana*, this might imply that the soldiers of each had been dispersed to their homes in the fall of 213 and ordered to reassemble at Bovianum the following spring, at the beginning of the new consular year.[80] But it is equally possible that each army had wintered where it had served, that the consuls had only replacements with them at Bovianum, and that Fulvius summoned his legions from the *castra Claudiana*, a day's march from Beneventum on the *via Appia*, to mount his attack on Hanno.[81] This would explain why Fulvius's colleague Claudius only arrived a few days after the battle: his legions had been at Arpi and so had farther to come.[82]

In the same year these same two consuls also began the siege of Capua. The work of circumvallation seems to have gotten under way late in the campaigning season and remained incomplete at the end of their year of office the following spring.[83] The massiveness of the siegeworks with their wide circuit and double wall, the strong counterattacks the enemy mounted during their construction, and the consuls' vigorous efforts to ensure an ample food supply for the besiegers over the winter of 212/11 all make it highly unlikely that many Roman soldiers got to go home that fall.[84] Although nothing can be said of the winter of 211/10, in the following year the consul who held the province of Italy, M. Claudius Marcellus, refused to return to Rome to conduct elections for his successors. Writing to the senate, he informed them that he was pressing Hannibal hard at that point and to break off his pursuit would be against the public interest.[85] Livy's chronological placement of the elections is imprecise: his phrase *Iam aestas in exitu erat* might imply a date at the end of summer or perhaps early fall. Yet consular elections, as best we can date them during the war, usually took place between January and early March. They apparently did so in this case as well for the senate feared that the new year would open without magistrates in office if the other consul was not summoned from Sicily soon to conduct elections.[86] Clearly if the season were late summer and Marcellus would soon dismiss his men to plant their fields, the *patres* would have had no cause for urgency. Possibly by this date the Roman year had begun to run a month or two in advance of the seasons, but whether elections for 209 were held close to a seasonal date of February or the end of consular 210 fell a month or two earlier in seasonal time—that is, in seasonal December or January—the fact that Marcellus was then in active pursuit of Hannibal suggests that he had not furloughed his soldiers that fall. Nor had Marcellus's pursuit only begun in the early spring, after such a furlough had been completed. He set off after Hannibal in the aftermath of the Roman defeat at Herdonia that summer and had apparently been at it for some time when consular 210 drew to a close.[87] Marcellus's soldiers were in winter quarters for a time after this but ready to resume the offensive quite early in consular 209, again suggesting that they had not dispersed to their homes for the season.[88]

Although little can be established about how the consular legions spent the fall and winter of the following year, Marcellus, now a proconsul, fought and lost a pitched battle against Hannibal. Thereafter, he withdrew his men into Campania and billeted them in winter quarters in the middle of summer.[89] For this his enemies in Rome roundly criticized him shortly before the time of the consular elections for 208.[90] Whether his troops remained in camp during the fall and winter or went home to their farms is, however, unclear. Yet when Marcellus returned to Rome to answer the complaints about his withdrawal of his forces from the field, he left legates in charge of his army and camp, suggesting that a substantial body of men was present there.[91] Note, too, that when the senate sent Marcellus as consul-elect north to investigate disaffection in Etruria, it could instruct him to summon forces from Apulia if necessary. Obviously at this point—after elections but before the start of consular 208—the legionaries were with the standards and ready to march if circumstances required.[92]

Information is scarce for the remainder of the war. One of the consuls of 207, M. Livius Salinator, brought his army to Rome, celebrated a triumph for his victory over Hasdrubal at the Metaurus in June or July of that year, conducted elections for the next year's consuls, and then dismissed his troops.[93] Again the chronology is unclear. The section of the *fasti* that would have recorded the triumph does not survive. Livy refers to *extremo aestatis* but, as indicated earlier, his use of the term cannot be pressed.[94] Elections during the war regularly fell at the end of the consular year, and it is difficult to see why in this case they would have taken place in September or October.[95] If the triumph and the elections occurred close together around January or February, it suggests that the legions had been kept together up to that point. Nor would a lengthy delay between an important victory and a triumph have been unprecedented: Fabius Rullianus entered the city in triumph in September of 295 for his great success at Sentium in April of that same year.[96] Yet even if the triumph and discharge took place early in the fall, note that at the same time the senate explicitly ordered the troops of Livius's colleague, C. Claudius Nero, to remain in their province facing Hannibal.[97] During the winter of 205/4, again around the time of the consular elections and after Hannibal had withdrawn into winter quarters, P. Licinius Crassus, one of the consuls of 205 wrote to the senate that the proconsul Q. Caecilius Metellus's army should be dismissed. He made the request however not to allow his men to attend to their farms, but to avoid further contagion from an outbreak of disease that was then decimating their ranks.[98] Nothing was said of dismissing Crassus's legions, so presumably they remained under arms watching the enemy. One of the consuls of the following year, M. Cornelius Cethegus, returned to his army in Etruria after conducting elections for his successors.[99] In this case, too, these troops probably had remained under arms with the consul up to that point. Cn. Servilius Caepio, consul in 203, skirmished with Hannibal and then, when Hannibal left Italy to return to Carthage, crossed to Sicily in an attempt to pursue him to Africa. Hannibal's departure occurred in the late autumn, following the breakdown of peace negotiations between Rome and Carthage.[100] Caepio was in Sicily around the same time.[101] His plan at that point to cross to Africa to continue the struggle against Hannibal makes it highly unlikely that in the meantime he had allowed his soldiers to go on leave to do their planting.

Nothing in the Hannibalic War requires the assumption that Roman soldiers regularly obtained furloughs during the fall and winter to work their farms. Consequently, it affords no basis for retrojecting such a practice into the preceding decades. However, the situation during the first third of the second century in Italy might have been different. Rome's principal enemies there in this period, the Gauls and Ligurians, were farmers, too, with their own fields to tend. That convergence of interests might have led to mutual armistices in the fall when agricultural imperatives urged a temporary suspension of hostilities. We know that in the fall of 170 an embassy returned from Greece complaining that the consuls had been lavish in granting leaves to the men serving there. In this case, the *patres* were highly critical of the consuls' actions, a reaction difficult to understand if furloughing most soldiers at some point in the fall had been customary.[102] Still, the practice might have occurred regularly in Italy. If so, it could furnish grounds for believing that the pattern of warfare was similar prior to the Hannibalic War when Rome's enemies were often farmers, too. But although several cases of armies disbanded in the summer or fall are on record, few of these would support the view that they had anything to do with the needs of agriculture. In 191 the consul P. Scipio Nasica won a decisive victory over the Boii, and before returning to Rome he dismissed his army with orders to reassemble at the triumph he expected to receive for his achievement. This apparently took place in midsummer.[103] Yet Nasica's point here was to emphasize the magnitude of his victory, which had ended the war against the Boii. They had made a formal surrender, given hostages, been deprived of half their land, and agreed to receive a Roman colony.[104] He had completely pacified the enemy, there was nothing further for the soldiers to do, and their dismissal underscored that fact. Significantly, however, a tribune of the plebs harshly criticized

Nasica's action, claiming that the consul should have led his forces into Liguria to assist the proconsul conducting operations there. The tribune demanded that the senate order Nasica to reassemble his troops and continue fighting; he could celebrate his triumph as proconsul.[105] Despite the opposition, the senate voted Nasica his triumph at once, but the tribune's demand for a delay was completely out of keeping with any need the soldiers may have had to return to their fields that fall. Nasica's year as consul would have ended at a seasonal date of mid-November in this period, well past the date when much of the planting for the following year ought to have been completed.

Other cases are no more indicative of a concern to return the men to their agricultural duties. In 186 the consul Q. Marcius Philippus dismissed his forces apparently in midsummer, but only to conceal the extent of the defeat he had suffered.[106] Three years later the consul M. Claudius Marcellus disbanded his legions and returned to Rome to hold elections. If, as was usual in this period, these fell toward the end of the consular year, then Marcellus sent his men home around mid-October or early November, rather late in the planting season.[107] In a curious episode in 180, a military tribune dismissed the second legion during his month in command. Much here is unclear, but this probably occurred over the winter of 181/0, between December and March.[108] If so, however, the reaction to it shows that it was anything but expected or customary. The consul Postumius raced to Pisa and vigorously pursued the soldiers in an effort to round up those he could catch. The senate ordered the others to return to the standards. If they did not, they were to be seized and sold along with their property. The senate punished the rest by depriving them of half a year's pay, even though they had only followed the tribune's orders.[109] The severity of the consul's and the senate's reaction is difficult to comprehend if furloughing the legionaries during the winter months was standard practice. In 170 the consuls dismissed their legions sixty days after taking the field, but only because there was nothing for them to do: the Ligurians had refused to take up arms.[110]

On the other hand, references to troops spending the winter in camp crop up with some frequency in the sources for this period. While the consuls of 170 sent their Roman troops home early in the campaigning season, the Latin allies went into winter quarters.[111] Something similar happened two years later inadvertently when the consul improperly proclaimed a day for his legions to assemble in Gaul. Consequently they remained at Rome, and he was compelled to spend the winter encamped with the Latin allies alone.¹¹² And in 182, as the consular elections were drawing near, the senate ordered one of the consuls to dismiss his legions and return to Rome, but instructed the other to winter with his forces at Pisa, because rumors of trouble in Gaul were circulating, and the possibility of an invasion loomed.¹¹³ For these soldiers, then, there would be no plowing that fall. Likewise for the consular legions of 178. Late that summer A. Manlius Vulso led his forces against the Histrians. The enemy humiliatingly routed the Romans from their camp, after which Vulso's colleague M. Iunius Brutus joined him with his legions. The two consuls spent that winter in camp whence they led their forces out against the enemy early the following spring.¹¹⁴ It is difficult to believe that after Vulso's defeat the consuls then sent the bulk of their legionaries and allies home and prepared to defend their encampment that winter against the possibility of another attack with only the skeleton force that remained. Note, too, the legions of the consul of 173, M. Popillius Laenas, which were sent into winter quarters after a victory over the Statelliates.¹¹⁵

The senate's frequent practice of ordering the new year's consuls at the beginning of their year in office to dismiss the previous year's legions and enroll new forces is also suggestive in this context.¹¹⁶ If soldiers usually obtained furloughs to return to their farms in the fall, this pattern meant that they made the journey home and then, after completing their planting, traveled back to the legions in the winter or early spring only to be sent home again shortly thereafter. The problems inherent in such a practice are obvious, not the least of which would have been convincing soldiers to return at all if they believed there was a good chance they were going to be dismissed as soon as they arrived. The seasonal date of the Ides of March between the years 190 and 168 ranged from November 19, 191, to January 5, 168, and so in many cases the senate would already have made its legionary dispositions for the coming year by the time the soldiers had finished planting and been starting their journey back to camp.¹¹⁷ One might therefore have expected the senate to have made its decision about enrolling new forces earlier in the fall and so saved the furloughed soldiers the bother of returning to their units when these were going to be disbanded anyway.

To be sure, Livy often notes that commanders ordered their legions to assemble at a particular time and place at the beginning of the campaigning year. But it must not be thought that such notices reflect the furloughing of soldiers over the winter with orders to return to the standards by a specific date. This was how Romans' customarily moved newly levied troops to their intended theater of operations.[118] Other cases, though clearly referring to legions from the previous year that the senate had assigned to a new commander, simply reflect the Romans' standard method of moving legions through friendly territory from one place to another.[119] This was how, for example, the consul Sempronius, who had been summoned with his army from Sicily to Gaul in September 218 to deal with Hannibal's invasion, managed the task: he had his troops at Regium swear an oath to be at Ariminum on a date forty days hence.[120] Likewise, the consul of 193 ordered the prior year's urban legions to be at Arretium in ten days. But these men were at Rome, not dispersed among their farms, for the veterans among them appealed to the tribunes for exemptions from service.[121] And when in 168 the consul Licinius inauspiciously proclaimed a day for the urban legions to assemble in Gaul, the troops refused to obey the order. But these legionaries spent the year at Rome, not in their homes, as one would expect if they had been dismissed for the winter.[122]

The evidence of the annalistic tradition for the second century, like that for the Hannibalic War, therefore gives no reason to believe that after 218 generals' dismissals of their legions in the fall regularly solved the problem that the conflicting demands of war and agriculture presented for Rome's soldiers. This finding, in turn, is critical, for it means that the same type of conflict that armies faced in the later fourth and third centuries—indicated by the evidence for overwintering in the literary tradition and the preponderance of late dates for triumphs preserved in the *fasti*—cannot be explained away on the assumption that furloughs in the autumn obviated the problem. Soldiers before 218 as afterward often if not usually remained with the standards throughout the fall and into the winter until they either triumphed or renewed their operations under new commanders. Extended terms of military service, whether in Italy or abroad, posed the same dilemma for small farmers before Hannibal's invasion as they did afterward.

Moreover, general considerations militate strongly against the notion that commanders would ever have regularly sent their soldiers home on leave with orders to report back to their legions' winter encampments once their planting was done. Offering agricultural furloughs to soldiers on campaign was a formula for organizational chaos. In the years when Rome's territory was fairly small and the Romans warred principally against near neighbors, the times for plowing and planting might be roughly the same for all the citizens, so that their mobilization and demobilization could be fairly easily coordinated with the times when they would be needed on their farms. The matter became much more complex however when Rome drew manpower for the legions from citizens and allies from the entire peninsula, from Apulia and Lucania to northern Italy. Generally, farmers in colder, wetter regions had to begin their work earlier in the fall, whereas those in warmer, dryer areas could wait longer. Planting time correlates not only with latitude but with elevation as well. The climate in mountainous regions differs sharply from that in adjacent plains. Simply arranging military obligations so that they coincided with agricultural seasons in Latium would be too late for some soldiers while others might return home weeks before they were really needed. But topography and the elements introduced further complexities. Partible inheritances, dowries for daughters, and a tradition of fairly modest holdings to begin with may in many cases have combined to break up a farmer's land into a number of small, scattered parcels.[123] Variations in the soil qualities of each and their specific microclimates will have led to corresponding variations in the times when each field needed to be plowed and sown. More unpredictable was the weather. Wheat and barley need water to germinate; thus sowing ideally follows the first rains of the fall that supply the necessary moisture to the soil. However, too much rain could asphyxiate the seed or young plants or wash them away altogether.[124] Hence successful sowing was to a significant extent a matter of guessing what the weather would do, and this also introduced a measure of uncertainty into the timing of the sowing. Similarly, wheat needs a minimum of about 38–39 degrees Fahrenheit (3–4 degrees Celsius) to germinate. Hence "sowing times have to be regulated by anticipation of the onset of winter temperatures."[125] Again, more guesswork was required of the farmer. The tendency was to sow early, therefore, but sowing too early could lead to too much plant growth before the onset of winter and hence damage by cold weather.[126] The uncertainty of successful planting also meant that farmers had to be prepared to replant; a second crop might have to be sown later that fall or even in the spring. Finally, no subsistence cultivator could afford to grow only one crop since its failure would mean starvation. Instead, small farmers normally sought security in a variety of crops.[127] Emmer wheat (*far*) probably formed the basis of most Italians' diets in the third and second centuries, but smallholders would have cultivated a mix of field crops that included some combination of barley, millet, panic, oats, and spelt as well as various legumes in addition to emmer and possibly other types of wheat. All of these might require somewhat different culture: barley for example was sown before wheat and emmer before unhulled types of wheats.[128]

Thus although the total number of days of work involved in plowing and sowing their fall crops might for some smallholders be quite modest in the aggregate, these could extend over several weeks or months because of the topographical characteristics of various pieces of land, the requirements of the crops being sown, and the vagaries of the weather.[129] Accommodating the agricultural responsibilities of his men therefore did not simply mean a Roman commander could dismiss them for a couple of weeks while they did their planting and returned, even if he imagined that they all needed to perform this task at about the same time. No smallholder could reasonably expect to get his crops in and ensure food for his family for the coming year if he was kept away from his fields most of the fall. They all will have wanted to return home as early as possible in order to be ready to plant at the right moment and to stay there as long as they could to be sure that their crops were well established before they returned to their winter encampments.

To accommodate the planting schedules of an army drawn from nearly the whole of Italy, a general could have dismissed piecemeal the various contingents from different regions one after another at whatever times seemed appropriate for plowing and sowing there (if these could be predicted with any degree of accuracy), thus dribbling away his manpower in reliance on nothing more than the hope that no delay would supervene to prevent a timely return. The soldiers would naturally have been bound by oath to return to the army by a specific date, but no commander could feel much confidence that men faced with a choice between completing their work in the fields and thereby ensuring that their families would have food for the following year or returning to the army and leaving their families to face possible starvation would elect the latter course, particularly when loopholes abounded.[130] Few generals, to say nothing of the senate, would have relished the prospect of seeing a large proportion of troops go on a kind of indefinite leave with the onset of the fall (even if the difficulty of keeping track of all their comings and goings could have been overcome). The *patres* were dismayed at rumors in 170 that large numbers of soldiers were absent from the Macedonian army *incertis commeatibus*.[131] The problem was compounded by the fact that the Roman portion of a general's forces fought not in an undifferentiated mass but in small, highly organized maniples. Because the members of these were not all drawn from the same geographic areas, allowing soldiers to go home when they felt that it was time to plant in their particular regions would have led to a situation in which the strength of the maniples at any given time would have varied widely, with some close to their full complement and others depleted of manpower. That in turn would have made their effectiveness in combat highly uneven. Even among the allied cohorts, which were conscripted on a regional basis, variations in local topography and microclimate will have meant that not every member was ready either to leave or return at the same time.

The alternative was simply to dismiss the entire army with the onset of fall, a course that seems a formula not only for enormous confusion but for strategic paralysis as well. As discussed earlier, the principal target in much of ancient warfare was an enemy's agricultural base, particularly the grain crops. These were particularly vulnerable not only at the time of harvest but also during the sowing. To deny an enemy the opportunity to plant was tantamount to destroying the crop at the point of harvest. But although farmers might be reluctant to plant late, sowing could in fact take place over a fairly lengthy span of time. This fact had made preventing sowing altogether a difficult task for armies made up of soldiers who paid their own expenses on campaign; as noted already, these men could not afford to stay away from their homes long enough to deny their enemies complete access to their fields during the planting season. However, the advent of payment for military service had solved this problem. Armies in fourth-century Greece, which relied heavily on mercenaries for their manpower, regularly campaigned through the fall and into the winter. Their greater staying power in turn gave to the wars they waged a greatly increased destructiveness directed chiefly against agriculture. At this point in Greece, interruption of planting became an easily available strategic option for generals as did inflicting serious damage on trees, vines, and farm buildings.[132]

Rome rarely in this period fought its wars with mercenaries.[133] But the republic did pay its citizen-soldiers in the third and second centuries, enabling them to buy food, and it also provided rations for the allied contingents. This subsidized commissariat and the sophisticated logistical system that supported it gave the Romans a far greater ability to pillage and destroy an enemy's agricultural establishment than the hoplite forces of the early republic had had.[134] The manipular legions of this era were particularly well adapted for ravaging and pursuit: more than a quarter of a legion's infantry manpower was light-armed. We do not know what percentage of soldiers in allied cohorts was light-armed, but if the proportion approximated that of a legion's, then this combined force along with the 600 Roman and 1,800 allied cavalry of a two-legion consular army gave a Roman commander the ability both to harry an enemy population attempting to work its fields and to pillage its territory.[135] That the Romans realized this potential seems clear from the discipline and effectiveness with which they carried out these tasks.[136] The clearest example of the legions' capabilities in these types of operations is to be found in the devastation that Roman armies operating in Campania between 215 and 213 inflicted. By 212 the Capuans had become unable to feed themselves because they had been prevented from sowing, and they were reduced to begging food from Hannibal.[137] Although these events took place late in the third century, no reason exists to suppose that earlier armies, which could be sustained for months in an enemy's homeland, would not have been capable of laying it waste on a similar scale. Sending the soldiers home in the fall to attend to their own farms, therefore, will have meant forgoing the ability to prevent an enemy from sowing its crops.[138] To do so, given the labor and difficulty of damaging other agricultural elements, will in many cases only have increased the duration and arduousness of campaigning and made it easier for the enemy to avoid marching out and fighting a pitched battle to protect its food supply. The republic's generals, therefore, had every reason to avoid a hiatus in their operations during the fall.

When Rome confronted mercenary armies, as in the war against Hannibal in Italy (or for that matter a generation earlier against the Carthaginians in Sicily or even earlier against Pyrrhus), dismissing troops for extended periods became impossible, for even though Carthaginian forces regularly went into winter quarters, their soldiers had no farms to return to. They remained ready to take the field if an opportunity arose, and for the Romans to have furloughed their forces on a large scale would have been to court disaster by conceding the strategic initiative to their opponents.[139] Such a decision would have exposed the Romans and their allies to attacks precisely when their own agricultural base was particularly vulnerable, during plowing and sowing for the following year's crops, and raised the specter of food shortages in the coming year. Conversely, they will have forgone the opportunity to visit similar attacks on their enemies.

Finally, amid all the preoccupation with the availability of manpower for planting in the fall, one must not lose sight of the other critical period in the agricultural cycle, the harvest. This was unquestionably the most labor-intensive task in a farmer's year: with the greatest part of his family's food supply for the coming year on the line, there was an understandable desire to get the crops in as quickly as possible once they were ripe.[140] The harvest therefore generated the year's peak labor demand, and all hands turned out to help.[141] Urgency over this task in classical Greece prior to the rise of mercenary service in the fourth century had represented an important constraint on the rhythms of warfare. Invaders' eagerness to return home to bring in their own ripening crops could limit the duration of incursions, while preoccupation with completing the harvest could delay the start of a campaign.[142] Again, there is no reason to presume that the situation was significantly different in the conduct of Rome's wars during the fifth and most of the fourth centuries. By the third century, however, military service regularly took men away from their farms during this crucial season. In 263 four legions were in Sicily at the height of the harvest.[143] With the outbreak of hostilities against Hannibal in 218, the consuls' armies appear to have been levied and ready to march by June or July of that year, when a revolt in Gaul forced the consul Scipio to dispatch one of his legions north to deal with the emergency.[144] Likewise, Hannibal destroyed the consul Flaminius's army at Trasimene on June 21 of the following year.[145] In 216 when Hannibal left his winter quarters, the wheat was ready to harvest in Apulia. Up until that point, the Roman army had remained encamped opposite the Carthaginians forces. When news reached Rome, the senate resolved to give battle but ordered the proconsuls not to engage Hannibal until the consuls arrived with additional troops.[146] And in 201 Gauls attacked a Roman force bent upon gathering in the enemy's ripe grain.[147]

In all the evidence that we possess, there is nothing to indicate that the exigencies of war ever yielded to the demands of agriculture for labor to bring in the harvest. For how could they? Although the duration of the harvest season in Italy is shorter than that for planting, it still can extend for a considerable length of time, from early June to early August, depending on the region.[148] In addition, variations in the weather could retard or advance the ripening of wheat and other crops from year to year. Even within a particular region, individual farms comprising several scattered fields may not have seen all their crops ready to harvest at once. Indeed, staggered harvest times can be an important mechanism for coping with a limited labor supply at a time of maximum demand. Yet the strategy of devastation required that farmers leave their fields in order to attack the enemy's. Trying to assemble an army from across the whole of Italy for this task while allowing farmers time to deal with their own harvests would have presented insurmountable problems. Again, the demands of war and agriculture lay at cross purposes.

The findings of the preceding discussion pose a serious challenge to the theory that the rise in landlessness evident at the time of Tiberius Gracchus's tribunate resulted from the introduction of year-round military service in the late third or early second century. No evidence supports the notion that before the Hannibalic War soldiers usually served for only a few months each year during the summer slack season on their farms. Year-round warfare had become the norm by the later fourth century. Nothing warrants the presumption that furloughs in the fall and winter accommodated the soldiers' need to return to their farms to see to the autumn planting. Still less did the republic make any attempt to allow men to return for the harvest in the early summer. And both the strategic design of Roman warfare with its emphasis on agricultural devastation and the geographical distribution of the coalition upon which the city drew for its military manpower argue strongly against such practices. If all this is so, then the wars the republic undertook at the close of the third century when Hannibal invaded Italy and subsequently in its conquest of the Mediterranean did not impose on small-scale agriculture any *qualitatively* different burden from that which Roman expansion had previously entailed. To be sure, the number of men under arms after 218 was often dramatically higher than had been the case earlier, and the implications of this increase are examined in the chapter that follows. Yet in the conventional view this change did not make the critical difference for farmers. Instead, where and for how long soldiers fought during and after the Hannibalic War are supposed to have created new and onerous obstacles to the viability of Italy's small farms. But that notion is illusory. Continuities rather than change mark the relationship between war and agriculture from at least the late fourth century down through the second.

The conflict between the imperatives of Roman warfare and the needs of small farms therefore arose much earlier and endured far longer than is generally supposed, and its appearance at least two centuries before agrarian problems became acute dramatically weakens the case for any causal link between them. And absent that temporal conjunction, little supports a connection between the two. No direct evidence, after all, explicitly states that the conscription of smallholders for extended periods of military service made it impossible for those left behind to work their farms. Certainly several ancient authors mention military service among the grievances of the poor that led Tiberius Gracchus and other reformers to act. Sallust reports that the *populus* was weighed down by *militia atque inopia*; Plutarch notes that those expelled from their land no longer presented themselves for the levy; and Appian states that the Italians were worn down by "poverty, taxation, and military service."[149] But only within the framework of the conventional claims about the deleterious impact of long-term conscription on the operation of small farms do these passages acquire probative value for this hypothesis. When they are read with that idea already in mind they naturally confirm it. But in and of themselves they do not require that hypothesis in order to be comprehensible. Other interpretations are equally if not more attractive. By 133 military service over the preceding two decades certainly had grievously burdened Romans and Italians, not because their farms were failing in their absence but because of the nature of the wars soldiers were being sent to fight. Combat in Spain since 153 had been hard and unprofitable. Defeats were frequent, and the loss of life there often heavy. Conscripts were notoriously reluctant to go, just as Plutarch reports, and morale in the army was at rock bottom.[150] The immediate military context, not a general structural flaw in the Roman military system, adequately explains these references.

A handful of passages referring to conditions in the fifth and fourth centuries is sometimes taken to reflect the dire consequences of long stints of military service overseas for small-scale agriculture in the second and first centuries, which annalists writing at that date then anachronistically wove into their accounts of the early republic.[151] Yet these texts do not bear out the claims made upon them.[152] Rarely do they attest the failure of farms because their cultivators were away at war. They focus instead on the problem of debts incurred by soldiers as a result of having to pay the *tributum*.[153] That tax, however, was suspended in 168 following the Third Macedonian War and not reimposed officially until 43.[154] Consequently, it can have played no part in the economic difficulties smallholders were experiencing in the thirty-five years prior to the tribunate of Tiberius Gracchus. Given the generally successful run of wars in the preceding thirty-two years, it is difficult to believe that the financial burdens that the republic's smallholders had to sustain in that period had plunged many of them deeply into debt. The tax was not levied every year, and it sometimes could be repaid if a war brought money into the treasury. Nor would the tax have been particularly onerous for the lowest census categories even if the usually accepted rate of one *as* per thousand is correct. That amounts to a mere 4 *asses* for an *assiduus* in the fifth class with property rated at 4,000 *asses*; 25 *asses* for a citizen in the fourth class worth 25,000. For the sake of comparison, a legionary's pay in the second century was 3 *asses* per day.[155] More important, the *tributum* was a kind of transfer payment owed by those *assidui* who did not serve to those who did.[156] Military service, in other words, exempted conscripts from the obligation to pay the tax. Hence, reports in our sources that soldiers in the fourth and fifth centuries were falling into debt because they could not pay the *tributum* correspond to nothing in the second century. Possibly these accounts preserve some echo of the situation during the worst of the Hannibalic War, when the republic's military crisis was exacerbated by a financial one.[157] But if so, the latter eased considerably after 210, and the sixty years following the end of the war ought to have provided ample time for smallholders to recover.[158]

Only once does a text explicitly link military service with the inability of soldiers to cultivate their fields, and that is in connection with the war against Veii in the late fifth century.[159] As Romans collectively recalled this event much later, the siege of Veii involved a continuous blockade of that city lasting as long as ten years.[160] Given this tradition, one might suspect that some annalist simply assumed that agriculture must have suffered in consequence. He would scarcely have had to have witnessed this occurring among his contemporaries to have reached such a conclusion. Because Livy makes reference to untended fields only in passing, in a passage reporting an inflammatory speech by tribunes of the plebs complaining of the hardships the commons were enduring on account of this and other wars, this passage may represent the historian's own contribution. And as one might expect, the plebeians' ills arise primarily from debt; agricultural decay is hardly a central theme in the speech. On the other hand, mention of the latter problem need not be pure invention. The war against Veii seems to have presented a far greater challenge to the Roman Republic than any it had previously faced. Quite possibly, the city undertook a major restructuring of its military forces in response, altering the basis of recruitment. Pay, too, may have been instituted for the soldiers at this point along with a tax to fund it in view of the length of time they were now expected to spend at war.[161] The conflict thus is likely to have placed new and different burdens upon the citizens, and for that reason complaints reported in the sources about the hardships that these entailed need not be dismissed out of hand. By the same token, difficulties among the soldiers in making the transition from brief campaigns to conducting a long drawn-out struggle would hardly be surprising. Without much prior ex-

War and Agriculture

perience in conducting operations of this type, problems were bound to arise as men sought to reconcile their agricultural responsibilities with the longer terms of service that the city was now asking of them. The important point is that no evidence furnishes any reason to see such problems, if they occurred, as anything more than the temporary result of a period of transition and adjustment.

Still, despite the lack of evidence, one might suppose that any time the household on a farm sent a man to war those left behind must have fallen into debt because they would have had to borrow money simply to live until the soldier returned.[162] However, that assumption begs the very large question of whether the remaining members of the family would have been able to keep a farm going for several years in the soldier's absence. As the following chapter shows, a negative presumption is unwarranted, at least in the great majority of cases. To be sure, common sense might make this contention seem self-evidently wrongheaded, but common sense misleads here. What comparative evidence can be mustered, indeed, points to the opposite conclusion. Athens in 415 sent some 1,500 of its citizen hoplites to Sicily for what was clearly expected to be a lengthy campaign. In the extensive debate Thucydides composed to represent the pros and cons of undertaking the Sicilian expedition, neither Nicias nor Alcibiades utters one word alluding to any economic hardship that the departure of these soldiers for many months might entail for their farms.[163] Alexander in 334 took half of the Macedonian heavy infantry—smallholders all—with him for an expedition that was to last years, yet there is no indication that their farms went to wrack and ruin in their absence.[164]

It remains possible that some other change in the conditions of military service during or after the Second Punic War led small farms to fail in large numbers. One might imagine, for example, that while year-round campaigning was usual before the Second Punic War, farmers in that era went to war only for a year or two at a time. After that their continuity of service ceased, and when they returned to their farms, conditions there had not deteriorated to the point where they could not easily be made good, particularly since soldiers often came home with money in their purses from booty and donatives to pay off debts and fund restoration.[165] But the practice of enrolling soldiers for far lengthier hitches, begun during the war with Hannibal and continued afterward, caused farms to decline far beyond the ability of soldiers to restore them upon discharge, even with the profits of a successful war. Thus the duration of the time that Rome took conscripts away from their farms rather than the conflict itself between military and agricultural rhythms constituted the critical change that turned a relatively benign situation prior to 218 into a crisis in the second century.

However, nothing in the record of Rome's third-century wars indicates that prior to Hannibal's invasion soldiers served for only a year or two and then returned to civilian life for a number of years before being called up again. The senate's practice in the second century of keeping legions in being for several years at a time before discharge would lead one to expect that a similar pattern obtained in the third.[166] Magistrates in the second century at times fought to avoid having to take over an army of *tirones*, being well aware of the greater military effectiveness of seasoned troops, and we have no reason to suppose that those who held office a century earlier would have been ignorant of this fact or any less eager to lead veteran troops into battle.[167] Even when armies were discharged in the second century, veterans were often immediately conscripted to serve in new ones being formed.[168] Again, in the absence of explicit evidence to the contrary, the presumption must be that recruiters in the third century likewise preferred to levy men who had served before. Consequently, a conscript in the third century would almost certainly have faced the prospect of several years of continuous service rather than a brief episode at war followed by a lengthy return to civilian life.

A more critical change is sometimes thought to lie in the recruitment of soldiers from a significantly poorer class of farmers after 215 than before. In this reconstruction, only about half of all *iuniores*, that is, men between the ages of seventeen and forty-six, prior to that date were wealthy enough to qualify as *assidui*. The remainder were *proletarii* and so exempt from legionary service, although they could be drafted as rowers for the fleet.[169] But the shortage of recruits that Rome faced during its long duel with Hannibal compelled the republic to lower the threshold for enrollment among the *assidui*, and the farms from which these new legionaries came were far less able to get along without their labor than the larger establishments that had sent men to the republic's armies prior to the Second Punic War.[170] But serious problems vitiate this hypothesis. Certainly, the censors of 214 found only 2,000 men on the rolls of the *iuniores* who had failed to present themselves for service in the legions prior to that date without a legitimate exemption or the excuse of ill-health. This fact, however, does not require us to conclude that because Rome had conscripted about 98,000 men between 218 and 215 to serve in the legions out of a citizen population of men under forty-six numbering about 210,000, the remaining 110,000 must have been *proletarii*.[171] *Proletarii* to man Rome's warships as well as *assidui* for its legions were in desperately short supply throughout the Second Punic War, something difficult to account for if half of all citizens were eligible for naval service.[172] Nor does Livy say that those who failed to serve had been exempted because they were *proletarii*, only that they had been granted *vacationes* or were excused because of their health. The most natural conclusion to draw, therefore, is that these reasons, not their poverty, allowed the men in question to avoid conscription.[173]

In addition, the senate's reduction in the minimum amount of wealth required to be enrolled among the *assidui* (which more likely dates to 212 or 211 than 215) probably dropped the threshold only about 27 percent in real terms, from 11,000 to 8,000 ounces of bronze.[174] Men who found themselves eligible for the draft in the wake of this change would not have been a great deal poorer than those conscripted before it, if they were in fact poorer at all. For the monetary value of an individual's property was to a considerable extent notional.[175] No Italy-wide market in small parcels of land such as a subsistence farmer might have owned is likely to have existed in the third century or for most of the second to serve as a mechanism for setting prices. And it is inconceivable that the censors — like modern tax assessors in the United States — would have determined the monetary value of such properties objectively and independently. Rather, a *pater familias* declared his property to the censors, who then rated him in one of the census classes.[176] A subsistence farmer, whose land represented nearly the entirety of his wealth, thus enjoyed a certain leeway in the matter — for who would bother to check, after all? — while the censors' decision about the monetary value of the property he declared must have been to some degree arbitrary as well.[177]

What monetary wealth in this amount represented in real terms is not easy to say; however, those from whom the republic drew its *assidui* prior to 218 by no means represented only the most affluent of its citizens. Although no source provides an explicit statement of the minimum amount of land that qualified its owner for enrollment in the fifth census class, a number of texts show quite plainly that this threshold could have been met by possession of a small farm of no more than seven *iugera* (about 1.75 hectares or 4.4 acres) and sometimes as few as two.[178] Citizens who participated in maritime colonies founded before the Second Punic War received allotments of that size and yet enjoyed a formal exemption from military service. This can only mean that other Romans with comparable farms regularly served in the army, for otherwise the colonists would have been *proletarii*, and no exemption would have been necessary.[179] Farms on this scale are subsistence operations, no more, and their owners can scarcely be described as rich. Slave ownership, which one might suppose enabled citizens in many cases to absent themselves from the work of running their farms for long periods at war, certainly did not extend to such men. Our evidence for the extent of slaveholding in this period is meager, to be sure, but the one text that bears directly on the problem identifies only those in the third census class and higher as possessing slaves in 214.[180] Indeed, the very notion that a large farm is better able to bear the loss of its cultivator than a small one is implausible on the face of it, for the former naturally requires overall a greater input of labor than the latter and makes its replacement that much more difficult.[181] Slaves may often have substituted for a master when he went off to war, but it is well to remember that unfree laborers require supervision. The only other slave society to go to war on a scale comparable with the Roman wars in the late third and second centuries was the American Confederacy in 1861. It mobilized perhaps 75 percent of its free adult male population during its attempt to win independence, and without question slave labor was crucial to sustaining its war effort.[182] Among the very few categories of exemption granted to white males was the one allowed for owners of twenty or more slaves, on the grounds that a free man's supervision was necessary to keep them in line. None was given to smallholders who were the sole supporters of their families.[183] Slave ownership and large properties, in other words, may often have made removing a man from his farm more difficult, not less.

If Rome was drawing its conscripts from the same class of small, subsistence farmers in the later fourth and third centuries that it recruited them from in the second, and if multiyear terms of year-round warfare were the norm both before and after the Hannibalic War, then the absence of an agrarian crisis prior to 218 poses a serious obstacle to the claim that these practices were at the root of the one that developed afterward. Admittedly, this is an argument from silence, but widespread landlessness and its attendant social and political tensions ought to have left some mark even in our sources, meager though they be for much of the third century. However, a critic might counter that in fact just such problems were occurring. Agitation for viritane distributions (that is, distributions to citizens individually rather than as members of a colony) of recently conquered lands in the *ager Gallicus* occurred in 232, and one might suppose that among those clamoring for allotments were many former soldiers whose years with the legions had deprived them of their farms.[184] Perhaps, therefore, the reason that we otherwise see so little evidence of the hardships that long-term military service was causing among Italy's smallholders in this period is because the city's extensive colonization in these years kept a lid on the problem. Distributions of recently conquered land would have enabled farmers ruined by conscription to make a fresh start and so prevented their distress from giving rise to serious unrest. The pivotal change came, however, once the safety valve that colonization offered ceased to operate around 181. The far greater numbers of men that Rome was levying for its conquests in that era transformed what had been a manageable problem of landlessness arising out of military service into a crisis.[185]

However, this reconstruction raises more questions than it answers. The republic's whole system of conscription was predicated upon the willing compliance of recruits and scarcely workable without it.[186] No preindustrial state possessed the elaborate police force and other instruments of coercion that would have been required to compel unwilling citizens to come forward for a levy, and Rome was no exception.[187] Yet the senate dispatched colonies infrequently and largely in response to the demands of military security.[188] Latin colonies, which account for the bulk of land distributed through colonization, number only thirteen in the eighty years down to the eve of the Hannibalic War. Four viritane distributions of land occurred during the same period.[189] Consequently, a ruined soldier and his family might wait several years before being able to obtain an allotment of land. In the meantime they surely will have been hard pressed in many cases to make ends meet. And the remedy that colonization afforded will have been available only *after* a farmer had completed his military service, whereas the shortage of labor this created meant hunger in his household immediately and continuously while he was away with the standards.[190] Thus, in exchange for his family's present hardship and possible starvation, a man called to war could solace himself only with a prospect of redress some years down the road. Why, then, would a prosperous small farmer willingly go off to war for several years if this entailed the very real possibility of economic catastrophe during his absence, even if he viewed his impoverishment as only temporary, to be reversed in a few years by a new parcel of land and a fresh start in a distant colony? Why should a man want to put himself and his family through such an ordeal?

One would expect considerable resistance to conscription and a reluctance among ordinary Romans to undertake expansionist wars if they saw any prospect that they would wind up landless and in poverty as a result. Yet nearly everything we know about attitudes toward war in the third century and indeed throughout much of the second indicates exactly the reverse: Rome's citizen-soldiers were generally willing and often enthusiastic participants in their city's imperial adventures. In 264, for example, voters in the *comitia* eagerly accepted the Mamertines' call for aid because they hoped for booty from the war despite, Polybius tells us, being "worn out . . . by recent wars and in need of any and every kind of restorative."[191] Likewise, in 200, voters in the *comitia centuriata* were ultimately persuaded to declare war on Macedon despite their exhaustion after eighteen years of war with Hannibal.[192] Those few cases in which ordinary Romans evinced any reluctance to serve were due to the hard, unremunerative combat they anticipated, not to any fears that their farms would fail and their families starve in their absence.[193] There is good reason to doubt, therefore, that colonization could have made workable a military system that was regularly impoverishing large numbers of its participants.

Further, it cannot be maintained that we hear nothing of the landlessness that third-century military service was causing because the extent of the problem was quite limited. One might estimate the number of Romans who served in the legions between the outbreak of the Third Samnite War in 298, when year-round warfare seems to have become regular (if it was not so before), and the eve of the war against Hannibal in 218 at roughly 240,000.[194] We do not know how many Italian allies Rome called upon to fight alongside its citizens in these years, but we may guess that their number will not have been much smaller than that of the Romans for a total of about 480,000 men. How many settlers Rome dispatched both to Roman and Latin colonies is likewise a matter of conjecture for the most part, but estimates are in the range of 62,600.[195] We are equally ill-informed about viritane assignments of land in this period, but a reasonable guess might be on the order of 18,000 to 27,000.[196] The resulting total, 80,600 to 89,600 settlers, seems scarcely sufficient to accommodate the soldiers who will have needed land if year-round military service prior to 218 was in any serious way impeding the successful operation of their farms and causing a high proportion of them to fail. Colonial and viritane distributions will have had to absorb not only these ruined soldiers but also men rendered landless or left with inadequate holdings from any other cause, including recently conquered *socii*, portions of whose territory Rome had confiscated, the poor and indigent both at Rome and among the rest of the allies, sons whose inheritance portions might be inadequate to support a family, and even Roman freedmen.¹⁹⁷ So at most only a few ex-soldiers can have availed themselves of this remedy. Consequently, if colonial allotments were palliating the landlessness that long service with the legions was causing, this would imply that very few farms were failing to begin with.

Colonization unquestionably played a vital role in Roman imperialism, building the city's *propugnacula imperii* in Italy, helping to reconcile *socii* to Rome's hegemony, and in some cases compensating them for lands taken from them as the price of defeat. It mitigated economic distress both in allied communities and within the republic itself and so obviated the political conflicts to which this otherwise might have led. But, as a way of salvaging the consensus view, the hypothesis that colonization in the third century was what prevented military service from causing the kind of economic hardship among smallholders it did subsequently entails serious difficulties. It cannot account for the willingness of citizens and allies to participate in warfare that posed a serious risk of financial ruin and starvation for their families in their absence. Within the structure of Roman imperialism, colonization only makes sense as part of what motivated conscripts to fight if they saw going to war as something that would better their lot, not cause it to deteriorate. If one explains their willingness to go to war despite these risks by claiming that in practice they resulted in the impoverishment of only a small minority of those who served, then one is at a loss to understand how most managed to escape these consequences if war and agriculture were so fundamentally at odds.

These problems do not categorically rule out the end of colonization as the key change that turned the consequences of Roman imperialism for farmers into the agrarian crisis of the later second century. But they do invite a crucial question that the consensus view takes for granted: were midrepublican warfare and subsistence agriculture necessarily in conflict? Would Rome's manpower demands in these years have drained vital labor from small farms, as common sense might lead one to suppose? Or is it possible to develop a different model of the relationship between year-round military service and small-scale farming, one that can explain how the Roman Republic could mobilize its citizens and allies so regularly for so long and ultimately on such a vast scale without causing widespread economic ruin among those recruited into its armies? And if such a model can be constructed, does this account for the willingness of recruits to serve and the absence of evidence for discontent over landlessness in the third century and even through much of the second better than the hypothesis that colonization suppressed any problems as long as it lasted, with its attendant difficulties? To these questions the following chapter must turn.

Chapter 3

War and the Life Cycles of Families: Three Models

Mid-republican Rome was by no means the first Mediterranean power to wage war year-round. Conflicts in Greece had broken free of the constraints that the agricultural cycle imposed well over a century earlier, and it is instructive to consider how classical and Hellenistic states solved the problem of finding men to fight wars of this kind. As is well known, Athenian sea power depended largely upon the city's *thētes*, poor citizens without land. Men like these had few if any agricultural responsibilities, and utilizing *thētes* enabled Athens not only to launch naval expeditions at any time within the sailing season but to conduct lengthy sieges abroad.[1] Mercenary oarsmen, too, made a critical contribution to the naval strength of Athens, and mercenary soldiers likewise were at the forefront of the "military revolution" of late fifth- and fourth-century Greece.[2] Such men, often exiles from their homelands, had no farms that needed their labor. Their exclusion from the subsistence economy coupled with the wages they earned from war enabled them to take the field any time a paymaster beckoned.[3] And if the campaigns they fought took them far from home for years on end, that was no matter. Consequently, as the classical age gave way to the Hellenistic, military power on the Greek mainland, in the Near East, and throughout Sicily and Magna Graecia came increasingly to depend on mercenary armies and on the economic strength to fund them.

But a variety of factors foreclosed this option for Rome. Although the city's wealth in the fourth century has often been unduly minimized, it could scarcely match the commercial power of an Athens, a Syracuse, or a Carthage. Yet even these states often strained to raise the funds they needed to meet their military obligations. Nor did Rome in this period possess the vast financial resources of a great Hellenistic kingdom, which could draw upon taxes extracted from an extensive subject population to pay its soldiers. On the contrary, Rome had built up its limited hegemony in Italy to that point through cooperative alliances, eschewing the imposition of money contributions upon its *socii*. Lacking subjects to tax, therefore, the Roman Republic would have had no choice but to turn to its own citizens for the money to hire mercenary armies, and here a host of potential problems loomed. Political power at Rome was vested in the hands of a landowning elite whose ethos disdained commerce.[4] Yet the need for ready cash to finance wars might easily have led to the elevation of the commercial class upon whose wealth the city's defense would have to depend. Avoiding that prospect would have forced the owners of large estates to shoulder the costs of going to war. However, over the course of the fourth century a large part of the solution that the republic had gradually reached to the Struggle of the Orders involved the emancipation of plebeian *nexi* who worked as dependent laborers on the estates of the rich. But basing the republic's military strength upon the wealth that these enterprises could create would have required even greater productivity from them and so necessitated increasing the exploitation of dependent plebeians rather than ameliorating their lot. Likewise, this same need for increased productivity would have pressed the owners of large estates to expand their holdings at the expense of the poor, whose demands for land had constituted another cause of social and political unrest during the early republic. Moreover, the oligarchic character of the city's government made its aristocrats deeply suspicious of any single figure who might rise to undue prominence and power among them. Command of a mercenary army however had enabled more than one would-be tyrant to overthrow an oligarchy. Paid soldiers as the foundation of the city's military power therefore can scarcely have appealed to those of its governing class with an interest in perpetuating their collective rule. Lastly, the constant, severe foreign threats that the republic had faced during most of its first 200 years had by the late fourth century brought about the development of a strong military ethos among both its aristocracy and commoners that made a citizen army seem the obvious and only way to defend the city's interests.[5]

Transforming a citizen militia that fought brief campaigns close to home into an army capable of waging war year-round in distant theaters first of all required that the legions acquire staying power. As the preceding chapter noted, in the late fourth century, if not before, the city began regularly to subsidize its soldiers' costs while on campaign, paying its citizens a small sum out of which they could purchase food and supplying grain to allied contingents gratis.[6] Once the soldiers' dependence on their own resources to feed themselves ceased, the republic's forces could remain in the field almost indefinitely, at least in theory. But in an overwhelmingly agrarian economy made up principally of subsistence cultivators, this was only part of the solution. As the role of mercenaries in Greece shows, to wage war year-round required mobilizing men who could be removed from the civilian economy for the lengthy periods necessary without causing serious damage to it. Several scholars, noting that underemployment is characteristic of the subsistence sectors of agrarian economies, have seen in this a key factor in Rome's ability to mobilize so large a proportion of its citizens for war. But underemployment is not unemployment. Even though the labor required of smallholders for considerable stretches of the year might be light, at certain periods those demands skyrocketed and the work involved became critical to their survival. The kind of year-round warfare that Rome began regularly to wage in the late fourth or early third century took men from their fields not just during those times when their labor was not needed, but also at others, like the fall planting season, when it was vital. Consequently, the transformation of Roman warfare from an activity that fitted into the agricultural year's slack time to a year-round pursuit necessitated transforming the farmers who fought these wars from part-time warriors into full-time soldiers.

What, then, would have been the consequences for the farms they left? Would year-round warfare have had to lead to shortages of labor—and hence to hunger, crisis, and abandonment—on many of the farms from which Rome conscripted its soldiers, as conventional wisdom has long maintained? The preceding chapter has argued that little in the admittedly sketchy testimony for the republic's domestic affairs during the third century supports an affirmative answer to this question; hence the a priori assumption that year-round military service in the second century would have entailed dire consequences lacks foundation. But the same scarcity of evidence renders an argument from silence on this point inconclusive. In seeking a solution to this problem it is necessary to find some way to move beyond the limitations imposed by the paucity of literary texts that bear upon it. The chapter that follows attempts to do so by developing a model of small-scale agriculture capable initially of testing the hypothesis that the demands of Rome's year-round campaigning caused the small farms from which the city drew its soldiers to fail for want of their labor and, in the event of a negative answer, of explaining how the removal of so much manpower for extended periods could avoid this result.

One approach to the task of constructing a model of subsistence agriculture capable of illuminating these issues would be, first, to determine how much food the members of a typical farming family needed to raise in order to survive. One would then establish how much work they needed to perform in order to grow it, next approximate the total labor potential available to them, and finally try to ascertain what would happen when the state's demand for soldiers removed some of that manpower for extended periods of time. Put in these terms, it should at once be obvious that no simple answer to these questions can be offered. A great deal depended on both the number and sorts of people in the household, who represented both its labor potential as well as its consumption demand, and the productivity of the land they cultivated. Each of these will often have varied significantly from case to case. Moreover, as Chapter 2 noted, farming practices in ancient Italy could be quite diverse, owing to the varied soils, topography, and climates throughout the peninsula.[7] Differences in methods of cultivation therefore may also have affected both the sizes of the harvests produced and the labor inputs they required. The more one seeks for definitive answers to questions about how Roman and Italian smallholders farmed, the more elusive they become. Fortunately, however, the problem at hand does not demand them. A solution instead requires developing some way of ascertaining the point at which the removal of the labor potential represented by one or more able-bodied young men would become a threat to the economic viability of a farming family. To do so, we can examine various hypothetical models of families on subsistence farms to see which would experience a critical shortage of labor if it were to lose a man to the legions for a period of years. Various modifications to the models can then be considered, reflecting the real conditions under which small farms might have operated, in order to consider how these changes might affect the conclusions to which the models point.

Consider the following hypothetical family: a father, age fifty, mother, age forty, and three children, two sons, twenty and fifteen, and a daughter, ten. Their food requirements can be fairly well approximated based on the recommendations of the 1985 FAO/WHO/UNU report, *Energy and Protein Requirements*.[8] On the basis of these, a man between eighteen and thirty years of age and weighing sixty-five kilograms (143 pounds) would need between 2,990 and 3,530 calories per day depending upon his level of activity. The lower figure assumes "moderate work" and the higher "heavy work," two standards of activity that should encompass most of the tasks involved in subsistence farming.[9] An adult male between thirty and sixty years of age and weighing sixty-five kilograms[10] would require between 2,900 and 3,380 calories per day, whereas a woman aged from thirty to sixty and weighing fifty-five kilograms (121 pounds) would need between

2,170 and 2,400 calories per day.[11] Estimating the food requirements of children and adolescents is somewhat more complex, but for purposes of this exercise, that for a male fifteen years old can be set between 2,650 and 2,900 calories per day and for a girl aged ten, between 1,950 and 2,350.[12] The yearly caloric requirements of this hypothetical family would then be between 4,620,900 and 5,314,400 calories per year.[13] This estimate probably overstates somewhat their food requirements because the weights assigned to the family members—which are critical in the formula for determining a person's daily energy needs—may be too high for antiquity.[14] That fact, however, should not matter for the present purposes, for we are looking for a worst-case scenario: the point of this exercise is to create a model of subsistence farming that can accommodate a loss of manpower to military service for extended periods of time. Consequently, using a higher demand for food and therefore a higher demand for labor will only increase its validity, for if actual weights ran lower, less food will have been required and so less labor to produce it, and the farm will have been that much more able to get by with the loss of one or more sons to the legions.

Determining how much labor such a family would have had to expend to produce the food it needed is a complex matter, but for present purposes we can make a couple of simplifying assumptions. First, let us suppose what is unlikely to have been the case in practice, that the family's diet consisted entirely of wheat. The consequences of using a more realistic diet for the model are considered later. The average caloric content of wheat is about 3,340 calories per kilogram.[15] A family that needed 4,620,900 to 5,314,400 calories per year to sustain itself therefore would have required 1,384 to 1,591 kilograms of wheat per year.[16] How much work, then, would have been required to produce this amount of wheat? An answer involves estimating both the productivity of the land they farmed and the amount of labor required to cultivate it. The only figures we have from Roman antiquity on yields of wheat are expressed in ratios of grain sown to grain harvested, and these are highly problematic.[17] However, because we are searching for the point at which removal of a significant portion of the family's labor supply would render it unable to sustain itself, a worst-case scenario can again be postulated, one that assumes a fairly heavy demand for labor under normal circumstances and so places the model closer to that point to begin with. Therefore, let us make a second simplifying assumption and use a 1:3 yield ratio, since Columella, who offers the most pessimistic of the yield estimates, claims that, "We can scarcely remember a time when over the greater part of Italy wheat returned a fourfold yield."[18] The standard sowing rate given in the agricultural writers for wheat is five *modi* per *iugerum*, and one *modus* of wheat weighs, on average, about 6.65 kilograms.[19] Thus this sowing rate works out to 33.25 kilograms of wheat per *iugerum*.[20] If one assumes a 1:3 yield, 33.25 kilograms of seed sown produces 99.75 kilograms of wheat per *iugerum*. However, seed grain for the following year's crop must be deducted from this; consequently the actual yield is only 66.50 kilograms per *iugerum*.[21] Thus, if this hypothetical family required 1,384 to 1,591 kilograms of wheat per year for food, it will have had to work 20.8 to 23.9 *iugera* of land.

Next, how many days of work would have been required to bring in this crop of wheat? Columella is the only source who preserves figures upon which such a calculation can be made. He gives a total of 9.5 to 10.5 man-days per *iugerum* of wheat: four days for plowing, three days for hoeing, one day for weeding, and one and a half days for reaping, with perhaps an additional day for harrowing after sowing.[22] Unfortunately, considerable variation seems to have existed among Roman agricultural writers on plowing times per *iugerum*; even Columella himself elsewhere gives ranges from 3.25 to 6.5 man-days per *iugerum* depending on whether the soil was difficult or easy to work.[23] Therefore, let us assume the highest plowing time, six and a half days per *iugerum*, another for harrowing, three more for hoeing, one for weeding, and two for reaping, for a total of thirteen and a half man-days per *iugerum* of wheat.[24] Columella, however, does not provide a figure for threshing, but one may estimate that this task added another two days to the total.[25] In addition, Columella mentions several other tasks associated with cereal cultivation but fails to take them into account in his total. If the four days these require are added in, the total labor required under this model is about nineteen and a half days per *iugerum*.[26] If the family needed to cultivate 20.8 to 23.9 *iugera* to raise the wheat it needed to meet its food requirements for the year, the total labor requirement is between about 405.5 and 466 man-days per year.

Finally, we may set against this the total labor available to the family. Unfortunately no standard method exists for computing the labor potential on a subsistence farm. I follow Gallant, therefore, who assigns a value of one man-day of work potential to an adult male, nine-tenths to an adolescent male, seven-tenths to adult and adolescent women, and one-half to children and elderly adults.[27] Columella estimates the average number of working days in an agricultural year at 290, assuming 45 days per year lost to rainy weather and holidays and allowing his slave plowmen an additional 30 days of rest following the sowing.[28] These last, however, a family of subsistence farmers might not be in a position to allow itself. Hence the agricultural working year of the hypothetical family postulated above ought to fall somewhere between 290 and 320 days, yielding a total labor potential of between 1,189 and 1,312 man-days per year.[29] Clearly this family would have had vastly more labor at its disposal than its subsistence needs required, and it should be obvious that removing the 290 to 320 man-days of work per year from the farm that the twenty-year-old son's labor represented would have made no difference in the family's ability to produce the wheat it needed to grow under this model, particularly since its overall food requirements would have dropped by the amount that the son consumed per year, between 326.75 and 367.8 kilograms of wheat.[30] His absence meant that the remaining family members will have had to cultivate between 4.9 and 5.5 fewer *iugera* of land, which will have correspondingly reduced the total amount of labor required of them by 95.5 to 107 man-days per year. Even drafting the second son is not likely to have brought the family's labor potential anywhere near a dangerously low level as far as simple survival was concerned.

Of course, the demand for labor on a subsistence farm was not constant year-round; it peaked in the summer with the need to get the harvest in quickly.[31] A crop of wheat between 20.8 and 23.9 *iugera* will have required between forty-one and a half and forty-eight days to reap.[32] Our hypothetical family could call upon 4.1 man-days of labor per day, meaning that it would take them between ten and a little less than twelve days to bring in the harvest. What would the effect of the loss of a son to the legions have been on this critical operation? The family would have now disposed of only 3.1 man-days of labor per day, but had to bring in only 15.3 to 19 *iugera* of crops, meaning 30.6 to 38 man-days of work. With 3.1 man-days per day of labor, the total time spent on reaping would have been between a little less than ten and slightly more than twelve, scarcely a significant difference.

This model is highly simplified, and the realities of small farming households were certainly more complex, particularly in the matter of diet. No one could survive on wheat alone, nor would anyone want to. The question is, however, what effect would a more varied diet have on the labor requirements of a farm family? Would a more realistic mix of crops alter the balance in such a way that the family would experience a serious shortage of labor if one or both sons went off to war? As noted in the previous chapter, subsistence cultivators normally grew a variety of crops both to guard against the failure of any one of them and in order to provide themselves with a more palatable diet.[33] The most obvious additional staple to a diet principally comprised of wheat was legumes, especially beans.[34] Their role in the archaic diet in Latium is secure, and they undoubtedly continued to compose a significant part of what smallholders and their families consumed throughout the third and second centuries B.C.[35] Beans were grown as field crops and required no more labor to cultivate than wheat, so devoting a portion of the family's acreage to them will not have entailed any increase in the overall total of labor output.[36] Other potential field crops will not have required substantially more days of work than wheat and in some cases needed fewer.[37] More troublesome is the task of estimating the amount of time required to cultivate a kitchen garden, vines, fruit trees, and olives. That the vegetables from the garden as well as wine, olives, and other fruit composed important parts of an ordinary Roman's or Italian's diet seems certain; how significant a part, in terms of the percentage of the required calories they supplied, is more difficult to ascertain.[38] However, we may again adopt a worst-case scenario and assume that two *iugera* of garden, vineyard, and orchard were cultivated *in addition* to the fields of grain and legumes already incorporated into the model.[39] Gallant estimates the labor requirements for vegetables, vines, and olives at 175 to 200 man-days per hectare, thus about 44 to 50 man-days per *iugerum*.[40] Clearly, an additional 100 man-days per year would scarcely tax the labor resources of our hypothetical family, even allowing for the removal of one or both sons.[41] Finally, the model assumes that animals supplied traction for plowing. Tibiletti reproduces figures suggesting that one hectare of intensively cultivated pasture could support between one and two oxen.[42] We may suppose, then, that this hypothetical family's two oxen would have entailed cultivation of an additional eight *iugera* for their pasture. This area would have required ten days per *iugerum* for cultivation, thus adding eighty man-days to the family's working year, for a total annual requirement of 585 to 648 man-days.[43] Yet even so, the departure of one or both sons for extended military service will not have brought this family's farm anywhere close to the point where its demand for labor outstripped the capacities of the work force it could muster. (See Table 1 and Figure 2.)

Admittedly, the foregoing is entirely hypothetical and cannot claim to represent the actual experience of any particular family of Italian smallholders. The purpose of the exercise, rather, has been heuristic: to es-

War and Life Cycles

TABLE 1. *Available vs. Required Labor in a Hypothetical Family of Smallholders*

Total Available Labor of Household in Man-Days

Father, Mother, Two Sons, Daughter 4.1@290 workdays/ year 1,189	Father, Mother, Two Sons, Daughter 4.1@320 workdays/ year 1,312	Father, Mother, One Son, Daughter 3.1@290 workdays/ year 899	Father Mother, One Son, Daughter 3.1@320 workdays/ year 992	Father, Mother, Daughter 2.2@290 workdays/ year 638	Father, Mother, Daughter 2.2@320 workdays/ year 704

Labor Required to Meet Nutritional Demands, in Man-Days, All Calories from Wheat Only

Level of Activity: Moderate Work	Level of Activity: Heavy Work	Level of Activity: Moderate Work	Level of Activity: Heavy Work	Level of Activity: Moderate Work	Level of Activity: Heavy Work
405	466	310	359	214	252

Wheat, plus Legumes and Garden

505	566	409	459	314	352

Wheat, plus Legumes and Garden, plus Two Oxen

582	648	489	539	394	432

Wheat, plus Legumes and Garden, plus Two Oxen, plus 20% Additional Labor

702	777	587	647	473	518

Wheat, plus Legumes and Garden, plus Two Oxen, plus Sharecropping

952	1,073	841	939	620	693

Wheat, plus Legumes and Garden, plus Two Oxen, plus 20% Additional Labor, plus Sharecropping

1,142	1,288	1,009	1,127	744	832

tablish parameters within which to assess the effect of long-term military service upon the labor requirements of Italy's small farms. The model presented here rests upon many, many assumptions, each of which is open to challenge. However, the critical point is not the degree of speculation that has gone into its creation but whether the assumptions that under-

War and Life Cycles

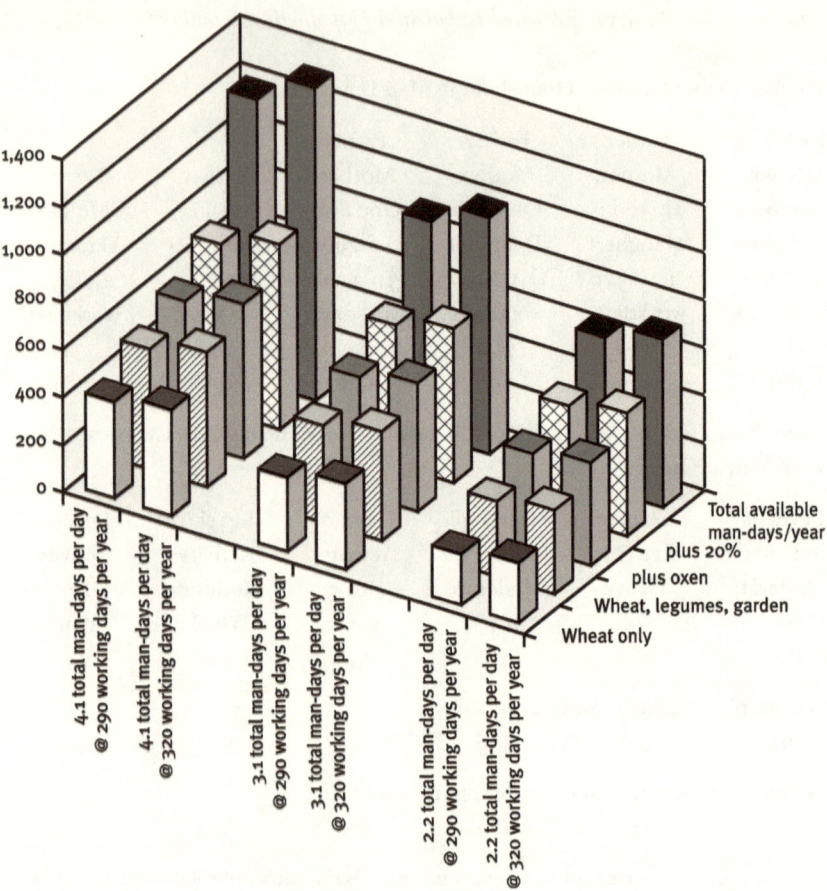

FIGURE 2. *Household Labor Potential vs. Household Subsistence Demand for Labor in a Hypothetical Family of Smallholders*

lie it are plausible. Put another way, to challenge the conclusion to which the model points, that conscription of a son for long-term military service will not necessarily have left a family of this sort on a subsistence farm dangerously short of manpower, one would have to make other assumptions about how such a farm operated that both put it near or past the point at which a shortage of labor occurred when its men were drafted and which are *more plausible* than those used here.[44] Yet as the preceding discussion has stressed, a worst-case scenario has been assumed throughout, one that is based on highly unfavorable assumptions. The reality is likely to have been significantly more benign. So, for example, to argue that the food requirements would have been significantly greater than those used here, one would have to assume that the family members were a good deal heavier. Yet the model already assumes weights that seem generous for people living in a subsistence economy who consumed little animal protein and fat; how plausible would it be to claim that people usually weighed even more?[45]

Similarly, one could push the hypothetical family closer to a labor shortage in the event that one or both sons were drafted by postulating that its per *iugerum* labor requirements were higher, but how likely would that have been? The highest average plowing times have already been built into the model, and, in practice, wheat could be grown if necessary in fields that had received minimal preparation.[46] Thus in many cases plowing times would have been substantially less, particularly if farms were too small to allow biannual fallowing.[47] Columella's average plowing times after all represent the sums of three separate cultivations, the first and second of which took place between midwinter and late spring on land not under crops sown the preceding fall.[48] Such practices could be carried out only on large establishments that possessed enough land to allow a substantial part of it to lie under fallow. These fields could be plowed under several times prior to planting in the fall.[49] Where farms were small, all or nearly all of the grain fields often had to be sown every year in order to raise enough food. Their cultivators may only have been able to work the soil once after the harvest, shortly before they began their sowing. Consequently, far fewer man-days per *iugerum* will have been needed to bring in a crop of wheat on these farms than the number assumed in the model.

Again, the eighty man-days required to tend the pastures assume that the family used oxen to pull its plows; in light soils a single donkey or even a cow could do the job, and a donkey, at least, required significantly less fodder than cattle, while a cow contributed milk and cheese to the family's diet.[50] Possibly, too, plows and oxen were shared among families who otherwise would not have been able to keep a team employed year-round, thus diminishing the amount of land and labor required of each family to maintain them.[51] On the other hand, it is not impossible that some families dispensed with animal power altogether and tilled their fields by hand.[52] The labor required would have been significantly greater but might have been offset by an equally significant increase in yields and hence decreased the total amount of land that had to be worked.[53] More plausibly, because the family had a surplus of labor under the model outlined here, it could afford to invest some portion of it in more *intensive* forms of cultivation.[54] Weeds, for example, represent a significant constraint on wheat yields because they compete for moisture, light, growing space, and soil nutrients.[55] Hence the greater attention to hoeing and weeding that a family of subsistence farmers could devote to their fields, compared with workers on the sorts of slave estates the Roman agricultural writers were concerned with, could have increased yields dramatically and hence lowered the overall amount of land required for subsistence.[56] Travel to and from distant fields, too, might in some cases have shortened the working day, but in general smallholders seem to have lived dispersed in the countryside and so much closer to their fields than later peasants in medieval Italy, whose residences in agro-towns often required a lengthy walk out to their farms each day.[57]

Yield ratios represent a more critical variable. Reducing the average yield from 3:1 to 2:1 would put the family right on the margins of its available labor to begin with, while the loss of even one son would push them into deficit.[58] Yet is it likely that yields this low were prevalent in this period? Although Columella's statement that yields as high as 4:1 were rare might suggest 2:1 as a reasonable average, his remark constitutes a case of special pleading for viticulture as the only sensible form of agricultural investment.[59] Comparative data from Italy in the medieval and early modern periods offer little support for the notion that yields as low as 2:1 were common.[60] Modern data collected from Greece in the first half of the twentieth century likewise do not justify placing yields this low.[61] Despite the pessimism of some scholars about the ability of small farms to maintain soil quality in light of the scarcity of manure on Italian farms in antiquity, there is good reason to believe that Italian farmers practiced wheat-legume rotation strategies to restore nutrients to the soil.[62] Hence reductions in average yields below the 3:1 ratio used here are implausible; more likely, Italian farmers could usually count on higher yields, although these will naturally have varied—sometimes considerably—from year to year.[63]

However, the introduction of sharecropping into the model would have much the same effect as reducing the overall yield ratio. If our hypothetical family owned only a small amount of land and had to depend on some type of sharecropping arrangement to grow the balance of the crops needed to meet its food requirements, the conscription of one or both sons could have proved disastrous.[64] Nearly everything about how a sharecropping arrangement would have operated in this period must be a matter of speculation.⁶⁵ Let us assume, however, that the family owned only 7 *iugera*, that it farmed the remainder of the land it needed as tenants, and that it obtained only half of the produce of its rented land, the other half going to its owner. If 2 of the family's own 7 *iugera* were given over to its garden, vines, and orchard, and 1 to buildings, the family obtained 100 percent of the produce of the 4 that remained, 2 of which, let us further assume, were fallowed every year. The family needed therefore to cultivate 37.6 to 43.8 additional *iugera* since it netted only 33.25 kilograms of wheat from each after deducting rent and seed for the following year. The labor required to cultivate this area will have been 733 to 854 man-days. One may add 39 man-days for the family's own 2 *iugera*, 100 man-days for the garden, vineyard, and orchard, and 80 for care of pastures for a total of 952 to 1,073, still within the family's maximum of 1,312 man-days per year. However, the loss of one son to the legions will have put them at 841 to 938.5 man-days of labor required as against an available maximum of 992.⁶⁶ With both sons gone, the family would be all but in deficit: 620 to 693 man-days required and 638 to 704 available.⁶⁷

But the critical question in evaluating the impact of conscription in such circumstances is how common sharecropping was in this period. The degree to which small farmers in the third and second centuries were dependent upon this or some other type of tenancy arrangement for access to land is a notorious crux. As is well known, the sources portray an ordinary Roman's farm in the early and middle republic as quite small, on the order of seven *iugera* (1.75 hectares) or smaller.⁶⁸ Yet few scholars imagine that their cultivators could have made such tiny plots yield enough to meet the food requirements of a family.⁶⁹ On the other hand, because of an understandable reluctance to jettison the tradition of small farmsteads, some scholars have claimed that many smallholders worked as tenants or sharecroppers on the estates of their wealthy neighbors in order to obtain access to the additional acreage necessary to support themselves and their families.⁷⁰ But De Neeve's detailed study of tenancy raises serious obstacles to such a hypothesis.⁷¹ What evidence can be gathered suggests that the institution of tenancy developed only toward the end of the second century B.C., while sharecropping is not specifically attested until the principate.⁷² Furthermore, the notion that tenancy or sharecropping was widespread in the later fourth, third, and second centuries ill accords with the republic's militarism during this period. As the previous chapter has shown, Rome's wars in the middle republic regularly kept soldiers from making any significant contribution to the running of a farm. This conflict would have existed whether the land in question was their own or someone else's. Yet citizens liable to conscription could be men of quite modest means, possessing in some cases little more than two *iugera* of land.[73] They were, in other words, often men without enough land to support their families and so precisely the sorts of men who we might suppose would have had to become the tenants or sharecroppers of their richer neighbors in order to survive.

If this conjecture is true, then we must further suppose that senators and other Romans wealthy enough to exploit their estates in this way were at the same time supporting an expansionist foreign policy that would often have required the conscription of their tenants. At a minimum, the loss of dependent workers would put landlords to the trouble of replacing their labor; at worst, it could seriously jeopardize their ability to profit from their estates. That the republic's elite regularly sacrificed its own convenience and financial well-being to further Rome's conquests seems inherently unlikely. While the few senators who commanded armies and their officers in any given year might have expected to enrich themselves from booty, those not with the legions would have lacked that compensation for the conscription of their tenants. Well-to-do citizens also composed the higher census classes, whose votes dominated the *comitia centuriata* where wars were declared. We might expect their farms, too, in many if not most cases to have required some form of dependent labor. And even though soldiers in the richer census categories might profit from the republic's victories, such men usually served in the infantry or, at best, the cavalry, and the shares that such soldiers could expect from booty or donatives were not lavish.[74] It is difficult to understand the assembly's regular ratification of declarations of war therefore if the richer voters, too, stood to lose members of their work force to the legions. Indeed, Appian says explicitly that one of the reasons for the distress of the poor in the period prior to Tiberius Gracchus was the preference of the wealthy for slaves rather than freemen to work their estates because of the latter's liability to the draft.[75] Such self-interest seems only natural. Much later, in the fourth century A.D., aristocrats displayed a deep reluctance to part with their tenants despite the army's pressing need for recruits.[76] One may compare, too, early modern Sweden, where the nobility supported a system of conscription only because their tenants enjoyed exemptions, which, because of the great unpopularity of military service, afforded aristocrats an important means of controlling their dependents as well as ensuring their labor.[77] Moreover, throughout the middle republic conquest usually led to colonization or viritane distributions that made land available to the poor. Had the rich depended on sharecroppers or tenants to farm their holdings, these initiatives, undertaken at the behest of the senate, would have reduced the pool of labor available to landlords and so placed them in a weaker bargaining position vis-à-vis those tenants who remained.[78] Neither the *patres*' support of continuous, year-round warfare, therefore, nor their regular dispatch of colonies provides any reason to believe that tenancy and sharecropping were widespread during this period.

Instead, slave rather than free labor composed the permanent agricultural work force on large farms during the middle republic. Rome was enslaving its defeated enemies on a large scale as early as 297–293 B.C. A few years earlier, passage of the *lex Poetilia* abolished the institution of *nexum*, which had provided the principal means of tying dependent labor to the domains of the wealthy during the early republic. These developments make it possible to explain the absence of sharecropping and tenancy in the middle republic by positing that the rich typically began to replace the labor of *nexi* with that of slaves in the latter part of the fourth century, a practice that the great slave hauls of the third century only accelerated.[79]

The foregoing therefore provides little reason to believe that Roman subsistence farmers would have found any additional cropland they needed on the estates of the rich. Rather, they are far more likely to have looked to the *ager publicus* for supplementary acreage.[80] Such an arrangement certainly helps make sense out of the vastly greater tracts of land that colonists received in some second-century Latin foundations as opposed to the much smaller allotments given to Roman colonists: fifty *iugera* went to each settler at Latin Bononia in 189 or Aquileia in 181, for example, while tiny five-*iugera* parcels were handed out at Roman Mutina and Graviscae at about the same time.[81] Although the size of the allotments in Latin colonies is usually explained as compensation for Roman participants' surrender of their citizenship or the dangers of living on the frontier, one may suspect that the greater size of Latin colonists' allotments also arose from the fact that these represented the total amount of land the founders expected them to cultivate, while Roman colonists would have access to *ager publicus* in addition to their individual allotments.[82] Its critical importance to smallholders seems evident as early as 367, when a *lex Licinia* was passed limiting the maximum amount of *ager publicus* that one individual could occupy.[83] Plebeian support for the measure makes little sense unless the *plebes* saw in the question of access to public land an issue vital to their independence and the economic viability of their farms. From that point on, if not before, the use of the *ager publicus* to supplement inadequate holdings began to constitute one of the mainstays of small-scale agriculture at Rome.

One need not, however, rule out the existence of sharecropping altogether.[84] In an era before widespread monetization of the economy, such arrangements may have formed the basis of exchanges in which small farmers contributed their surplus labor to large landholders at critical times during the agricultural cycle and received in return a portion of the crop or perhaps access to equipment or products that they could not afford to maintain or produce on their own—the use of a team of oxen, for example, or olive presses or a share of olive oil from the harvest.[85]

Postulating the *ager publicus* as many smallholders' source of supplemental cropland, however, raises the issues of whether this hypothetical family would have paid a tax on any public land it worked and, if so, how much. Scholars debate the first of these questions, but fortunately reaching a decision on it is not necessary for present purposes; we are interested only in considering a worst-case scenario.[86] Therefore, let us assume that the tax was both required and, equally important, regularly collected. On the second question, scholars also divide. Some believe that the tax due for use of the *ager publicus* represented only a token payment intended simply to acknowledge the republic's ownership of the land by its occupier. The sums, on this view, would have been quite nominal, on the order of the one *as* per *iugerum* annually that the city's creditors were required to pay to use the *ager publicus* in the vicinity of Rome in 200.[87] On the other hand, Nicolet has argued that the measure in 200 was exceptional. He points instead to a passage in Appian giving much higher rates—a tenth of a grain crop and a fifth of the yield of an orchard—and asserts on the basis of this text that the *vectigal* constituted an important source of income for the treasury.[88] Hence, apart from cases in which public lands were fraudulently claimed to be private, these taxes would not have been "forgotten" as some scholars have assumed but regularly collected by the *publicani*. Let us therefore once again assume the worst, that Nicolet is correct and our hypothetical family had to pay a tenth of the grain crops it raised on the *ager publicus* to the treasury. Even so, this charge would scarcely have overburdened its resources. If the family had to raise the whole of the wheat it consumed on public land, it would have had to produce 1,538 to 1,768 kilograms in all in order to yield the 1,384 to 1,591 kilograms per year it required after paying 10 percent of its harvest in rent. The difference, 154 to 177 kilograms, represents the yield of between 2.3 and 2.7 additional *iugera* at a 1:3 yield ratio and so between 45 and 53 additional man-days of labor. Adding them to the figures developed previously of 585 to 648 man-days of work the family would have had to expend to grow the food it needed to survive does not come close to its maximum labor potential of 1,189 to 1,312 man-days per year. The same holds true even if the labor of one or both sons is subtracted from this total, as Figure 2 demonstrates.[89]

The supposition that this family would have worked a substantial amount of *ager publicus* raises a further consideration. The model presented earlier has assumed that the size of its farm was large enough to grow enough food to feed the household—that the available land, in other words, was always more or less in balance with total subsistence needs. Obviously, this assumption is open to question for land is rarely an unlimited resource. Conflict between the rich and poor over access to the *ager publicus* forms a leitmotif in the history of the early republic, and a drive to acquire more land, which was unquestionably a crucial stimulus to Roman expansion as well as a key step in resolving the Struggle of the Orders, must owe part of its origins to a shortage of farmland.[90] Hence, the question arises of how altering the model's assumptions about the adequacy of this hypothetical family's farm would affect the conclusions that one might draw concerning the effects of conscription on it. Clearly, a household without access to enough land to feed itself or one close to the margins would only have gained by having one or more sons drafted for a lengthy hitch with the legions. With one fewer member to support, the productive capacity of its land would come into closer balance with its subsistence requirements. Far from causing a labor deficit, we can expect that in at least some and perhaps many cases, warfare mitigated the threat to the survival of small farming families arising out of a scarcity of land. The expropriation and distribution of conquered territory not only enlarged the area families of Roman citizens could exploit, but even before any expansion of the *ager Romanus* had taken place, military service eased the economic pressure on them by taking over the burden of supporting their sons while they were away at war. This linkage has obvious implications, too, for the development of Rome's hegemony in Italy. The republic often mulcted a conquered enemy of a substantial part of its territory and then forced it to supply contingents to fight alongside the legions. In many cases, these two steps must have been necessary complements of one another: by drawing off a large number of young men from a population suddenly deprived of much of the land that had formerly enabled it to feed itself, Rome brought its victims' agricultural resources more into balance with the demands placed upon them. In this way, Rome palliated at least somewhat the impact of conquest upon the agrarian economies of its victims.

The margins of subsistence necessitate one final refinement to this model. No family of small-scale cultivators could afford to ignore the ever present risks that the high interannual variability of Mediterranean rainfall imposed on farming in that region.[91] In some years, the crops would simply fail for lack of water. Beyond that lay the very real possibility of sickness, accident, or some other misfortune that could threaten a family's survival. Making provisions for hard times was essential, and these generally took the form of stored food and favors done for others or surpluses shared with them to be reciprocated when trouble struck.[92] Other strategies aimed at the acquisition of assets such as jewelry or livestock that might easily be sold in case of need.[93] It is impossible to know how much additional labor, if any, the creation of such reserves might have entailed. Food was generally stockpiled in good years when harvests surpassed the household's requirements, so that in these cases, at least, stores would have entailed little additional labor.[94] Similarly, passing along surpluses to neighbors would not have much increased the family's work for the year. Entering into larger redistributive networks, such as those that required marketing produce in order to buy additional food or other goods to lay aside, involved time spent in travel and transport. And a household might have invested a considerable portion of its labor in handicrafts or work for wages on wealthier neighbors' farms or in exchange for items that it could not produce for itself or simply as favors done for others in the expectation of future reciprocation.[95] But even if one were to suppose that 15 or 20 percent of a family's total annual working time might have been directed toward these ends, adding this amount would not push it into deficit.

One important additional element in strategies to minimize risk was diversification—not putting all a family's subsistence eggs into one basket, as it were.[96] As noted already, no small farmer ever gambled on a single crop; commonly a variety of grains and other foods were grown to ensure at least something to eat in case one or more of the others failed. But diversification extended well beyond cultivators' crop selections. Livestock, wild plants, and even the choice of whether to invest time in farming or other kinds of work could enter into the picture, and among the latter one ought to count military service.[97] The republic not only fed the sons of farming families it conscripted but paid its citizens a wage (as its allies did their contingents) and offered them the prospect of booty and donatives in the event of victory.[98] Although it may be morally distasteful to think of plundering the wealth of those Rome conquered and often enslaving them to boot as just one more form of the redistribution of resources, in ecological terms this is precisely what these activities represented. Likewise, the payment Roman soldiers received transferred wealth from members of the community who did not serve to those who did. From this perspective, sending sons to war constituted a means to diversify further a household's overall survival strategy. It presented it with a different way of investing the labor of one or more members that had the effect of buffering the risk inherent in farming. Hence to think simply in terms of conscription "costing" a household some amount of work is misleading. Within a family's domestic economy, military service did not cause a net reduction in the labor force it could deploy in its struggle to survive. War rather allowed it to redeploy a portion of its labor assets into areas outside of agriculture that could bring substantial benefits when set against the possible adversities that farming could encounter (although as the following chapter shows, it was a strategy that entailed very real risks as well).

As stressed repeatedly, the foregoing model has not been intended to represent a "typical" family but only to provide a basis for assessing the effects of conscripting one or more of the members of a family of subsistence farmers configured along the lines indicated. The model shows quite plainly that such a family disposed of a significant surplus of labor. It possessed more than enough to compensate for the loss of its sons to the military, particularly since consumption declines more or less in step with labor. Pushing this hypothetical family into a labor deficit that would threaten its survival and cause its members to abandon their farm is unrealistic since the assumptions upon which the model has been developed already incorporate a very high demand for labor, and significant increases are not plausible.

Instead, the alteration that would make the biggest difference in the conclusions to which the model points come on the supply side, through a reconfiguration of the family's composition in such a way as to change drastically the potential labor left on the farm in the event of one or more male members being called away to serve in the legions. Therefore, let us suppose instead a very young family, one comprising a husband, a wife, and their three children all under the age of five. It is not necessary to go through all the calculations discussed earlier to realize that in this case the husband's departure on military service for several years would all too likely have proved disastrous for those he left behind. A solitary woman with small, dependent children could not have devoted seven-tenths of her working day, on average, to agriculture; children so young would not have been able to work in the fields; and so in the absence of kin or neighbors willing to lend a hand to make up the farm's labor shortfall, this soldier's wife and children would very likely have faced severe hardship if not outright starvation in his absence.[99]

Yet one may justifiably wonder how often such a tragedy will have been played out. As noted earlier, the Roman levy depended to a significant degree on the voluntary compliance of draftees.[100] Unlike a modern industrial state, Rome lacked the kind of elaborate police infrastructure that could enforce orders to report for duty in the face of widespread unwillingness to serve. It is difficult to imagine that the senate could have regularly compelled men to abandon their families to starvation. More critically, the dire consequences that this scenario foresees assumes, along with many scholars, that married men were regularly called upon to fight Rome's wars in this period.[101] But despite a few scattered references to soldiers' wives and children in the sources, the age at which Roman men usually married would have made the conscription of fathers of young families rare.[102] Roman men became eligible for military service at age seventeen, and one must presume that more or less the same held true for the *socii* as well.[103] Saller has argued, however, that during the imperial era men typically began to get married in large numbers in their middle or late twenties. The median age at first marriage according to Saller's data occurred around thirty, and many men postponed marriage even later.[104] Although evidence for this conclusion dates from the first, second, and third centuries of the Roman Empire and includes inscriptions from provinces other than Italy, there is little reason to think that the situation was significantly different in Italy during the third and second centuries B.C.[105] Certainly, a few mid-republican aristocrats are on record who married much earlier than Saller's data would suggest—Scipio Africanus, for example, seems to have become a husband in his teens—but isolated counterexamples do not refute a statistical average.[106] Exceptions will be found to any norm derived from a great number of individual cases. For aristocratic families in particular, political and social considerations probably led to a tendency to marry early in many cases.[107] On the other hand, the phenomenon of men delaying their marriages into their late twenties is widespread throughout western Europe and the Mediterranean over the

War and Life Cycles

course of many centuries. It stands in distinct contrast, for example, to a regular pattern of early marriage for men in Greece and the Balkans.[108] In the absence of any direct evidence, to argue that Italy in the middle republic was quite different, that ordinary men then married in their late teens or early twenties, one would need to account for an upward shift in men's age at first marriage by as much as a decade by the first century A.D., when our epigraphic evidence begins, as well as for why it remained in the late twenties to early thirties for centuries thereafter. And these are not easy developments to explain.

One might suppose, for instance, that the burdens of military service account for the change. On this assumption, so many men were being conscripted for long periods in the late third and second centuries and delaying marriage until after they had completed their terms of service that the average age at first marriage rose dramatically.[109] But this hypothesis fails to account for the fact that the marriage age for men remained high for centuries thereafter. Augustus's reforms of the military after 31 B.C. made war the business of professionals. After his reign, Italians had little expectation of conscription, and in the decades that followed fewer and fewer elected to join the army.[110] Therefore, if the duty of carrying out Rome's conquests in the third and second centuries had been causing Italian men to delay their marriages until their late twenties or early thirties, then once that burden was lifted, their age at first marriage ought to have dropped significantly. Saller's data for Italy, however, demonstrate that it did not.[111] One might also adduce a progressive impoverishment of Italian smallholders to explain why men increasingly delayed marriage between the second century B.C. and the first century A.D. and continued to do so thereafter. But again such a claim lacks cogency.[112] If men were too poor to marry in their early twenties, how had they accumulated enough money to do so by their early thirties? Some, of course, may have had to wait until their fathers died before they could become owners of a farm and thereby obtain the means to support a family. But the fathers of a substantial proportion of all men, perhaps as large as half, will have died before their sons reached the age of twenty, which would have allowed them to inherit at least a portion of their family's farm.[113] Yet Saller found virtually no husband below the age of twenty-five whose memorial was erected by his wife, and this fact strongly suggests that the lack of financial means to support a family cannot be a satisfactory answer for the practice of late male marriage under the empire. Moreover, those who during the empire commemorated deceased sons and husbands clearly were not poor;

if anything, Saller's sample is skewed toward the better-off segments of the population, families that could afford the cost of putting up a stone.[114] Those commemorated, therefore, were scarcely debarred from marriage by poverty, and yet they usually elected to put off marriage until their mid-twenties at the earliest. If poverty cannot explain why men married late during the first three centuries A.D., then it is unlikely to have caused the age of marriage to rise by a decade between the middle republic and the Augustan era. The cultural preference, it appears, for late male marriage was not a response to some short-term development but had probably existed in Italy for many centuries by the time the subjects of Saller's study began erecting their inscriptions, and it continued to influence the ages at which men married for centuries thereafter.[115]

Confirmation of a sort is to be found in Augustus's legislation forbidding Roman soldiers to contract legally valid marriages.[116] This regulation takes its place beside the emperor's other attempts at social engineering—most notably the legislation on marriage and adultery—as part of a broader, ideologically driven program to shore up the *mos maiorum*.[117] Its thrust was to underscore the restoration of the traditional *disciplina militaris* and to emphasize in particular the status of military service as an exclusively male preserve. But like those other initiatives, this rule did not so much innovate as endorse and codify what had been (or was believed to have been) traditional practice. The marriage ban was thus an affirmation of the republic's long-standing custom of drawing its legionaries principally from Roman and Italian bachelors. But the ban also had more immediate, practical ends in view. The regulation is best understood as part of a larger codification of conditions of military service, including the establishment of sixteen years as the usual term of service for ordinary soldiers.[118] In fixing this limit however, Augustus will have recognized that military service was henceforth going to extend well into the period when Roman and Italian men began to marry in significant numbers, during their late twenties. Men signing up to be soldiers at seventeen would normally expect to have wed by age thirty-three instead of just beginning to look for a wife. Hence, if terms were to be as long as sixteen years, steps would have to be taken to prevent soldiers from doing toward the end of their enlistments what their civilian counterparts in their late twenties or early thirties would ordinarily be doing.[119]

A late age for men at first marriage in turn greatly affects any understanding of the impact that military service would have had on Roman family life. Although the law held Roman men liable for conscription until the age of forty-six, recruiters in practice usually seem to have adhered to a much lower maximum age. In the crisis that followed Flaminius's disaster at Lake Trasimene in 217, for example, a levy of freedmen was held at Rome to supply marines for the fleet. But only those under thirty-five went aboard the ships; older men remained in the city as a garrison.[120] The passage strongly suggests that, a fortiori, citizens thirty-five and older rarely saw active duty, an impression confirmed at the close of the second century by the consul Rutilius's edict in 105 preventing anyone under thirty-five from leaving Italy during the emergency that followed the catastrophe at Arausio.[121] Even at moments of dire need, therefore, the republic did not expect to draft soldiers older than their early thirties.[122]

Moreover, the frequency with which older men normally will have been called to serve was limited by the fact that a Roman legion was structured by age-groups whose numbers were weighted to place the lightest burden on older men. The youngest and poorest recruits formed the corps of *velites*, who usually numbered 1,200 in the third century. Those slightly older composed the *hastati*, whereas the *principes* were men in the prime of life. Both of these groups were also 1,200 strong. The oldest men, veterans of proven courage, formed the legion's 600 *triarii*.[123] Assigning age ranges to these groups is somewhat conjectural; Meyer assumes that the *principes* were between twenty-four and thirty, and in broad terms this is likely to be correct.[124] Polybius describes the *hastati* as "those next after" (6.27.7) the youngest recruits and as "second according to age" (6.23.1), while Livy calls them the *florem iuvenum pubescentium* (8.8.6), making it difficult to believe that they were men much older than twenty-three or twenty-four if the youngest soldiers were in their late teens.[125] Polybius terms the *principes* τοὺς δ' ἀκμαιοτάτους ταῖς ἡλικίαις, "those in the prime of life" (6.21.7), and Livy describes the age of these maniples as *robustior* (8.8.6). At what age, then, did a man become a candidate for the *triarii*? Probably no formal minimum or maximum limits existed for this or any of the other categories, but as a practical matter extending the upper age range of the *principes* into the mid-thirties and selecting *triarii* from men older than that would have meant preferring men in their thirties to men in their twenties as *principes* and men in their late thirties or forties to men in their late twenties or early thirties as *triarii*. This is because if most *assidui* began their service at seventeen or eighteen and then, after spending two or three years as *velites*, served as *hastati* while they were in their early twenties, only part of them will have been able to go on to serve as *principes* if those already serving in that capacity had a tenure of as long as ten years.

For in order to keep the current *principes* on for those extra years, recruiters will have had to let a significant proportion of men serving as *hastati* go home in order to make room for older soldiers from among the *velites* who were graduating into the *hastati*, whose departure in turn was making way for newly recruited seventeen-year-olds to enter the ranks of the *velites*. Such a practice would be quite surprising. Those recruiting armies, given the choice, usually prefer younger veterans to older men, and the Romans seem to have made their preference clear in limiting the oldest soldiers within each legion, the *triarii*, to only a small proportion of the total.

If therefore the *triarii* were men in their late twenties or older, then the actual number of men of marriageable age called to serve in each legion would really have been quite small compared with the number of men in their early and middle twenties and teens: 600 as against 3,600 or a mere 14 percent of the total number of infantry. Even after 200, when the usual infantry complement of a legion was raised from 4,200 to about 5,200, this may not have resulted in an increase in the number of *triarii*. Polybius, in describing Roman military practices during the Hannibalic War, notes that even when the Romans enrolled more than 4,200 men in a legion, the number of *triarii* remained constant at 600.[126] The additional 1,000 men enrolled in a legion after 200 then may have augmented only the maniples of the *hastati* and *principes* and perhaps the *velites* as well. Yet even if we assume that the number of *triarii* increased after 200 in proportion to the overall increase in a legion's strength, this brings their total number only to 740 men per legion. On any plausible assumption about the age structure of the Roman population in this era, the military burden placed on men of marriageable age and above was far less than the proportion of the population as a whole that their numbers represented. If we assume, for example, a life expectancy for males of $e_0 = 25$ and a rate of reproduction of $r = 0.0$ (Coale-Demeny[2] Model West 3 Female), roughly 21 percent of the male population will have been between the ages of thirty and forty-five while about 23 percent was between seventeen and thirty.[127] Yet from the latter group the republic drew about 86 percent of the infantry for a legion. One cannot escape the impression that recruitment patterns intentionally minimized the contribution of men precisely at the point in their lives when they were likely to marry and take on the responsibilities of supporting a family.

Some notion of how this practice would have affected men of marriageable age can be gained by considering the following hypothetical examples. During much of the third century, the Romans typically fielded four legions of, probably, 4,200 infantry each, of whom 600 were *triarii*. Thus generally 2,400 *triarii* saw service each year. If we assume what is unlikely to have been the case in fact—namely, that Rome drew its *triarii* exclusively from citizens between the ages of thirty and forty-six—then over the sixteen years any cohort of thirty-year-olds remained liable to the draft, 38,400 man-years of service will have been required of them to meet the city's need for *triarii* during that period. Yet, if we again assume an average life expectancy at birth of twenty-five and a stable population where the rate of growth is nil (r = 0.0), out of a citizen population of about 300,000, about 100,000 men will have been between thirty and forty-six, meaning that if each of them served only one year as a *triarius*, only about 38 percent of them will ever have had to serve. If each *triarius* was conscripted for three years, then the proportion of men thirty and older who would ever have had to serve drops to only about 12.66 percent.[128] After the Second Punic War, the population was lower and the demand for military manpower greater. Census returns in the decades after the war are problematic, but let us suppose a total citizen population of 280,000, and, further, that during the first third of the second century nine legions on average were in the field during any one year, requiring the service of 6,660 *triarii*.[129] Over sixteen years, then, these nine legions will have necessitated 106,560 man-years of service from *triarii*; thus, if each *triarius* served three years, far fewer than half of all men between thirty and forty-six will at some point have had to serve.[130] Even during the height of the Hannibalic War, when Rome's manpower demands were at their peak, the burden on men thirty and older need not have increased beyond this point. In 212 Rome fielded twenty-three legions of citizens.[131] Brunt estimates that these contained about 72,000 men or roughly 3,200 each.[132] These figures presumably include cavalry and so, if of a legion of 4,500 men, including cavalry, the *triarii* composed 13.3 percent, then these would have made up about 426 men of a 3,200-man legion and totaled about 9,798 in twenty-three legions. If all were thirty or older and the total citizen body stood at about 230,000, then the *triarii* serving in 212 represented about 13 percent of all Roman men between thirty and forty-six.[133] If over the next eleven years of the war the senate kept an average of twenty legions of about 3,200 men each in the field, then 93,720 man-years of service on the part of *triarii* will have been necessary. Again, if each *triarius* served three years on average, then 31,240 men will have had to serve at some point during those eleven years, or about 41 percent of all men between thirty and forty-six.[134]

These figures, of course, must be considered purely heuristic and representative of only general orders of magnitude.[135] What they suggest, however, is that with a median marriage age of thirty and an average length of service of three years for men serving in the oldest class of legionaries, the republic need not have conscripted very many smallholders with young, dependent families during the third and second centuries in order to fill the ranks of the *triarii*. Many soldiers in this category were probably under thirty and as yet unmarried, while, as Saller's data suggest, as many as half of all men past that age would not yet have found a wife. For conscripts from among this latter group, a delay in marriage for two or three years while they completed their military service would have put them not much beyond the average age for beginning their families. Moreover, not every man would wed. Some would remain lifelong bachelors, although their numbers are impossible to determine.[136] But to the extent they existed, such men constituted a group whose members' conscription as *triarii* for periods considerably longer than three years would have had little or no impact on the viability of a young family. Among those conscripts who had already married or did so during their last years of military service, some will have been sufficiently well-off to make hired or servile labor economically feasible in their absence.[137] Others may have been able to serve knowing that their wives and any children would be cared for by their kin while they were away.[138] And some older men with adolescent sons not yet of draft age could have left the farms in their hands while they went on campaign. In one way or another, the republic could have avoided taking most fathers of young families without leaving its legions shorthanded.[139]

These limited military burdens in turn were critical in preventing the republic's demands for manpower, even at their peak, from removing essential labor from Roman (and, one must presume, Italian) small farms at a particularly vulnerable point early in a family's life cycle, when the arrival of young children would often have placed nearly the entire responsibility for agricultural work on their father. Consequently, we must look elsewhere for ways to alter a family's configuration that might bring about a critical labor shortage in the event of one or more men being drafted. Although the hypothetical mature family adduced earlier—a father and mother, two sons in their late teens and an adolescent daughter—certainly represents one possible mature family—that is, one entering the later stages of its existence before the deaths of the parents and the departures of the children to begin families of their own would dissolve it—

its makeup probably did not typify a majority of cases. As noted already, Roman and Italian men married in their late twenties or early thirties, and given the most plausible assumptions about life expectancy in this period, their late marriage ages meant that a significant number of these men would have died by the time their sons reached adulthood. As many as 50 percent of all twenty-year-olds in the computer simulation developed by Saller had already lost their fathers while fewer than a third of all thirty-year-old men still had a living father.[140] These results suggest that we cannot assume that in a majority of cases young men conscripted at seventeen would have left behind a father to carry much of the burden of working the farm. Moreover, family sizes tended to be small. The same computer simulation suggests that a substantial minority of men aged seventeen to thirty would have been only children.[141] Yet, as noted, Roman men between the ages of seventeen and about thirty contributed the bulk of the legions' infantry: a little more than 85 percent of the total prior to the second century and the same or perhaps slightly higher thereafter. Very often, therefore, men in precisely the group that bore the heaviest burden in fighting Rome's wars will also have constituted the principal labor force on their families' farms. Their departure for an extended period of military service raised the possibility of a serious conflict between war and the survival of the families they left behind.

Still, the republic's demands for soldiers in the third century down to 218 were fairly moderate, and so a recruiting pattern that primarily conscripted men aged seventeen to thirty could still have accommodated the dependence of a great many farms on these same men's labor by taking only those whose fathers survived or who had another older adult relative to work the land in their absence. Those who were the sole support of their families would have been allowed to remain civilians. Exemptions for only sons of widows and older brothers with younger, dependent siblings are not without parallels in other systems of conscription.[142] And the censors' finding in 214 regarding the extent of *vacationes* granted to that point in the war suggests that exemptions might have been awarded to some unmarried men younger than thirty as well as to those older men who had already wed.[143] On the assumption that Rome's total citizen population in 225 B.C. was about 300,000, about 107,500 of them would have been between the ages of seventeen and thirty.[144] To field four legions each year prior to the Hannibalic War required about 14,400 men from this age-cohort to serve as *velites*, *hastati*, and *principes*. Over any thirteen year period in the later third century down to 218, then, four legions will have required a total of 187,200 man-years of service, and if each solder served an average of six years, 31,200 of the men aged seventeen through twenty-nine, or about 29 percent in all, will have had to serve at some point. This proportion naturally diminishes as the length of service increases, so that if the average length of service was, for example, eight years, only about 22 percent of this age-group would have to have been drafted.[145]

The republic's efforts to stave off defeat by adopting a strategy of attrition following Cannae, however, dramatically increased the burden on men in this age-group, a burden that the end of the war did little to diminish. If the twenty-three citizen legions fielded in 212 were understrength and numbered perhaps only 3,200 men each, including cavalry, then in a legion of that size about 2,560 men will have formed the maniples of *velites*, *hastati*, and *principes*, where men aged seventeen to thirty principally served.[146] Consequently twenty-three legions will have required about 58,880 men. We might also once again estimate the male citizen population in this year at about 230,000.[147] If so, then *iuniores* between the ages of seventeen and thirty numbered about 84,000, and twenty-three understrength legions might have required conscripting about 70 percent of them.[148] If Rome fielded an average of twenty legions of about this size for the duration of the war, then during those eleven years the legions will have required a total of 563,200 man-years of service from men in this age-group. If each man drafted served an average of ten years, then 56,320 men in all, or about 67 percent, would have had to serve. If the average length of service was longer, the percentage having to serve naturally decreases while if the number of men enrolled in each legion was larger than Brunt assumed, the proportion who ever served will conversely rise. One should bear in mind, too, that while the number of citizens between the ages of seventeen and thirty includes *proletarii*, the number of men conscripted for the legions does not. Yet the war at sea had required nearly all of the former to be pressed into service as rowers.[149] Consequently, the total proportion of citizens in this age-group who were serving was in fact greater than that represented by legionaries alone. The burden that conscription imposed during the war's middle years was enormous—crushing in many cases, as the flat refusal of twelve Latin colonies in 210 to furnish any more soldiers for Rome's armies demonstrates.[150] The republic must have had to draft nearly every able-bodied citizen (as well as many who undoubtedly were not) under the age of thirty to sustain this effort, and in these circumstances it probably could not have avoided taking some older, married men from their families as well.[151]

Following the conclusion of the war, the maniples of *velites*, *hastati*, or *principes* in each legion contained about 4,460 men in all on the assumption that *triarii* made up 600 of the 5,200-man legions (excluding cavalry) levied after circa 200. If, on average, Rome kept nine legions in the field in these years, over the course of the thirteen years when a man was between seventeen and thirty the city's manpower requirements from his age-cohort will have been 538,200 man-years of service. Rome's citizen population, as noted earlier, is uncertain in the first quarter of the second century, but let us again use 280,000, which would imply a total male population of about 444,360.[152] If men between the ages seventeen and thirty represented 23 percent of this total, then they will have numbered about 102,000. If, as appears likely, recruits averaged ten years of service as a *veles*, *hastatus*, or *princeps*, the legions will have taken only 57,980 or 53 percent of all men in these age-cohorts; if the average was as long as twelve years, then 44,850 men or about 44 percent of them will have had to serve.[153]

If these estimates are correct, then the question is, What would their implications have been for mature families where the father had died or was unable to carry the main burden of work on the farm? In the decades after 201, as prior to 218, the republic could have exempted a substantial number of men between seventeen and thirty from the draft, so that in many cases, those who were the mainstays of their farms and whose departures would create serious hardship did not serve at all.[154] But exemptions may not have covered every case, and the republic's desperate straits during the Hannibalic War can have allowed few if any men under thirty to avoid the draft and probably not every married man over that age either. When exemptions were not forthcoming, would those left behind have been able to muster the labor they needed to survive? And if not, what alternate sources of manpower could small-farming families have turned to?

Not all farms, of course, were small. Wealthier families often must have been able to rely on slaves and overseers when a son went off to war.[155] Among smallholding families unable to afford slaves, on the other hand, younger brothers in their middle teens might in some cases have been able to take on the main burdens of working the farm in conjunction with other family members. However, this solution would at best have represented only a temporary expedient, because eventually these boys would turn seventeen and become liable to conscription. If not exempted, they, too, would depart, leaving the family in the same fix. Some type of reciprocal arrangement might have allowed a family to call upon its neighbors' help, at least occasionally, when additional hands were needed, especially if a store of earlier favors existed on the requesting family's side.[156] Kinsmen might help out, but one should not exaggerate their numbers.[157] In a stable population or one growing only slowly, family size on average is small. The children surviving to the age of reproduction generally only replace their parents or perhaps surpass their numbers by just a bit. At Rome, therefore, under the assumptions normally used in estimating the structure of the population at this time, the universe of a widow's kinsmen would ordinarily have been small. Of women between the ages of forty and sixty, when sons born twenty years or so earlier might be away with the legions, on average only around half of them might have had a surviving brother.[158] About the same number will have had surviving sisters, but many of their husbands, too, will have died.[159] More important, nephews, like sons, will often have been away in the army, although husbands of nieces may have formed a more promising source of labor for a farm. Some households may have comprised unrelated adults as well as family members other than parents and children, although the frequency of such arrangements is impossible to discern.[160] But to the extent that they existed, such households, too, obviously contained additional workers who might plug whatever gaps in the work force that conscription caused.

Wage labor in some instances might have made up shortfalls, particularly since the *stipendium* soldiers received allowed them, at least potentially, to send home money with which to pay a hired hand.[161] But the availability of these sorts of workers would have depended to some extent on the relative scarcity or abundance of public land in the vicinity. The most likely source of hired labor would have been neighbors: the men between the ages of seventeen and thirty who had not been drafted, those thirty and older who had completed their military service and were now raising their families, and those older fathers of families who remained vigorous enough to do a full day's work. But Chayanov has argued that subsistence farmers tend to work only as much as they have to in order to feed their families. Beyond that point the drudgery of the labor tends, in the eyes of the family members, to outweigh the value of whatever a family might gain by that extra work.[162] In an agrarian economy where smallholdings on the order of five or ten *iugera* predominated, access to extra land was essential to meet a family's basic subsistence needs. If *ager publicus* was available to exploit, then one might expect that families in many cases would have labored only as much as they had to in order to cultivate whatever amount of land their needs required and generally to have been unwilling to do much work beyond that point.[163] Consequently, while a large pool of manpower may have been available within any neighborhood, Chayanov's "drudgery factor" may have limited the ability in practice of a family whose son had gone off to war to draw upon it for extra labor. However, if subsistence farmers' access to *ager publicus* came to be restricted, as might have happened for example in some regions of Italy during the second century where population densities were high relative to an area's agricultural productivity, then the need for additional land to cultivate beyond their own holdings may well have driven the men in an area to seek to trade their surplus labor for access to land whose cultivators had been drafted.[164]

Yet one may doubt whether some type of sharecropping arrangement would in most cases have offered a solution to the problems created for a widow when her son went off to war. For in order to provide enough food to feed both the family that owned the land and also to meet some of the substance requirements of the sharecropper's family, a substantial amount of land would have been required, far more than the five- or ten-*iugera* plots most families owned. So by way of illustration, consider the situation of the first hypothetical family after it had suffered the death of the father and the departure of its two sons for military service. The mother and teenage daughter remaining on the farm will have needed between roughly 463 and 539 kilograms of wheat a year in order to survive.[165] If one assumes a yield of 66.50 kilograms per *iugerum* after seed grain has been deducted, they will have had to farm between about seven and eight *iugera* to raise this amount. If half or even a quarter went to a sharecropper, however, then the area cultivated will have had to increase by between half again and double, and it seems quite unlikely that most subsistence farmers will have owned this much land. Assuming a better seed-to-crop ratio than the 1:3 figure used here, of course, would bring the amount of land needed down into the range of five to ten *iugera*, and so it would be unwise to exclude sharecropping altogether as a potential strategy for meeting a labor shortage.[166]

An equally if not more promising solution to the problem, however, might come through the marriage of a daughter to a man who would replace some or all of a departing son's labor on the farm. At roughly the age when sons began to be called away to serve in Rome's armies, daughters began to acquire husbands.[167] They and their families will have sought prospective spouses from men in their late twenties or early thirties, precisely the point at which the latter had completed their normal military obligations and were ready once again to enter the agricultural work force. Because Roman, and presumably Italian, inheritances were partible among all children, a man's patrimony might be only half of his parents' farm or even less if he had more than one surviving sibling, and a marriage that brought access to land would in many cases have been essential to the economic survival of the family he hoped to start. Roman daughters generally inherited equally with their male siblings when their fathers died.[168] In most smallholding families, farmland must have constituted the bulk of the patrimony, and consequently most daughters whose fathers were deceased will have brought their portion of this with them when they wed.

The degree to which marriage could have served as a means for a family to replace the labor of a son lost through conscription depends, obviously, on its having a daughter available to wed a prospective son-in-law in the first place. Some have suggested, however, that the practice of exposing unwanted infants, especially girls, would have meant that the number of potential brides would have fallen considerably short of the number of men seeking them.[169] Yet little evidence supports the assumption that infanticide, much less femicide, was widespread during the republic.[170] The passage in Dio Cassius sometimes cited in support of the latter does not say that there were more men than women "among the free-born population" during the reign of Augustus. The statement refers only to the upper classes.[171] A very low average age of Roman brides at first marriage has also been held to demonstrate a shortage of women. On this view, men were obliged to secure wives, at least in the upper classes, as soon as the latter came of age since prospective spouses were scarce. Only female infanticide, it is therefore claimed, could account for a shortage of women.[172] This argument, however, can be stood on its head: a very young age for brides might equally indicate that husbands were at a premium and girls' families were therefore compelled to marry them off as soon as possible.[173] But any argument along these lines is irrelevant to the problem. Women's age at first marriage, at least among ordinary Italians, was much higher than earlier scholars supposed.[174]

One point worth bearing in mind in regard to the question of femicide is that when families were generally small and the chances high that the draft would at some point take any sons away from it for an extended period of time, the value of daughters to families, and hence the incentive to raise them, ought to have increased. Conscription could not remove their labor, and, as already noted, their marriages represented a potential strategy for securing a replacement to carry on at least some of the work formerly done by the son. Although Roman women left their natal families when they married, they could retain strong legal and affective ties to them. If they married *sine manu*, they stayed in the *potestas* of their father or, if he had died or emancipated them before their marriage, they remained *sui iuris*.[175] The fact that legal authority over a wife — and hence formal title to any property she had brought to the marriage — might remain with her natal family or with the wife herself would in many cases have created an incentive for her husband to help his in-laws meet their need for supplementary labor. Although the evidence for relations between Roman mothers and their daughters is limited and largely confined to the upper classes, it does suggest that mutual affection and support characterized them as well as a significant degree of moral authority of mothers over their daughters.[176] There is no reason to think, therefore, that a married daughter would have failed to do whatever she could to prevent her mother and siblings from starving to death, or that she would have lacked leverage to induce her husband to cooperate in assisting them.

Of course, not all families were blessed with daughters of marriageable age through whom they could replace the labor of sons gone off to war. Where sisters were younger than their brothers, they might not acquire a husband for several years. Even families with daughters to wed, however, may have found that the supply of potential spouses fell short of demand. Because men married a decade later than women, ordinary mortality will have meant that fewer thirty-year-old potential husbands existed than twenty-year-old prospective brides. The substantially higher mortality among men serving in the legions than their civilian counterparts would only have added to this imbalance.[177] So what would happen if, in the end, no replacement for the sons the legions took could be found by marriage or by hire or among kin? Would a woman and her children have disposed of enough labor to carry out the work of the farm by themselves? In the hypothetical family being considered here, the mother and her teenaged daughter would have had to work between seven and eight *iugera* in order to grow only the wheat they would need to survive. If each *iugerum* required 19.5 man-days of work to bring in a crop, the two women will have had to labor for a total of between 136.5 and 156 man-days. They would, in addition, have cultivated a garden and orchard, but smaller, certainly, than would have been required for the hypothetical family of five originally discussed. Therefore, one may estimate the time that these required at fifty man-days per year. Finally, pasture for draft animals, if this had been cultivated, will have absorbed still more labor, but again, less than estimated earlier, for the two women will not have needed two oxen to plow a mere eight *iugera*. Let us assume, therefore, they kept only one and again halve the man-days required for its feeding to forty. The total labor requirement, then, would be between 226.5 and 246.5 man-days.

Estimating these women's total labor potential is somewhat more complex for, as noted earlier, the factor of seven-tenths of a man-day for an adult or adolescent woman used in prior calculations is not the product of assumptions about a woman's physical capabilities compared with a man's but of assumptions about the additional demands on an adult or adolescent woman's time under ordinary circumstances: cooking, housekeeping, and other domestic tasks in addition to fieldwork.[178] But the circumstances under consideration cannot be considered ordinary. The women's survival was at stake, and we would expect nearly every other activity to have given way to the exigencies this imperative imposed. We might, therefore, justifiably assume that at least one of the women would have worked nearly full time in the fields and rate her labor potential at or at least near that of an adult or adolescent male, that is, at one or nine-tenths. Yet let us again adopt a worst-case scenario and assume that the daughter was capable of only half the exertion of a man and the mother of only seven-tenths. With a combined potential of 1.2 man-days, if the two women worked 290 days per year, they generated 348 man-days of labor and 384 man-days if they worked 320 days, more than enough to meet their needs in this case. Introducing young children into this hypothetical example would, of course, increase the food required and hence the acreage that the family had to till, but the corresponding increase in available labor would more than offset any increase in demand. Nor would a crisis result, on the other hand, on the assumption that the daughter's labor was unavailable and that, when a son left to join the legions, his mother was on her own. She will have needed to cultivate about three and a half or four *iugera* of wheat and legumes on the assumptions outlined earlier, and this will have entailed between 67 and 78 man-days of work per year. If we assume, again, 50 man-days for a garden, vines, and orchard and 40 man-days for cultivation of pasture for an ox, the total labor required of her will have been between 157 and 168 man-days per year. If we rate her potential labor at seven-tenths, then she will have disposed of between 203 and 224 man-days per year. The situation would only become critical if the mother was incapable of doing more than half a man-day of work due to age or illness.[179]

Still, some have doubted that a woman would have been capable of bearing the physical burdens that fieldwork imposed, particularly plowing.[180] However such claims are not persuasive.[181] It is true, of course, that a plow of the sole-ard type was not necessarily an easy thing to use.[182] And certainly, the work on a subsistence farm is physically very demanding.[183] But to posit that women on small farms in this period were typically weak or unable to undertake strenuous physical labor is quite anachronistic. Comparative evidence suggests nothing of the sort.[184] Emily Burke observed the fieldwork of female slaves in the southern United States during the antebellum period and remarked: "In those large fields of which I have previously spoken, thirty and forty men and women promiscuously run their ploughs side by side, and day after day, till the colter has passed over the whole, and as far as I was able to learn, the part the women sustained in this masculine employment was quite as efficient as that of the more athletic sex."[185] Basque women in the nineteenth century, too, showed themselves no whit inferior to men in farm work, including plowing.[186] In the ancient agricultural writers, of course, only men work in the fields, but these authors are discussing slave estates where the owner could control who was available to perform the work simply by purchase. And doubtless on Italy's small farms as well men usually did the heavy fieldwork, all other things being equal.[187] But when no men were present and hunger was the alternative, there is little reason to think that a woman in her forties or fifties and in relatively good health or her adult daughters would have been unable to work the land in order to survive. The principal deterrents to women working in the fields are generally cultural rather than physical.[188] As Thomas Jefferson noted when he visited the Champagne region of France in 1787, "I observe women and children carrying heavy burthens, and labouring with the hough. This is an unequivocal indication of extreme poverty. Men, in a civilised country, never expose their wives and children to labour above their force or sex, as long as their own labour can protect them from it."[189] In the nineteenth century, an unwillingness to expose themselves to the sexual advances of unrelated men or bring dishonor on their husbands kept the women of rural southern Italy at home.[190] Where expectations differ, however, as in many parts of the non-Western world, women form the primary agricultural work force.[191] Biological differences enter into the picture principally in connection with childbearing. The later stages of pregnancy or the care and nursing of infants are serious impediments to heavy physical labor like fieldwork.[192] But for widows in mature families, pregnancy would usually not have been an issue. And when survival was at stake, one must presume that cultural constraints like a concern to avoid dishonor or breaching the boundaries that defined appropriate feminine behavior would have taken a back seat to growing the food women needed to feed their families and themselves.[193]

The family farms of the Confederacy during the American Civil War offer confirmation on this score. In the course of its struggle against the Union in the period 1861–65, the Confederate States mobilized an extraordinarily high proportion of its population for the war: nearly 75 percent of the free white male population between the ages of seventeen and fifty served, comparable with the highest rates of conscription at Rome during the Hannibalic War.[194] The southern economy, like Rome's, was largely agrarian, and consequently the Confederate army drew most of its recruits and volunteers from farms of one sort or another.[195] Slavery permitted the South, as it permitted the Roman Republic, to remove so high a proportion of free men from its agricultural economy.[196] However, although unfree labor could grow the food necessary to supply the army and the cities, on tens of thousands of small southern farms, drafting the men who had traditionally done most of the fieldwork in many cases left free women and children to fend for themselves. As a result, wives and mothers in large numbers went into the fields to plow, sow, harvest, and thresh.[197] In a majority of cases, they survived. Their example is particularly revealing since expectations of the "proper" role for yeomen's wives in the antebellum south were premised upon a clear division of labor according to which men worked in the fields and women mainly in the home.[198] But in the crisis that the war forced upon many farming families, the need for survival took precedence, and many women proved themselves physically capable of performing work traditionally marked off as the preserve of men.

However, the absence of widespread death from starvation in the southern states during the Civil War does not mean that the burdens of working their farms in the absence of their men did not lead to terrible hardships for many families. Humans can tolerate extended periods of severe malnutrition before they die.[199] Hunger and privation were widespread among the civilian population in the southern states.[200] We have no way of gauging the degree to which a Roman or Italian widow who was forced to undertake the primary burden of running her farm when a son was conscripted would have suffered similarly. As discussed, however, Rome would not have had to draft every seventeen- to thirty-year-old to fill the ranks of its armies. During the first third of the second century, about half of all men in this age range could have been exempted, perhaps not coincidentally roughly the same proportion that in Saller's computer simulation might be expected to have lost their fathers. Where one or both parents or the others who would be left behind were incapable of supporting themselves in a son's absence, recruiters could have granted deferments. The Hannibalic War however placed much greater demands upon this cohort. There could be few exemptions when upward of 70 percent were called to serve, and for those mothers and siblings who found themselves unequal to the burdens of running a farm by themselves, the consequences may in many cases have been grim.

But several points of contrast caution against assuming without qualification that the experiences of rural southern families during the Civil War paralleled closely those Roman and Italian woman and children during the struggle against Hannibal. Although both Confederate and Roman and Italian soldiers received payment for their military service and so were able to send money home to their families, in the former case very severe inflation during the war greatly eroded its purchasing power. Consequently, much of the hardship that southern families faced arose because military pay did not allow wives to purchase either food or additional necessities, particularly the salt required to preserve meat.[201] Our understanding of the monetary history of the middle republic is quite limited, but it seems unlikely that it ever experienced a period of comparable hyperinflation. During the Second Punic War shortages of money were chronic and led to a debasement of the coinage between 217 and 212 that might have brought with it a corresponding rise in prices. But the introduction of the denarius in 212 or 211 stabilized the currency and so ought to have mitigated this pressure.[202] In addition, the effect of inflation upon the Confederacy's civilian population was all the more severe because of the degree to which southern rural families participated in a market economy. In the United States before the war, at least half the agrarian population consisted of specialized as opposed to subsistence farmers.[203] Even self-sufficient yeomen in the up-country portions of the South bought much of what they needed to survive.[204] Although the degree to which Italian smallholders were integrated into the market may have been greater than usually thought, there can be no question that American farming families depended to a far greater degree upon purchase to secure life's necessities than did small farmers in antiquity.[205] Hence, the restriction on southern families' access to the market that inflation caused resulted in much greater hardship for them than inflation would have imposed upon families much less dependent upon purchase to meet their needs.

Finally, a crucial difference between the experience of Confederate civilians and those in republican Italy lies in the simple fact that, save for four very difficult years between 216 and 212, Rome was generally successful in its struggle against Hannibal, although progress was often slow and uneven. The Confederacy, on the other hand, lost its war for independence, and a strong case can be made that its failure was due in no small measure to the fact that losses on the battlefield led its civilian population to see no point in the sacrifices it was making.[206] Significantly, civilian morale in the Confederacy seems to have been strong despite all the hardships until the last stages of the war, when a series of defeats and, more important, a Union strategy of bringing the war home to noncombatants —exemplified most terrifyingly by Sherman's march through Georgia— caused civilian morale, and particularly women's, to falter.[207] A sense of hopelessness, in other words, led to a failure of will that made privations seem far less bearable than if victories had demonstrated their value to those making them. Losing the support of those for whom they were fighting in turn powerfully affected morale among the soldiers at precisely the time they were facing overwhelming challenges on the battlefield. The situation in Italy during the middle and later stages of the war against Hannibal was quite different. Rome's strategy minimized the possibility of suffering defeats and the enemy's ability to harm civilians. Key victories at Syracuse, Capua, and Tarentum and in Spain gave meaning to the sacrifices the republic was calling upon its women and children to make. While the hardships themselves may have remained great, therefore, the civilian population's perception of the extent to which they could be borne would have been quite different than the Confederacy's. And the belief that they could—and, more important, should—endure them can only have served to strengthen Roman women's determination to remain on their land.

The foregoing discussion affords little reason to assume a priori that the conscription of a widow's adult son would inevitably have left her bereft of the means to support herself and any remaining children. A variety of substitutes for his labor were available. Failing that, she and her children could have worked the farm themselves, although this may have entailed hardship in some cases. One cannot rule out the possibility that sometimes the difficulties and privations became great enough to drive a widow from her farm or even cause her death, but these need not have constituted a majority of cases or even many. But again, let us assume the worst and consider the consequences of such a development or, perhaps a more frequent occurrence, what might happen when a crisis for the family arose not upon a son's recruitment but only later, as its condition altered. Soldiers' average terms of service in the late third and second centuries were long, and much could change over that time. In at least some cases, healthy, vigorous parents or the others who worked the land in a son's absence would fall ill, grown weak with age, or even die in the meantime. In cases like these, where a family after several years found itself no longer equal to the task of keeping the farm running and lacked a source of supplementary labor, whether hired or servile, what would the impact have been upon the farm and, more critically, upon the ability of sons who returned after completing their time in the legions to resume life as subsistence farmers and begin families?

Two obvious possibilities suggest themselves. In one, the land would lie fallow for many years, trees and vines would suffer neglect, livestock would vanish, age might render many tools and other agricultural implements useless if they were not stolen, and houses and other buildings would decay. The task a returning soldier faced in resurrecting an abandoned farm and restoring it to productivity might have been a daunting one, therefore. But two other considerations need to be borne in mind, factors that would in many cases have significantly mitigated the situation. The first is that many soldiers returned with cash in their purses from booty, donatives, and perhaps savings from their pay. To be sure, one must not overestimate the amounts—soldiers did not often get rich from war—but one should not unduly discount them, either. Soldiers expected to make some money from their conquests, and this expectation was a powerful incentive in securing their willing participation in the city's wars in the third and second centuries.[208] A man who came home with a bit of cash in his purse might be able to survive for some time without an income, therefore. And time was perhaps the most essential element in restoring a neglected farm to productivity, time to prune trees and vines, clear fields, and repair buildings. Established trees and vines that go untended do not usually die; fields unworked for several years do not become infertile. The latter may require heavy labor to clear them of weeds, brush, and trees, but on the other hand a long period of fallow can make fields *more* productive than land under continuous cropping or even biannual rotation, at least temporarily.[209] Thus a man who could support himself until he could bring in a crop might have been able to expect handsome returns for his hard work in returning fields to cultivation.

Second, our returning veteran will have been looking for a wife at just the time when many families with marriageable daughters were seeing their sons depart for legionary service and so stood in need of replacements for their labor. If marriage for such families represented a means to that end, then a veteran's chances of making an advantageous match would have been very good. A difference in average ages at first marriage of about ten years between men and women meant that, under the conditions of ordinary mortality generally assumed for ancient populations, prospective brides are likely to have outnumbered their potential husbands, while a higher rate of mortality among soldiers than for civilian men in the same age-groups will have further increased the imbalance.[210] Families of daughters, especially those concerned about a possible shortage of male labor, would often have been willing to be generous in the matter of dowries and other support to secure sons-in-law. Marriage, therefore, could have brought a returning soldier not only the land that would come as his wife's marriage portion but access to the tools, animals, shelter, and various other things of which the abandonment of his own family's farm might have left him in need.[211] A veteran's success in setting himself up as an independent farmer once again will in such cases have been less a matter of having immediately to restore a ruined farm to productivity than of taking over as a son-in-law part or all of the responsibility for running an established one and using that position as a base from which to undertake whatever remedial work his own family farm might require.

In the second scenario our veteran might return to find that death or abandonment had allowed an outsider to occupy his family's farm who was now unwilling to vacate the land. What will have been his recourse? If he did not have access to a local tribunal or the praetor's court in Rome in which to plead his case;[212] if he and his family had not enjoyed the patronage of a locally powerful family whose influence, legal assistance, or physical intimidation he could now enlist in his cause; and if he did not have other friends and relatives with whose help he could drive the interlopers off his land, then he will have been out of luck. Unquestionably, usurpations of this sort could have occurred in the countryside; the real issue is how often they did. Was it frequently enough to serve as the engine that drove great numbers of smallholders off their lands and created a large, landless rural proletariat? This is doubtful. Some farms—whose families had completely died off or abandoned them while sons were away in the army, where no arrangements had been made for their upkeep until a son returned, and where no kinsman was nearby to take charge in the meanwhile—will have been vulnerable to usurpation. But these cannot have been many. And only in those cases in which sons lacked the sorts of recourses just described will they have found themselves unable to reclaim land wrongfully occupied by others. Still, scholars commonly used to assume that because the numbers and size of large, slave-staffed estates were increasing dramatically in the early part of the second century, their expansion had to come at the expense of Italy's smallholders, and consequently in many cases when farms fell vacant during a son's absence at war, wealthier neighbors simply absorbed them into their holdings. Poor men would have been hard pressed to regain their farms in such cases. However, as argued here, the extent to which slave-based agriculture developed prior to the last decades of the second century has been greatly overestimated, while the value of preserving a network of neighboring clients to members of the Roman and Italian elite should not be minimized.[213] Consequently, this phenomenon is not likely to have been a major factor in agrarian developments during the early or middle portions of the century.

Problems with usurpation will also have arisen mainly when soldiers were unable to return home temporarily in order to make arrangements to have a farm looked after in their absence. Although Roman men served on average lengthy terms with the legions, these were not necessarily continuous. Legions were regularly discharged and new ones enrolled, and while commonly veterans may simply have been reenlisted, the discontinuity in service created opportunities for returning home.[214] Leaves were also possible. While Roman armies regularly remained in being year-round until discharged, that does not mean that they actively campaigned all that time. Those serving in Italy as well as abroad regularly went into winter quarters, and while most soldiers certainly remained in camp during this time, others with pressing personal business to attend to must often have obtained leave to return home. Soldiers from Flamininus's army in Greece during 196 traveled from their winter camp on various errands—possibly connected with the army, but possibly also on personal business.[215] Some may even have returned to Italy. Fifteen years later, during the winter of 170/69, many soldiers from Rome's forces in Greece were allowed to go on leave and many of them wound up in Italy, where the censors took it upon themselves to send them back to their legions.[216] One would expect then that men serving in Italy could likewise have traveled home during winter encampments. Spain may have been a different matter, but then again, it may not. A fit of anger reportedly drove the proconsul Q. Metellus Macedonicus, during his command there in 141, to grant leaves to any soldier who applied for one without examining the causes or setting any time limit.[217] This criticism clearly suggests that consuls allowed soldiers in some cases to absent themselves from their units for limited times and for specific reasons. The lavish leave-granting from the army in Greece over the winter of 170/69 was blamed on *ambitionem imperatorum*.[218] Commanders in Spain during most of the second century were usually of praetorian rank and so often cherished hopes of reaching the consulate. They had good reason therefore to gratify soldiers who sought to return to Italy temporarily to attend to pressing personal business. Leaves would certainly not have been practical for the army as a whole, but for a handful of men who could plead hardship they would have seemed a reasonable concession, particularly during the decades in the second century when the Romans waged little active warfare in Spain.

Consequently, there is little reason to think that even a soldier serving overseas could not at some point within a year or two have returned to his farm on leave and made whatever arrangements were necessary to ensure that he would not forfeit ownership in his absence. To be sure, in some cases a soldier might never receive word that his parents had died and his farm had fallen vacant. In others, provisions made on his behalf by relatives or others may have proved ineffective so that the soldier returned to find himself displaced from his family farm. But it seems highly unlikely that such a fate represented more than a handful of the tens of thousands of Roman and allied smallholders who fought Rome's wars in the middle republic. But ultimately, the question is not how many soldiers lost their farms while serving in the legions but how many veterans this would have kept from becoming farmers once again. Colonization offered obvious remediation, and when this was lacking, a surplus of nubile women over men could well have meant that, even in cases where a soldier returned to find his land gone with no hope of recovery, marriage served as his entrée back into the world of small-scale agriculture.

This chapter has offered three models of families in order to test the widespread presumption among scholars that year-round, long-term military service beginning in the late third or early second century created a critical shortage of labor on the small farms from which the Roman Republic drew its soldiers. As a result many families are supposed to have been forced to abandon their land and become part of a growing, landless rural proletariat. But the foregoing discussion reveals that no conflict need have existed between the viability of a small farm and military service provided that the latter accommodated the course of a family's life cycle. At certain stages, families disposed of a large surplus of labor, while at other times they had little or none. As long as the republic took men from their farms when their labor was superfluous and avoided removing them when it was essential, the impact of the city's warfare on their families would not have been injurious. The only potentially serious shortfalls revealed by these models would have occurred in some forms of mature families whose circumstances made them unable to replace a son's labor through marriage, hired help, slaves, or kinsmen. If the remaining family members were incapable of carrying the burden of working the farm, serious hardship could have ensued. But situations of this sort need not have been many. Farm women were generally capable of undertaking fieldwork if necessary. Where infirmity or illness prevented them from supporting themselves, Rome could have exempted *assidui* under thirty from military service when their labor was indispensable on their farms. To be sure, some cases must have arisen where these strategies and substitutions failed and those left behind when sons went off to war could not keep up the farms, causing the fields to lie fallow. Yet their numbers will have been few, and that fact alone will not in every instance have caused a returning veteran to lose his farm or forced him into the ranks of the landless poor.

Of course, no reconstruction of this sort is susceptible of proof, given the inadequacy of our evidence for rural life in the middle republic. It is only a hypothesis about the impact on Italy's subsistence agricultural economy of the republic's mobilization of an increasingly large portion of the peninsula's farming population—a mobilization for wars that ultimately enabled it to conquer the world over the course of more than two hundred years. To assess its validity and that of the models of small farms and family development on which it is based, one can ask various questions. Is it internally coherent? Are the assumptions that it incorporates more plausible than their alternatives, and do they accord with what few data we have on these topics? Finally, does the hypothesis have more explanatory power than its alternatives? In other words, if, as this chapter has argued, long-term, massive mobilization for war could *in theory* have coexisted successfully with subsistence agriculture, if the one did not necessarily spell the demise of the other, then we must choose between the hypothesis of a symbiosis, as it were, among war, small-scale agriculture, and family life cycles, on the one hand, and the consensus view that warfare ruined Italy's smallholders by depriving them of the labor that was vital to their survival, on the other.

The hypothesis offered here suggests that the republic's ability to exploit Italian marriage patterns for military ends made a vital and previously overlooked contribution to Rome's imperial success. The tendency of Roman and Italian men to marry for the first time around the age of thirty created a large pool of underemployed younger men on the small farms of Italy from whom Rome could draw the manpower necessary to carry out its conquests, even when these campaigns kept the soldiers away from their farms at crucial points in the agricultural cycle and for years on end. Once men reached the age of marriage, Rome's military system minimized the demands it placed upon them, returning soldiers to civilian life and keeping them with their young families when their labor was essential to their support. In view of the lack of any compelling evidence for a crisis in recruitment in the second century despite the fact that Rome conscripted an extraordinarily high proportion of its male population to carry out its conquests, in view of the overall willingness of Roman and Italian men to comply with the draft year after year (save in cases where the wars in prospect seemed hard, dangerous, or unprofitable) without elaborate mechanisms in place to compel them to serve, and in view of the absence of anything in the archaeological record that could demonstrate the widespread disappearance of small farms in the second century, there is every reason to prefer the reconstruction presented here to the conventional picture. Whatever was causing the landlessness and poverty that Tiberius Gracchus sought to remedy in 133, it was not abandonment of farms owing to families' inability to keep them going while their men were off at war.

Chapter 4

Mortality in War

If the reconstruction advanced in the preceding chapter can be accepted, it reveals much about the underpinnings of Roman power in the middle republic. The massive Italian manpower that formed the basis of Rome's military might and imperial success itself rested on a bedrock of converging economic, cultural, and sociopolitical strata. Because most farms were small and their labor requirements, at least as far as subsistence was concerned, lay well within the capabilities of most mature households, adult sons could usually be conscripted without imperiling a family's survival. The typically late age of first marriage for most men further yielded a large pool of potential recruits who lacked the obligations and emotional attachments that a wife and children entailed. The abolition of debt bondage had led Rome's upper classes to substitute chattel slavery for the labor of free citizens on their estates, and this shift in turn obviated the potential for a conflict pitting the republic's need for soldiers against the private concerns of landlords eager to keep sharecroppers and tenants at work in their fields. This combination of factors enabled the republic to mobilize its citizens and allies on a scale that was probably without precedent among complex societies to that date and certainly would not be equaled, in Europe and North America at least, until the efforts of the Confederate States to win their independence in the American Civil War.[1] Moreover, these same factors, coupled with the financial and organizational resources necessary to sustain a complex logistical system and offer payment for military service, allowed Rome to maintain its armies in the field year-round, far from home, and for years on end if need be. Staying power was critical, for it allowed the legions to visit widespread, systematic devastation on its enemies' territory, confronting them with a stark choice between surrender, accepting the destruction of their crops, or fighting to defend their land. In the latter event, Roman armies' long months together furnished the time required to develop the discipline, cohesion, and skill at arms that were essential if they were to prevail in a pitched battle.

Still, two critical questions remain: how did Rome's wars in the late third and second centuries affect the farms and families of the men who fought them? For even if, as the two previous chapters have argued, soldiers did not return to find their fields uncultivated, their families overwhelmed by debts or on the verge of starvation, and no hope of ever setting things right again as some scholars believe, the consequences of removing so many young men from civilian life and exposing them to the risks of war were far from negligible. And, second, what accounts for the social and economic conditions that formed the background to the tumultuous events of Tiberius Gracchus's tribunate? For in rejecting the conventional view of the effect that Rome's second-century wars had upon smallholders and on the revised chronology of the development of the "slave mode of production" in Italy, the argument presented so far leaves us without an adequate explanation for this upheaval. Indeed, the great strength of the conventional view has always been its ability to explain so convincingly its origins, and the challenge facing any critique is to produce an alternative hypothesis of comparable explanatory power.

This chapter and the next argue that military mortality, although little studied, had a major impact on Roman and Italian society in the generations prior to the tribunate of Tiberius Gracchus, particularly among the small farming families who supplied the bulk of the men that fought Rome's wars. That Rome won its struggle against Hannibal at a very heavy cost in lives, civilian as well as military, is of course well known.[2] What scholars have less clearly perceived is that just as the republic's military efforts subsequently, although diminished from their Second Punic War levels, remained quite high, so, too, did these wars, despite their victorious conclusions, continue to exact a substantial toll among conscripts. The republic's progress along the path toward a Mediterranean-wide hegemony was not without its setbacks, and heavy casualties often accompanied those defeats. But even the city's far more numerous successes could sometimes require a steep price in lives. And quite apart from immediate death in combat, wounds and disease also contribute to far higher rates of mortality for soldiers than among their civilian counterparts. All these deaths profoundly affected Roman and Italian smallholding families—not only individually but on a very general level and in broad terms—both for the worse and, perhaps surprisingly, in some cases for the better. But in order to gauge their effects it is first necessary to attempt to quantify insofar as our sources allow the extent of military mortality following the Hannibalic War. This chapter shows that even working within the parameters of a fairly conservative set of assumptions the numbers will have been

Mortality in War

far higher than Rome's many victories might lead one to expect. The next chapter attempts to gauge the impact of these deaths upon Italy's smallholders and, on this basis, to explain how they and other consequences of Rome's wars led to the crisis that Gracchus sought to solve.

Scholars occasionally have attempted to put the extent of the deaths that the republic's war making produced among its own soldiers into quantitative terms, but the lack any systematic treatment of the evidence limits the usefulness of these efforts.[3] The starting point for an investigation of the numbers of men Rome lost to warfare during the second century must be the figures our sources, mainly Livy, preserve for the first thirty-two years of the period, when our evidence is at its (admittedly limited) best.[4] These are listed in Table 2 along with estimates of the sizes of the armies involved.[5] They indicate that of the 627,800 soldiers who may be assumed to have taken part in these battles, 55,281 individuals or 8.8 percent died.[6]

This mortality rate cannot of course simply be taken at face value, for at least some if not all of the numbers on which it is based can be questioned. Brunt has argued that even contemporary observers would have had considerable difficulty in arriving at an accurate count of casualties. Roman generals would not in every case have taken the time and trouble to count the corpses on a battlefield, particularly in defeats. In other instances, survivors of a battle who had been captured or deserted might nonetheless appear as killed in action when they failed to return to their units.[7] Yet on balance, concerns on these points are probably overdrawn. Roman commanders certainly had, at least initially, a fairly clear idea of how many men they were leading since reckoning the size of their forces would have been a simple matter of having headcounts passed up the chain of command. Subtracting the number of those present afterward from the earlier total ought to have enabled a general to know how many men he had lost.[8] Moreover, commanders and their quaestors had to be prepared to account for the funds disbursed to them for pay and food for the troops under their command. They deposited their *rationes* in the *aerarium* and were expected to produce them for examination on demand.[9] Keeping track of the sums involved perforce entailed accurate counts of their recipients, particularly when their numbers changed, because command of armies also brought responsibly for their provisioning.[10] Therefore those in charge needed to know with some precision how much food to requisition. The monthly ration of 100 men represented almost three tons of wheat, to say nothing of other items; seemingly trivial errors in reckon-

TABLE 2. *Combat Deaths, 200–168 B.C., as Reported by the Ancient Sources*

Date	Number Killed	Number of Soldiers	Percentage Killed	Place	Result	Sources
200	f. 2,000	26,500	7.5	Gaul	Victory	Livy 31.22.2
199	> 6,700	26,500	25.3	Gaul	Defeat	Livy 32.7.5
197	f. 700[1]	26,500	2.6	Greece	Victory	Poly. 8.27.6; Livy 33.10.8
196	f. 3,000	26,500	11.3	Gaul	Defeat	Livy 33.36.4
194	5,000	26,500	18.8	Gaul	Draw	Livy 34.47.8
	(?)6,700	13,400	50.0	Spain	Defeat	Livy 35.1.2
193	> 5,000	26,500	18.8	Gaul	Victory	Livy 35.5.14
	73	13,400	0.5	Spain	Victory	Livy 35.1.10
191	200	26,800	0.7	Greece	Victory	Livy 36.19.12; App. *Syr.* 20
	< 300	26,600	1.1	Liguria	Victory	Livy 36.38.4
	1,484	26,600	5.5	Gaul	Victory	Livy 36.38.6
190	6,000	13,400	44.7	Spain	Defeat	Livy 37.46.7
	< 324	26,700	1.1	Asia Minor	Victory	Livy 37.44.2; App. *Syr.* 36[2]
186	4,000	26,600	15.0	Liguria	Defeat	Livy 39.20.7
185	f. 5,000	47,200	10.6	Spain	Defeat	Livy 39.30.6
	> 600	42,200[3]	1.4	Spain	Victory	Livy 39.31.15
181	> 1,030[4]	23,600	4.4	Spain	Victory	Livy 40.32.7
180	1,491[5]	23,600	6.3	Spain	Victory	Livy 40.40.13[6]
179	109	23,600	0.5	Spain	Victory	Livy 40.48.7
178	237	26,800	0.8	Histria	Defeat[7]	Livy 41.4.8
176	52	27,900	0.2	Liguria	Victory	Livy 41.18.13
173	> 3,000	21,600	13.8	Liguria	Victory	Livy 42.7.10
171	> 2,200[8]	29,400	7.5	Greece	Defeat	Livy 42.60.1; Plut. *Aem.* 9.2
168	80/< 100	29,400	0.3	Greece	Victory	Livy 44.42.8; Plut. *Aem.* 21.3
Totals	55,280	627,800	8.8			

1. Not all of these were Romans or Italians: see Chapter 4, n. 39.

2. Klotz, "Eine römische Verlustliste," argues that the figure of 324 Roman dead at Magnesia represents only Roman citizens, not citizen troops and *socii* together, on the grounds that Appian specifies that those killed were Ῥωμαίων ... ἐξ ἄστεος,

TABLE 2. *Continued*

"Romans from the city," while Livy merely calls them "Romans." Klotz assumes that Appian's identification of the dead in this way is unlikely to be his own addition. Instead, he explains it on the assumptions that Roman commanders' official reports would not have mentioned allied dead but only the number of citizens killed and that Appian here drew, like Livy, on a source that derived ultimately from the official report of the battle. But while Livy simply classified all those killed as *Romani*, Appian preserves the specification of these as Roman citizens contained in the general's original report. This argument is unconvincing, however, for there is simply no reason to believe that Roman generals would not have included casualties among the Italian allies in their dispatches to the senate. The republic's *socii* regularly contributed half or more of a Roman army's manpower, and the *patres* took care to keep allied as well as citizen contingents up to strength when they dispatched *supplementa*. The senate cannot have afforded to adopt the cavalier attitude toward allied casualties that Klotz suggests, nor could the republic's generals, who were responsible for provisioning Rome's *socii* as well as citizen troops. And other figures for casualties strongly suggest that official counts of the dead after battles did not omit those among the allies who fell. Livy reports, for example, that at Pydna not more than 100 died on the Roman side and that most of these were Paeligni, that is, *socii*: 44.42.8. Likewise, he breaks Roman casualties in the city's Spanish victories in 181 and 180 down into citizens and Italian allies. Rather, Appian's identification of those killed at Magnesia as "citizens" is certainly his own conclusion, the product of the fact that, when he wrote, Roman citizenship had been extended to the whole of Italy and the distinction between the allied and citizens components of Roman armies largely forgotten: compare the casualty figure that Appian reports at *Iber*. 45, where the Roman dead are identified as ἐξ ἄστεος, and *Iber*. 46, 56, and 58, where they are not. Yet the first figure is impossible to explain on the assumption that these dead were exclusively Roman citizens and that an equal or, more likely, far greater number of Italian (and Spanish) allied were killed in the same battle.

3. The combined armies of the praetors in Spain were first defeated and then victorious in this year. The size of the army that won the victory has been therefore determined on the assumption that the 5,000 soldiers killed in the earlier defeat were all Romans or Italians and included no native auxiliaries, although this is unlikely to have been the case. See Chapter 4, n. 39.

4. This figure is the sum of the slightly more than 200 Romans and 830 *socii* killed according to Livy but does not include the 2,400 Spanish auxiliaries who also fell.

5. The figure is the sum of the 472 Romans and 1,019 Latin allies who fell, according to Livy, but does not include the 3,000 Spanish auxiliaries who also died in this battle.

6. Walsh doubts the occurrence of this battle, believing that it is a doublet of that recorded at Livy 39.30; Walsh, *Livy Book XL*, 160.

7. The enemy drove the Romans from their camp in a surprise attack, but the legions subsequently recovered the camp in a counterattack the same morning.

8. The losses Livy records at 42.66.10 are omitted here on the grounds that they are unhistorical. See Chapter 4, n. 161.

ing up the amounts required could lead to substantial logistical consequences.[11] Individual magistrates also compiled detailed personal *commentarii* recording the events of their terms of office; these, too, perhaps added a further reason for commanders to keep accurate counts of their casualties.[12] But certainly the Romans' practice of collecting and cremating their dead, at least following victories, provided an occasion for reckoning their numbers.[13]

Military administration therefore simply cannot have been as haphazard as Brunt's suggestion that commanders were ignorant of their armies' casualties implies. The republic did not send out mobs of men on the warpath. Rather, its nearly continuous, successful warfare over so long a period implies an extraordinarily high degree of organization and systematization in its military operations.[14] Consequently, it would be highly surprising if these structures did not usually enable the Romans to know how many deaths an army had sustained in a battle. Even in the chaos following the catastrophe at Cannae, Varro was able to inform the senate of the number of men he had under his command at Venusia.[15] Nor was this an isolated instance. Generals' dispatches to the senate concerning their operations in the late republic regularly reported figures for both enemy and Roman dead, to judge by a tribunician law of 62 B.C. seeking to ensure the reliability of those numbers.[16] And no reason exists to believe that the practice was not common throughout the second century.

The number of enemies slain constituted an important element in the senate's deliberations over the award of a triumph, as indicated by the law, usually dated to the middle of the second century, stipulating that at least 5,000 of them had to have fallen.[17] Yet quantifying the extent of Rome's successes in this way naturally raised the question for the *patres* of the price at which they had been purchased, and more than once the size of the republic's losses was used against a commander seeking a triumph. In 197 tribunes criticized the heavy casualties that the troops of Q. Minucius Rufus had suffered.[18] And ten years later Cn. Manlius Vulso apparently had the losses his army sustained in a fight with Thracian bandits thrown in his teeth by his opponents in the senate.[19] Even when a general sought to hide the extent of the damage, the truth could come out: L. Cornelius Merula's hopes for a triumph in 193 were scuttled by his legate, M. Claudius Marcellus, when the latter wrote privately to many of the senators complaining, among other things, of heavy Roman casualties in Cornelius's victory over the Boii.[20] Three years previously, a letter read to the senate detailed the extent of the losses C. Sempronius Tuditanus's army had suffered in its defeat in Spain.[21] At other points, Livy records the numbers of military tribunes, prefects of allies, and centurions, often with their names, who had fallen in a particular battle along with the total number of men lost.[22] These reports again suggest official communications by commanders to the senate reporting the results of combat. Even where no request for a triumph was at issue, the senate regularly considered the question of levying *supplementa* for the legions at the beginning of each year when the new consuls took office.[23] Such discussions would have been otiose unless the *patres* possessed accurate information on the current strength of Rome's forces and, consequently, could estimate the losses suffered over the preceding year. These passages provide sufficient grounds to believe that many if not most of the figures for Roman casualties listed in Table 2 entered the historical tradition through reports of commanders and their officers in the field back to the *patres* in Rome.[24]

Generals, to be sure, sometimes may have had difficulty distinguishing those killed in battles from captives and deserters, yet this was not inevitable: Livy reports 600 captives when Persius defeated the Romans at Callinicus in 171.[25] More revealingly, the Romans included in the treaties of peace they imposed on their defeated enemies a requirement that all prisoners of war and deserters be returned to them.[26] Because the latter were put to death, the Romans must have been able reliably to separate them from prisoners of war without too much trouble.[27] Such discrimination suggests that commanders' ability to account for the fates of their troops in battles was not as limited as one might think. Yet whether soldiers died or deserted or became captives does not make a great deal of difference for present purposes. Some captives in this period will have been enslaved and sold, thereby removing them as permanently from their farms' labor force as if they had died in battle.[28] Similarly those who deserted are not likely to have returned home if execution awaited them upon discovery. There is little reason to think therefore that, even if the number of his soldiers a general reported killed in battle had been inflated by the inclusion of captives and deserters, this will have made much difference as far as the impact on their farms was concerned.

Brunt also suggests that serious inflation may have occurred at a later stage in the transmission of the casualty figures, when Roman annalists enlarged them, perhaps out of a desire to glorify the city's victories by exaggerating the magnitude of obstacles it had overcome to achieve them.[29] Yet here, again, one should probably not make too much of this possibility. While Livy sometimes records the various (and to his mind often exaggerated) numbers of enemies the Romans killed that he found in his sources, comparable discrepancies in the numbers the Romans lost are rare in this period.[30] For the Hannibalic War, in those few instances when variations in Roman casualties occur, Livy interestingly tends to prefer the lower figures. Polybius, clearly working from a Carthaginian source, claims the Romans lost 70,000 men at Cannae; Livy's total is 48,200.[31] Likewise, Livy reports 1,500 Romans killed at Zama, while Appian's figure is a thousand higher.[32] For the battle at Lake Trasimene, Livy notes that while some of his sources preserve totals for the Roman casualties higher than the 15,000 he records, he elected to follow Fabius Pictor's authority in this matter.[33] Polybius, on the other hand, records not only 15,000 Romans killed in this battle but another 15,000 captured, again probably following a Carthaginian account.[34] One can suspect, therefore, that when Livy indicates that one of his sources gave 13,000 Roman dead and another 7,000 for the second battle of Herdonia in 210, the higher figure is due not to patriotic amplification in the annalistic tradition but the tendency of Punic historians to magnify the extent of their victories.[35] However, if a similar tendency toward inflation was occurring for the casualties Rome sustained in its second-century battles, it has left little trace in our sources. In all probability, rather, variant figures are lacking because no broad "anti-Roman" tradition existed for these conflicts comparable with that which led the pro-Carthaginian chroniclers of the Hannibalic War to exaggerate their side's prowess in killing Romans.

Discrepancies in the figures for casualties in second-century battles are not, however, entirely absent. Plutarch found two different statements in his sources of how many men the Romans lost at the battle of Pydna: Scipio Nasica put the number at about 80, whereas Posidonius (not the later philosopher) claimed 100.[36] Both men were participants in the battle, and at first blush this case might appear to bear out Brunt's claim that even contemporary observers would have had difficulties arriving at an accurate count of the dead. Similar variations appear in reports of the numbers killed on the Roman side at the battle of Callinicus in 171: 2,200 according to Livy but 2,500 in Plutarch.[37] The latter figure also probably goes back to Posidonius, who wrote a history of Persius's reign from a sympathetic point of view to judge by his defense of the king's flight from Pydna.[38] Hence in each case the higher figures may be due to an anti-Roman source's eagerness to magnify the successes of the republic's enemies. But the discrepancies in question, 20 in the one case and 300 in the other, scarcely seem great enough to bestow much glory on the Macedonians. Instead, they may have arisen not from a desire to magnify Roman losses but because of differences in status among those who died. Rome made frequent use of local forces in its battles, particularly those it fought in Spain and Greece, and those few cases in which Livy breaks down the city's casualties by category show that these *auxilia* could comprise a substantial proportion of those who died in the fighting.[39] In these two cases, then, the different totals may reflect the inclusion of non-Italian allies in some instances but not others. In addition, one may suspect that this same factor has inflated at least some of the other casualty figures in Table 2, particularly those for combat in Spain, which are among the highest for this period.[40] Could the local auxiliaries be factored out, the mortality rates for Roman and Italian soldiers in such cases might drop significantly.

Still, the existence of some very high numbers of Romans and allies killed in a few victories in northern Italy and Spain might in and of themselves induce suspicion that inflation by Roman writers has occurred. One might suppose that the most reliable figures for casualties would come from Polybius, particularly when he is reporting battles that took place in Greece during his own lifetime, such as the 700 Roman dead he reports from Cynoscephalae, or from other contemporary participants, such as Scipio Nasica or Posidonius for Pydna. The losses they report are fairly small compared with the casualties reported in some of the victories the Romans won in northern Italy—in 193 and 173, for example—or Spain in 180. This contrast might in turn suggest that the latter are unreliable, the product of Roman annalists who exaggerated them to glorify Roman feats of arms. But in fact no dichotomy exists distinguishing small Polybian totals for casualties from supposedly inflated annalistic reports. For the numbers killed in other Roman victories over Spanish or Ligurian foes are quite low, and Livy almost certainly found these not in Polybius but in his annalistic sources. One cannot assume, in other words, that high casualty figures for Roman victories are a hallmark of a mendacious annalistic tradition and are therefore inherently suspect.

Nor should there be anything suspicious about a high casualty figure for any particular victory. Although the average loss in the victories listed in Table 2 is about 4.2 percent killed, any statistical construct may contain significant deviations from the mean.[41] On the contrary, in gauging the plausibility of higher figures, one must bear in mind that some of Rome's opponents were considerably more formidable than others. The Gauls, especially, had a reputation as fierce warriors in antiquity.[42] This trait is an integral part of their characterization in authors of the late republic and the Augustan age, and certainly had a firm basis in fact to judge by the senate's fear and frantic preparations at the news of an impending Gallic attack in 225.[43] Against a tough, stubborn foe there ought to be nothing surprising in an army suffering much heavier casualties than usual even when it conquered.[44] One may note, for the sake of comparison, Polybius's report that Hannibal lost about 5,700 men in his slaughter of Rome's forces at Cannae, more than 11 percent of his total force. Even the ambush at Lake Trasimene cost him about 1,500 casualties, equivalent to about 3 percent of his army, despite the fact that the Romans could offer no organized resistance to his attack.[45] There can be no question of annalistic inflation in these cases; Polybius certainly found these figures in the Greek historians of Hannibal he consulted, and they are quite plausible.[46] Desperate fighting by the Roman troops in both battles, notwithstanding the hopelessness of their situations, ensured heavy losses among Hannibal's men, even though they ultimately triumphed. Nor does the consul L. Cornelius Merula's costly victory in 193, in which almost 20 percent of his force died, afford any grounds for suspecting the work of annalistic fabricators. His losses occasioned severe criticism of the general on precisely this point by his legate Marcellus in the letter to the senate alluded to earlier.[47] The political conflict these losses touched off cost Merula his hopes of a triumph and confirms the veracity of the high figure that Livy reports. Heavy losses in some victories in Spain also seem quite defensible in view of the fears and reluctance to serve that reports of fighting there in the late 150s touched off among Roman recruits.[48] On the other hand, the limited number of casualties the Romans suffered in several of their victories in the Hellenistic East strongly suggests not that these ought to be taken as "typical" for all battles but that some measure of truth lay behind the Roman perception of the Greeks and other easterners as unwarlike foes over whom victories could be easily and bloodlessly won.[49]

On the contrary, some of the trifling figures for casualties that appear in Table 2 might suggest a tendency on the part of generals or perhaps later authors to minimize the extent of Rome's losses. However, generals' dispatches to the senate were heard by men who had long experience at war. Presumably, they had some idea of what figures were plausible, which must have limited to some extent the degree to which the number of deaths could be adjusted. And, as Marcellus's letter to the *patres* shows, a commander could not be certain that his attempts to cover up his losses would not be exposed by others with access to that information. For example, the consul of 186, Q. Marcius Philippus, suffered a serious defeat in Liguria.

Before returning to Rome he disbanded his army in order to conceal the extent of his losses. Yet Livy reports 4,000 men killed, information that can only have come from Philippus's officers.[50] And in view of this figure as well as the other very large losses that Table 2 reports, there is little reason to think that the annalists who transmitted these figures to Livy were systematically minimizing the casualties Rome suffered in these battles.

Finally, doubts about whether the numbers of Roman dead have any basis in fact at all might arise in view of Scheidel's remarkable study of sums of money reported in our sources for the imperial period. In this he argues that "between ninety and one hundred percent of all existing financial numerical data are merely conventional figures which cannot automatically be accepted as rough approximations or rounded variants of actual figures known to the authors (like 300,000 instead of 289,700 *vel sim.*)." He bases his case on the overrepresentation in such data of figures based on multiples of 10, 30, or 400—far beyond what is statistically probable—and concludes that when these sorts of financial figures appear in our sources, they "may be no more than indicatory of a certain order of magnitude."[51] Although Scheidel's study examines only reported sums of money, its pertinence is obvious to any evaluation of the trustworthiness of the casualty figures that Livy and other ancient sources preserve for the middle republic. Yet the figures given in Table 2 do not display the dramatic overrepresentation of any particular number that the sums of money Scheidel analyzed do. Most are certainly rounded approximations, as Livy himself often indicates when he introduces them with phrases like "less than," "more than," and "about."[52] Hence, the difference between the 80 Roman dead that Nasica reported and the 100 in Posidonius may represent nothing more than the latter rounding up the actual total (which Nasica, too, may have rounded off). Likewise the divergent casualties reported for the battle of Callinicus. A few multiples of 30 and 400 occur, but they scarcely constitute a majority or even a significant minority among the casualty figures.[53] On the other hand, some of the figures Livy preserves do not appear to be rounded approximations at all but quite specific numbers— for example, the 472 Romans and 1,019 *socii* killed in a clash in Spain in 180 or the 109 who died there in a victory the following year.[54] Such precision suggests, at least prima facie, that in these cases the figures go back to an official count following the battles, an impression strengthened, as mentioned earlier, by Livy's occasional inclusion of figures for the numbers of centurions, *praefecti sociorum,* and *tribuni militum* killed.[55] And what comparative evidence can be mustered from classical Greece offers fur-

Mortality in War

ther grounds for believing that the rest of the figures in Table 2 are not entirely fabricated. Krentz has calculated that the average rates of death in hoplite battles during the fifth and fourth centuries were about 5 percent for the victors and 14 percent for the defeated. As Krentz notes, the numbers on which his calculations are based have an inherent likelihood of being accurate owing to the care that Greek poleis took to recover and bury their dead. The information was known, therefore, and what we know of Thucydides' working methods makes it probable that he, at least, took the trouble to inform himself on this point from those who knew.[56] The figures for Roman battles from Table 2 are on average 4.2 percent killed in victories and 16 percent in defeats, about what one would expect on the basis of the Greek evidence.[57]

No grounds exist, therefore, for rejecting out of hand the casualty figures Livy and our other sources preserve. The foregoing discussion suggests on the contrary that, while reported figures for this period may be in some cases too high, they are not altogether worthless. If any adjustment needs to be made, it would be to revise our totals downward somewhat on the grounds that the inclusion of foreign *auxilia* has raised the count in some cases, that deserters who made their way home after battles or captives later repatriated did so in others, and that non-Roman historians for patriotic reasons added to the total in others still. Such adjustments might be offset to some degree, however, by a tendency among Roman generals to minimize their own casualties, so that in the end, the total number of deaths in Table 2 might change little if at all. Of course, a casualty rate of 8.8 percent or somewhat lower still does not provide an average annual rate of mortality in combat for the simple reason that Livy does not report Roman casualties for every battle that took place in every year. These missing data need to be taken into account for unquestionably if we had a full list of casualties for all the battles Rome fought during these thirty-three years the average annual rate of death in combat would drop significantly below 8.8 percent. The decrease stems from two factors. The first is that, as already noted, reported casualty rates for Rome's defeats reveal significantly more men killed in action than in its victories: 16 percent as against 4.2 percent. Yet, somewhat surprisingly, the proportion of all defeats for which our sources preserve the number of casualties is far higher than the proportion of victories. Every defeat Rome suffered in this period but two is included in Table 2, while of the thirty-eight pitched battles Livy calls victories, we have the number of Roman forces killed for only fifteen (sixteen counting the victory in Spain in 185).[58] Consequently, if we had the missing casualty numbers for Rome's victories in this period and could add them into the totals given in Table 2, the result would be to reduce significantly the average combat mortality rate below Table 2's overall rate of 8.8 percent.

Some idea of the extent of the reduction can be gained from Table 3, which lists the combat experience year by year for all legions in the period 200–168. Combat experience by legion has been adopted as the unit of reckoning here because some battles involved only a single legion with its complement of allies while full consular armies of two legions and their allied contingents and occasionally four legions and allies fought others. In addition, some legions fought more than one battle and hence suffered casualties more than once in a single year. Table 3 has been constructed in order to take these variations in troop strength and frequency of combat into account. Table 3 records a total of thirty-eight victories in pitched battles over these thirty-three years, which involved combat by a total of ninety-four "victorious legions."[59] Those involved in winning the fifteen victories listed in Table 2 numbered thirty-one. The "victorious legions" not represented in Table 2 therefore numbered sixty-three, and this figure affords a basis for estimating the total number of casualties that Rome's armies might have suffered in the engagements they fought. Only twice did a victory cost the republic as much as 10 percent of its forces; far more commonly (eleven of fifteen victories in Table 2) the figure was below 5 percent, several times falling below even 1 percent.[60] One might assume therefore that in the majority of the twenty-three victories for which no casualty figures are recorded, the Romans' losses might have been around the 4.2 percent average for those victories for which numbers have been preserved. If so (and on the further assumption that each of these legions in the year it won its victory or victories had a complement of 5,500 citizens and 6,700 *socii*), each "victorious legion" would have sustained an average loss of about 512 dead.[61] The total number of men killed in these sixty-three legionary combats would therefore have been 32,281.[62]

However, this figure is probably too high. The median percentage killed in the victories listed in Table 2, about 1.1 percent, is significantly lower than the mean of 4.2 percent. This divergence suggests that the latter does not afford a good indication of how many soldiers Rome might lose in a typical victory since the figures preserved in our sources include a few exceptionally large numbers of soldiers killed that pull the average up. A more usual figure is likely to lie somewhat below this, although probably

TABLE 3. Combat Outcomes, 200–168 B.C., by Year and Theater

Date	Gaul & Liguria	Etruria	Rome	Bruttium & Apulia	Sicily & Sardinia	Spain Cit./Ult.	Greece & Asia	Totals V/D/Mn/Mx/Nc/?
200	VV		NcNc			V Nc	NcNc	3/-/-/-/5/-
199	DD					Nc Nc	NcNc	-/2/-/-/4/-
198	NcNc NcNc					V Mn[1]	MnMn	1/-/3/-/4/-
197	VV MnMn					[2]	VV	4/-/2/-/-/-
196	3V3V DD[3]		NcNc			V[4] Nc[5]	NcNc	7/2/-/-/5/-
195	2V2V[6] NcNc[7]					VV[8] VV	MnMn	8/-/2/-/2/-
194	MxMx[9] NcNc					D[10] V[11]	NcNc	1/1/-/2/6/-
193	VV MnMn					Mn V		3/-/3/-/2/-
192	VV[12] MnMn NcNc		NcNc	NcNc		Mn 3V		5/-/3/-/6/-
191	VV VV		NcNc	NcNc		Nc Nc	VV	6/-/-/-/6/-
190	NcNc NcNc	Nc		NcNc		D[13] Nc	McNc VV	2/1/-/-/10/-
189	NcNc	Nc		NcNc	Nc	V Nc	2V2V[14] MnMn	5/-/2/-/7/-
188	NcNc NcNc[15]					Nc Nc	MnMn NcNc[16]	-/-/2/-/10/-
187	2V2VMnMn[17]					NcNc NcNc		4/-/2/-/4/-
186	DD NcNc		NcNc			MxMxVV VV[18]		4/2/-/2/4/-
185	MnMn MnMn[19]					2D2D2V2V[20]		4/4/4/-/-/-
184	NcNc NcNc					MnMn NcNc		-/-/2/-/6/-
183	NcNc NcNc					MnMn NcNc		-/-/2/-/6/-
182	MnMnMnMn[21]					VV[22] NcNc		2/-/4/-/4/-
181	NcNc NcNc VV[23]					MnMn VV		4/-/2/-/4/-

Year					V/D/Mn/Mx/Nc/?
180	MnMn MnMn		VV NcNc		2/-/4/-/2/-
179	VV NcNc		VV[24] 2V2V[25]		8/-/-/-/2/-
178	MxMx[26] NcNc	Nc	MnMn VV[27]		2/-/2/2/3/-
177	VV[28]	Nc			4/-/-/-/3/-
176	VV ??[29] MnMn[30]		VV		2/-/4/-/2/2
175	VV VV		MnMn[31]		5/-/-/-/2/-
174	NcNc ??[34]		Nc[32]		1/-/-/4/2
173	VV NcNc		V[33] Nc		3/-/-/4/-/-
172	NcNc NcNc		Nc		-/-/-/-/6/-
171	NcNc		Nc Nc	DD	-/2/-/-/8/-
170	NcNc		Nc Nc	DD	-/2/-/-/8/-
169	NcNc		Nc Nc	NcNc	-/-/-/-/8/-
168	NcNc		Nc Nc	VV VV	4/-/-/-/6/-
Totals	NcNcNcNc				94/16/43/6/153/4
	NcNcNcNc				
	NcNc				
	NcNc				

Note: Each symbol stands for a single legion; two symbols adjacent indicate two legions under a single commander. Victories are indicated by V; defeats by D. Where little or no combat occurred in a theater in a year, this is indicated by Nc. Mn stands for minor actions of various sorts—cities stormed, skirmishes, and other small-scale engagements. Battles in which the result was mixed—neither a clear-cut victory nor defeat—are indicated by Mx. Cases where there is no information on military events in a province are indicated by ?. Where a numeral precedes a V or D, this indicates that more than one pitched battle was won or lost by these legions in a particular year.

1. On Stertinius's activities in Spain, see Richardson, *Hispaniae*, 34.

2. Although no legions were present in the Spanish provinces in 197, each praetor commanded 8,000 allied infantry: Livy 32.28.11. One of these praetors, Sempronius, suffered a serious defeat with this force.

3. Marcellus's legions fought three battles, in one of which they were defeated, though they were victorious in the other two; his colleague Purpurio's legions participated in one of the victories: Livy 33.36.4–37.9. Hence in this year the four legions in Gaul had between them three victories and one defeat.

4. Late in this year or early in 195, Q. Minucius Thermus won a victory as proconsul before the arrival of his successor.

TABLE 3. Continued

5. Although no legionary combat is recorded in Hispania Ulterior in this year, the proconsul M. Helvius won a victory with a 6,000-man escort force accompanying him as he retired from his province.

6. Valerius's legions won a victory during the year of his consulate, 195, and another in the following year while he was proconsul before they apparently came under the command of one of the consuls of that year: Livy 34.22.1–3, 46.1, and compare Afzelius, *Römische Kriegsmacht*, 36–37.

7. I follow Afzelius in assigning the praetor Porcius two legions in this year, although Livy mentions only 10,000 infantry and 500 cavalry, not legions: Livy 33.43.9, and compare Afzelius, *Römische Kriegsmacht*, 36.

8. In addition to one major victory, Cato won several smaller engagements using a portion of his legionary forces.

9. Sempronius's legions fought a pitched battle with mixed results, although Livy claims more than twice as many Gauls fell as Romans: Livy 34.46.4–47.8.

10. The praetor Digitius sustained a number of defeats that Livy claims cost him half his forces: 35.1.1–3.

11. Scipio Nasica won a number of victories as praetor and then as proconsul in the following year; one of these was clearly a major battle: Livy 35.1.3–12.

12. Minucius's command at Pisa was prorogued where he won a victory; Livy records that one consul devastated the enemy's territory but fought no pitched battle in this year while the opponents of the other remained quiet: Livy 35.40.2–4.

13. Possibly Aemilius Paullus suffered two defeats in this year, as Richardson, *Hispaniae*, 98, suggests; compare Livy 37.46.7–8.

14. The legions of Manlius were involved in a number of minor operations in addition to fighting two pitched battles.

15. We are expressly told that the consul Valerius did nothing worthy of note in his province in this year: Livy 38.42.1; Livy, however, makes no mention of his colleague Salinator's activities in his province at all. It is assumed here that he, too, had an uneventful year and did little or no fighting.

16. Fulvius's legions apparently completed the siege of Same very early in 188: Walbank, *Commentary on Polybius*, 3:136, on the chronology.

17. The legions of the consul Lepidus seem to have prevailed in two pitched battles as well as several minor engagements; the legions of the other consul, Flaminius, won "many battles" according to Livy, but these appear to have been something less than full-scale engagements: Livy 39.2.1–11.

18. In Hither Spain, L. Manlius Acidinus fought one indecisive battle with the Celtiberians and then shortly thereafter won a victory over this same enemy; his colleague C. Atinius won a victory over the Lusitanians.

19. Again, although Livy mentions that the consul Claudius's legions fought several successful battles, these appear to have been minor actions against light-armed enemies rather than full-scale, set-piece battles.

20. Quinctius and Piso combined their legions in this year; these four legions suffered one defeat and won one victory.

21. Although Livy describes no battles fought by these legions in Liguria, the senate did vote a one-day *supplicatio* to the consuls *quia prospere ibi res gesserunt*: 40.16.4; compare 17.6–7.

22. Fulvius fought a number of hard battles and lost many men against a Celtiberian relieving force as he besieged a town: Livy 40.16.8.

23. The legions of the consuls did no fighting in this year: Livy 40.35.1. Early in the year those of the proconsul Paullus won a victory over the Ligurians and were subsequently disbanded. Toynbee, *Hannibal's Legacy*, table IV, n. 1, claims that four other legions were active in this year, those of Q. Fabius Labeo and M. Pinarius Rusca, which had been commanded in the previous year by Marcellus and Baebius, respectively. However, Marcellus's legions were dismissed at the close of 182; the troops he handed over to Fabius Buteo (not Labeo) consisted of a force of 7,400 Latin infantry and cavalry: Livy 39.56.3, 40.1.6, 26.2–3. The troops that Pinarius took from the army of Baebius at Pisa likewise consisted only of Latin allies: Livy 40.19.6–8. Although Baebius had been instructed to dismiss his army when he returned to Rome to conduct elections, he probably either delayed doing so until after the elections were over or, perhaps more likely, he dismissed the legions but kept the allies with him over the winter: compare Livy 43.9.3.

24. Gracchus's second major victory, described by Livy at 40.50.2–5, is doubted by Richardson, *Hispaniae*, 102.

25. Livy denies Albinus could have won the two victories ascribed to him by Livy's sources because he arrived in his province too late in the year, but see Chapter 4 at n. 159.

26. These legions were driven from their camp in a surprise attack at dawn but counterattacked later that morning and recaptured the camp. Albinus apparently won a victory over the Lusitani in this year: Livy, *Per*. 41; compare Richardson, *Hispaniae*, 102.

28. These two legions not only won a pitched battle against the Ligurians but stormed three Histrian towns: Livy 41.11.1–9, 12.7–10.

29. The legions of the consul Petillius fought a pitched battle in this year; those of his colleague Laevinus also campaigned against the same enemy, but a lacuna in Livy's text makes it unclear whether they also participated in this same battle: Livy 41.18.6–13.

30. The legions of the proconsul Claudius recaptured Mutina by storm: Livy 41.16.7–8.

31. Livy reports that Sempronius fought many successful battles, but whether these involved large-scale, pitched battles or minor operations is unclear. The claim that the Romans killed 15,000 Sardinians may be disregarded: Livy 41.17.1–4.

32. No activity in this province is reported in this year, but a lacuna in Livy's text prevents certainty on this point.

33. The proconsul M. Titinius Curvus's actions in Spain were sufficient to earn him a triumph, but the date of his victory is unclear.

34. One consul, whom Livy seems incorrectly to identify as M. Aemilius Lepidus, the consul of 175, checked a sedition in Patavium but otherwise did little: Livy 41.27.3–4. There is no information about the activities of his colleague.

not as low as the median. One may suspect that the Romans also tended to remember and record exceptionally low numbers of casualties, as at Pydna for example, as well as the very large ones. Putting the typical casualty rate midway between the median and the mean, at about 2.65 rather than 4.2 percent, reduces the total significantly, to 20,368 men, for those victories where no casualty figure is recorded.[63] The overall losses in all battles would thus be 75,649 and the average combat mortality rate for all these legions would be about 5.6 percent.[64] Adding in the two defeats involving legions for which we lack casualty reports would not raise this figure significantly since neither apparently involved a large number of deaths.[65]

No allowance has been made for losses in minor battles and skirmishes or those combats that ended in a draw. Although Livy or another source occasionally mentions such clashes and sometimes even provides a figure for casualties sustained in them, the majority undoubtedly go unrecorded.[66] How frequently they occurred and how many men the Romans usually lost in them must be a matter of conjecture. A reasonable guess might be that a legion's losses in minor actions over the course of a year would on average be about half of those incurred in a major battle. In addition to the ninety-six actual legions represented on Table 3 as winning one or more victories or being defeated, Livy reports another forty-three as having fought various sorts of minor actions.[67] Therefore, another 22,469 deaths ought to be added to those that resulted from major battles.[68] Finally, six legions fought with mixed success in this period. Four of these, two commanded by Ti. Sempronius Longus in 196 and two that served under A. Manlius Vulso in 178, are represented on Table 2, and their losses are included in that total. Only the deaths resulting from the mixed success of L. Manlius Acidinus's two legions in 186 need to be brought into the reckoning, therefore. A reasonable guess for these would put them midway between typical loss figures for Rome's victories and defeats, so at about 6.675 percent or 1,647 deaths. These additions increase the total number of deaths in combat to 99,765 over the entire period 200–168.[69]

Yet any rate of combat mortality that these deaths might suggest would need to be reduced still further owing to the second of the two factors referred to at the start of our discussion—namely, the strikingly large number of legions that fought no battle at all during any given year of service. For example, while the consul Marcellus reportedly lost 3,000 of his 26,500 man army—11.3 percent—in 196 when he was ambushed by the Boii in Gaul, that same year the proconsul Flamininus's 26,500 soldiers in Greece were involved in little or no fighting. Combining the two cases would result in a casualty rate of just over 5.5 percent for the year. In addition, the two legions stationed in Rome during 196 also represent another 26,500 men who saw no combat during 196, and including them in the calculation would drop the average even further, to under 4 percent. Table 3 also affords a sense of how great a reduction would have to be made in the overall proportion of conscripts who died in combat annually if the 155 legions that fought no battle in a year of service were added into the total. According to Afzelius's calculations, as modified slightly by Brunt, on average 8.7 legions were in the field during the period 200–168. If, once again, we assume that each of these was at full strength in every year, they represent a total of 3,847,140 man-years of service or 116,580 men per year.[70] Using these figures, the overall annual rate of mortality from combat among all men conscripted in this period can be reduced to 2.6 percent.[71] And if Afzelius is correct to hold that fairly sizable contingents of *socii* regularly garrisoned parts of southern Italy as well as Sicily and Sardinia, where they presumably saw little combat, then this rate would drop even further.[72]

Outright death in combat, however, is not the only factor involved in trying to form an impression of military mortality in this period. The effects of wounds and disease must also be considered, but unfortunately the evidence on these topics is even more limited and problematic than for deaths in battle. Some of those wounded in battles would later succumb to tetanus, gas gangrene, or septicemia, or would die from shock and loss of blood. But how many? The complete absence of quantitative data in the ancient evidence might seem to urge recourse to comparative material. For example, 17 percent of wounded soldiers in the Union army during the American Civil War died as a result of their injuries; the figure is said to have been 20 percent among the wounded in the British army during the Crimean War.[73] The rate among British troops during the Napoleonic Wars has been estimated at 12.5 percent.[74] Yet one might question the relevance of such data to the experience of combat in the middle republic. Firearms and artillery accounted for the overwhelming majority of the wounds that soldiers suffered in the eighteenth and nineteenth centuries. Around 70 percent of the French casualties admitted to Les Invalides in 1715 and in 1762 had suffered gunshot wounds; cannon fire accounted for about 10 percent; and swords for under 15 percent.[75] Among Union forces during the American Civil War, over 99 percent of all reported wounds were shot wounds.[76] The damage done by bullets may well have been considerably more deadly than that inflicted by swords and spears in close combat: "When a full-bore musket bullet penetrated with force, it frequently described an eccentric course as it skidded off the tissues inside the body, causing horrible damage on the way."[77] Certainly, the prospect for those who suffered such wounds before the twentieth century was grim. As one early modern general, Don Luis de Requensens, wrote of his casualties in 1575 after an engagement in the Low Countries, "Most of the wounds come from pikes or blows, and they will soon heal, although there are also many with gunshot wounds, and they will die."[78] About 90 percent of Union soldiers wounded in the abdomen or head died; for wounds to the chest the figure was about 60 percent.[79] Moreover, serious gunshot wounds to the extremities usually fractured the bone, which often led to amputation and thence, owing to the lack of sterile conditions in the operating rooms of the Union army, to infection, gangrene, and death.[80]

This is not to say that the sorts of weapons wielded in battles before the modern era never inflicted wounds that would ultimately prove fatal. Keegan notes of the French soldiers wounded by English archers at Agincourt in A.D. 1415 that "penetrating wounds . . . which had pierced the intestines, emptying its contents into the abdomen, were fatal: peritonitis was inevitable. Penetrations of the chest cavity, which had probably carried in fragments of dirty clothing, were almost as certain to lead to sepsis."[81] But significantly, no passage in the medical or the nonmedical literature from antiquity refers to the need to extract clothing or pieces of armor from a wound. This silence probably stems from the infrequency of such operations, compared with the procedures later required in the case of gunshot wounds because swords, spears, and arrows cut or pierced clothing and armor cleanly rather than carrying fragments of these into the wound, as musketballs did.[82] Hence infections from this source are likely to have occurred much less often than following the gunpowder revolution. Moreover, the deadliness of the wounds that Rome's enemies could typically inflict will have depended to a significant extent on the kinds of weapons they were wielding, and the damage these could do might vary widely. Polybius, in a well-known passage, remarks on the ineffectiveness of Gallic swords in the 220s. Able to wound only by slashing, they were not much good even for that, he claims, owing to their poor design: their first blow bent them into the shape of a strigil.[83] And until the first century B.C., when Rome went to war with the Parthians, whose military strength was based on mounted archers armed with recurved composite bows, Roman forces never faced bowmen with weapons comparable in striking power to an English longbow.[84] When a contingent of Cretan archers attempted to repel a Roman assault in 200, their arrows failed to penetrate the legionaries' shields.[85] To be sure, javelins, swords, and spears could inflict serious penetration wounds, and some slashing-type wounds could be quite horrible.[86] Livy's description of the carnage that Roman cavalrymen's "Spanish swords" inflicted on their Macedonian opponents in a skirmish in 199 is notorious.[87] But as the impotence of the Cretans' arrows shows, a Roman legionary, unlike his early modern or nineteenth-century counterpart, enjoyed the protection of effective defensive equipment: a *pectorale* (or *lorica* in the case of cavalrymen and infantry from the first census class), a metal helmet, and a large, thick shield.[88] The *scutum* especially will have reduced significantly a soldier's exposure to serious, penetrating wounds and debilitating blows, at least as long as he maintained his position within the line of battle.[89] The protection it afforded is perhaps most vividly evident in the remarkably small number of Romans and Italians killed and wounded at Pydna, despite facing antiquity's most formidable array of pikemen, the Macedonian phalanx.[90] One cannot assume therefore that mortality rates from the kinds of wounds characteristic of early modern or nineteenth-century warfare can be simply carried back to the middle republic, when the weaponry was very different. The frequency with which Romans suffered the deadliest kinds of wounds—those that penetrated deeply the abdomen or chest—was in all probability much more limited than that among soldiers facing gunpowder weapons and without the protection that shields and defensive armor afforded.

Nevertheless, tetanus, gas gangrene, and septicemia in the absence of modern antibiotics are fatal in the great majority of cases. Statistics from some wars in the nineteenth century as well as the First World War indicate that an average of about 5.5 percent of all casualties contracted tetanus, and in the latter conflict on average an additional 5 percent (or probably more) of wounds became gangrenous.[91] Yet here, too, caution is required in applying such figures to antiquity. Clostridium bacteria present in the soil cause gas gangrene and tetanus, and the connection is repeatedly drawn between the length of time and extent to which soil has been manured and the prevalence of these types of organisms in it.[92] We can be fairly sure that most fields in Roman Italy and elsewhere in the ancient Mediterranean had not been nearly as heavily manured for as long as, for example, those of northern France by the early twentieth century.[93] Septicemia has been claimed for an average of 1.7 percent of wounds, but this condition usually develops in wounds involving major veins and arteries, so again one wonders how the differences between the kinds of wounds typically inflicted in ancient and modern combat affect the application of such a figure to the problem at hand.[94]

Mortality from wounds is also dependent upon the availability and type of medical treatment a wounded person receives, and here, too, any attempt at retrojecting early modern or nineteenth-century experiences back to the middle republic raises a host of questions. Certainly no medical corps or hospitals accompanied Roman armies in the middle republic, unlike those of the nineteenth or twentieth century.[95] Yet the absence of the kind of medical attention these could provide does not necessarily mean that mortality rates will have been higher among the former than the latter, given the appalling lack of cleanliness in hospitals and the ignorance of sterile and antiseptic techniques within the medical profession in the eighteenth and nineteenth centuries. Wounds treated by the legionaries themselves may have been significantly less prone to infection that those benefiting from the ministrations of modern physicians prior to the twentieth century.[96] Certainly, one can point to the examples of much scarred veterans like M. Servilius Pulex Geminus, C. Marius, or the Elder Cato to show that not all wounds would prove fatal.[97]

But even if we could establish that an average of 10 or 20 percent of wounded soldiers would die as a result of complications, we would have to ask, 10 or 20 percent of how many? — and here the problem becomes nearly insurmountable. Livy offers only two references to specific numbers of wounded in books 21–44: at Pydna the Romans sustained somewhat more wounded than killed in action, so a few more than 80 or 100; and among the Romans and *socii* that Scipio led to victory in Spain in 206, 1,200 fell and another 3,000 were wounded.[98] Otherwise, the most one can say on the basis of the evidence he offers is that often the wounded outnumbered those killed, hardly an unexpected state of affairs. Including the limited data available from other battles merely leads to the conclusion that the number of men wounded in a battle could vary widely. Alexander the Great's Macedonian army, for example, sustained ratios of wounded to killed as high as 10:1 or 12:1 in some of its early victories, while ratios derived from Caesar's few explicit references to the numbers of casualties his legions suffered range from 8.5:1 to 1:2.[99] Moreover, we do not know how many of those whom the sources indicate were killed in battle in fact lived through the fighting but died of their wounds soon afterward. Good reason exists to suspect that this would have been especially the case when the Romans lost and were compelled to flee the battlefield. Unlike the Greeks, the Romans apparently never requested a truce following a defeat to recover their dead and other casualties.[100] Hence, most of those seriously wounded in defeats, who would have been the least able to escape under their own power, will have remained on the battlefield either to die where they lay or be killed by the enemy when they came to despoil the corpses.[101] From the point of view of a Roman commander, however, such unfortunates would simply have been counted as killed in action. Paradoxically, therefore, a general may have seen a higher proportion of his wounded recover after defeats than following victories simply because the kinds of injuries sustained by those among the wounded who had been able to flee the battlefield would usually not have been very serious.

Ultimately, the evidence for fatalities arising from wounds is simply insufficient to allow anything more than an educated guess about their numbers. Thus one may suppose that Roman and Italian wounded numbered about twice those killed. This is about half the ratio of those killed to wounded in battle for the British army during the period 1793–1815 and for Union forces during the American Civil War, but a lower figure can be defended on the grounds that soldiers facing gunpowder weapons and without defensive equipment would have been far more vulnerable than men who protected themselves behind heavy shields against swords and spears.[102] Of the wounded who would later succumb to their injuries, a plausible estimate would be about 5 percent. This figure, again considerably lower than those for more recent wars, is based not only on the presumption that the kinds of wounds that muscle-powered weapons could inflict were usually much less likely to lead to complications and death than those that the muskets and cannon of the late eighteenth and nineteenth centuries could cause but also on consideration of the obvious contrast between the heavy defensive equipment that protected the heads and torsos of ancient soldiers and their early modern and nineteenth-century counterparts' lack of it.[103] The use of shields and body armor, despite their weight and discomfort, suggests that they substantially reduced the likelihood of suffering the most life-threatening kinds of wounds in combat. Nothing similar availed soldiers after the advent of gunpowder until the flak jacket issued to American soldiers during the Vietnam War. Further, this lower estimate also assumes that the figures we have for combat fatalities in most cases included those who died soon afterward of serious wounds to the head, chest, or abdomen. Roman generals and their officers in this period had long experience at war and, like Don Luis de

Mortality in War

Requensens, they ought to have been able to predict with some accuracy what kinds of wounds were likely to prove fatal and consequently which of their wounded were likely to die. If it is true that the initial sources of the casualty totals that Livy and our other sources preserve lay in the armies' food and pay records, then it is likely that in reckoning up changes in the disbursements required for the army as a result of battles commanders and their quaestors would have made no distinction between reductions owing to outright death in battle and death from wounds occurring soon thereafter. Finally, the supposition that 5 percent of all wounded later died assumes that the medical care the wounded received from their comrades was usually not of such doubtful benefit as the treatment injured soldiers received at the hands of physicians in the eighteenth and nineteenth centuries, whose lack of sterile and antiseptic practices significantly increased the likelihood of infection. A ratio of killed to wounded of 1:2 and a 95 percent survival rate among the latter would add 9,977 deaths for a total of 109,742 if outright deaths in combat numbered 99,765.[104]

Evidence for mortality due to disease is even scarcer and in many ways more problematic. Livy occasionally mentions serious outbreaks of disease in Roman armies during the Hannibalic War and afterward: one occurred in 212 among the army besieging Syracuse; another in 205 devastated the legions facing Hannibal in Bruttium; in 178 armies in both Sardinia and Ariminum were affected.[105] Thirty years later Appian notes that disease broke out in the army of the consul Censorinus during the siege of Carthage and did so again in the winter encampments of the proconsul Pompeius's army during its siege of Numantia in 140, resulting in the deaths of a few of his men.[106] Additional, if indirect, evidence is to be found in the complaints voiced by Latin and allied representatives in 209: Rome was bleeding their cities white with its ceaseless demands for recruits, some of whom perished in combat, others from disease. One may note as well the serious illnesses that occasionally incapacitated commanders of armies, for example the praetor Q. Mucius Scaevola in Sardinia in 215, the consul-elect Valerius Laevinus in Greece in 211, Scipio Africanus in Spain in 206, and M. Helvius and P. Sempronius Longus, proconsuls in Spain in 195 and 183.[107] But although occasional outbreaks of disease in Roman armies are undeniable, the paucity of the evidence makes questions of their extent and frequency very difficult to answer, once again inviting recourse to analogies drawn from modern warfare. Here the facts are sobering. More than twice as many soldiers in the Union army during the American Civil War, the first for which we possess detailed statistics, died of disease as were lost in combat or from wounds, about the same ratio as among British troops during the Boer War.[108] These figures are on the lower end of the scale. Conventional wisdom in the late nineteenth century, prior to the introduction of modern antibiotics, apparently held that at least four soldiers would die of disease for every one the enemy killed.[109] Such data might suggest that disease was a far greater threat to the lives of Rome's soldiers than combat, yet losses on such a scale were not inevitable. The Japanese army introduced strict regulations for drinking water, food, and sanitation following its war with China in 1894–95, in which about 80 percent of Japanese deaths resulted from disease. As a result, when war with Russia broke out in 1905, Japan lost about 8 percent of its forces to enemy action and only 2 percent to disease.[110] Consequently, drawing conclusions based on modern rates of mortality from disease requires circumspection and caution. In estimating where the armies of the middle republic would have fallen within a range of potential rates of mortality, we cannot automatically presume that they would have been at the high end or even in the middle simply because they went to war in an earlier age.

In the first place, several of the epidemic diseases that commonly decimated early modern and modern populations until the twentieth century cannot be assumed to have been prevalent in antiquity. Measles and rubella were unknown, and while mumps is attested as early as 410 B.C., it appears to be endemic rather than epidemic at that point, and hence primarily a childhood disease.[111] Plague and cholera had not yet arrived in Italy.[112] And although earlier scholars often ascribed the great epidemic of 430 at Athens to an outbreak of typhus, this diagnosis now appears unlikely.[113] The earliest identifiable typhus epidemic in Europe occurred among the troops besieging Grenada in 1489–90.[114] Although nothing resembling smallpox is mentioned in either the Egyptian medical papyri or the Hippocratic corpus, some scholars have recently revived the case for smallpox as the cause of the great plague at Athens as well as a subsequent outbreak of disease among a Carthaginian army besieging Syracuse in 396.[115] But, if so, it never appears to have become endemic. Rather, smallpox probably died out because concentrations of people large enough to sustain an endemic form of the disease did not exist in antiquity until the second century A.D. when Marcus Aurelius's soldiers returning from Mesopotamia reintroduced it to the West as an epidemic.[116] We have no reason to suppose, therefore, that any of these diseases would have caused the very high mortality among soldiers that they did in the early modern and modern eras.[117]

The only modern epidemic diseases that might have seriously affected the rate of mortality among soldiers and whose antiquity is certain are dysentery and typhoid fever.[118] Greek and Roman medical writers describe clear cases of each.[119] Improper sanitation and contaminated water supplies are largely responsible for outbreaks of these diseases. During the American Civil War, for example, sanitation was appallingly bad in Union army camps. Few men bothered to use the slit trench latrines, and those who did often failed to shovel in dirt afterward. Water supplies were frequently unhealthy. Garbage, rats, and dead animals were everywhere. Poor nutrition among the soldiers only increased their susceptibility to disease.[120] Not surprisingly, typhoid was the leading single cause of death from disease among Union soldiers, and typhoid and diarrhea/dysentery occurring together accounted for about a third of the mortality attributable to sickness.[121] Significantly, of the five outbreaks of disease in Roman armies during the Hannibalic War and the second century, four occurred in the context of sieges or when an army had occupied an encampment for a lengthy period of time, circumstances that strongly suggest that smallpox or typhus was not at the root of these epidemics.[122] Lice or fleas spread typhus, whereas smallpox is a viral disease usually transmitted via airborne droplets that enter the body through the upper respiratory tract. Contagion occurs over a distance of no more than a few meters.[123] These vectors ought to have been operating no matter where the army was, even on the march, so long as the men were in close contact with one another, as men sharing tents in a camp undoubtedly were. On the other hand, lengthy sieges or encampments suggest precisely the sort of filthy, squalid conditions necessary to breed outbreaks of typhoid or dysentery.

Yet there is little reason to suppose that similarly poor sanitation usually obtained in the armies of the middle republic. A Roman soldier's diet was relatively good, probably better than that of most of his civilian counterparts.[124] Consequently, we cannot assume as a matter of course that soldiers were typically malnourished and hence more liable on that account to get sick. We know very little directly about hygiene in Roman camps in this period. Polybius does not mention latrines or the disposal of refuse in his well-known description of the layout of a legionary encampment. Schulten, who excavated several second-century camps in Spain, thought that three small buildings near barracks in an encampment he identified as that of Q. Fulvius Nobilior, consul in 153, might be latrines and noted a fragment (400M) from Lucilius, *qui in latrina languet,* that Marx connected with Scipio Aemilianus's activities in restoring discipline in the army at Numantia when he took command there in 134.[125] Certainly, by the time of the principate, latrines flushed with running water had made their appearance in the army's permanent camps while marching camps used "deep slit trenches fitted with wooden covers and removable buckets."[126] It is not difficult to believe that imperial armies had inherited this concern for sanitation, like the basic layout of their camps themselves, from mid-republican practices. As Schulten remarks, given the nature of these camps as fortified strongholds, allowing men to wander outside to answer nature's call would have been folly because of the threat of ambush and the possibility of desertion, while letting ten or twenty thousand men relieve themselves within the walls wherever they pleased would have been intolerable.[127] Certainly, the good order and discipline within Roman encampments that so struck contemporary observers, along with the care the Romans took to locate their camps near good supplies of fresh water, would make one suppose that sanitary practices were good, at least by premodern standards.[128]

Yet even if this were not the case, Roman armies were highly mobile and often pursued a strategy of devastation that required frequent shifts of encampment. This pattern of warfare will have therefore acted to minimize significantly the danger of outbreaks of typhoid or dysentery. In effect, Roman armies were regularly moving away from areas where water supplies had been contaminated by human wastes, and this practice, in the absence of modern sanitation and hygienic techniques, was probably the most effective prophylactic measure an army could undertake. So when, as noted earlier, disease broke out among Censorinus's soldiers during the siege of Carthage in 149, the consul simply moved his camp.[129] Even when their armies were not engaged in active campaigning, commanders probably shifted their encampments from time to time precisely because remaining in one place for too long contaminated the area.[130] Hence Roman armies in the middle republic did not typically find themselves in the kinds of situations that would have been likely to lead to serious deterioration in sanitary conditions. Long sieges were rare. Moreover, as Chapter 3 has demonstrated, Roman armies usually campaigned well into January or February. Consequently, while winter quarters were a regular feature of Roman warfare, their duration usually will have been no more than for two or three months, making the period of time during which sanitation in an encampment could become unhealthy shorter than would have been the case in northern Europe, where severe winters required far longer periods of inactivity.[131] Yet even in northern Europe, long sieges and winter encampments do not always seem to have resulted in high losses due to typhoid or dysentery among early modern armies, at least where medical records can be consulted. The war Spain's army waged in Flanders during the sixteenth and seventeenth centuries, for example, required numerous lengthy sieges, yet Parker's study of its archives suggests that venereal disease was the principal ailment among the soldiers. Corvisier records eye trouble and rheumatism along with chest problems as the complaints accounting for the largest percentages of admissions to Les Invalids in 1715–16 and 1762–63.[132]

Equally important are the places in which Roman armies were quartered for the winter or for other lengthy periods. The Romans apparently harbored a deep suspicion of the corrupting effects of urban luxury on soldiers to judge by Livy's claim that the quality of Hannibal's army deteriorated as a consequence of spending the winter of 216/15 in Capua as well as subsequent complaints about Roman commanders who ruined their armies' discipline by exposing them to the soft life in the cities of Asia Minor.[133] Hence, republican commanders seem to have only rarely billeted their troops in towns and cities, where the probability of exposure to the poor sanitation and host of diseases prevalent in ancient population centers would have been high.[134] Roman soldiers did sometimes serve as garrisons for towns and cities, yet the extent of the practice is difficult to determine.[135] Generally, the Romans did not use their legions as an army of occupation in the middle republic, at least in the conventional sense of garrisons placed in the urban centers of a conquered territory to keep its population in subjugation.[136] The deference that the Romans demanded from those within their *imperium* did not entail the sort of direct administrative control that would have necessitated the maintenance of standing armies among them.[137] Keeping the republic's *amici* in power along with the threat of reprisal was enough to ensure obedience. Spain is the only exception, but urbanization in this province was minimal.[138] Because the mechanics of the republic's imperial administration tended to keep its armies from long periods of contact with the residents of towns or cities, the applicability of some truly horrifying rates of mortality due to disease among early modern armies is quite doubtful. The experience of the parish of Bygdea in Sweden is the best attested example of the latter. It provided 230 men for service in Poland and Germany during the wars of Gustavus Adolphus between 1621 and 1639, of whom 215 did not return. Yet "[t]he greater part of them, 196 men, died from illness while perform-

Mortality in War

ing garrison duties on the eastern or southern shores of the Baltic," where typhus, plague, dysentery, and smallpox were rampant.[139] Such rates of death among soldiers simply cannot be carried back into antiquity with any confidence.

A simple answer to the problem of reckoning how many soldiers would have perished from disease during the first third of the second century does not emerge from a survey of comparative data. Ultimately, it is necessary to guess, and the question is whether to guess high or low. Should we suppose that four or five times more men on average died of disease every year than perished in battle, as the examples of some early modern or nineteenth-century armies might suggest, or should we imagine that "barring costly military engagements, mortality in the legions would not have been much higher, if at all higher, than in the civilian population in general," as Scheidel has argued was the case during the empire?[140] Where the comparative evidence can help is in determining at what point to come down between these two extremes. A high estimate would require rejecting those few indications we have about hygiene and sanitation in Roman armies during the middle republic, which tend to suggest that the situation in this regard would not have been too bad, at least compared with conditions in some nineteenth-century armies. We would have to assume instead that generals and their officers routinely allowed conditions to deteriorate to a level common in the camps of the Union army during the American Civil War. Putting the number high would mean, further, rejecting the obvious point that the Romans had, by the time of their struggle against Hannibal, long practice in the business of war and, consequently, ample time to develop procedures that would have minimized diseases like typhoid and dysentery as well as, and perhaps more important, the discipline necessary to enforce them.[141] The Union army, by contrast, was hastily levied by a nation with no long martial tradition and a small standing army up to that point. To ascribe a high toll to disease would also necessitate the assumption that diseases like typhus, smallpox, or plague regularly broke out in Rome's armies, despite the lack of firm evidence for their existence during this period and despite clear indications that even during the early modern era, when their presence cannot be doubted, they did not ravage every army. They were certainly a major cause of death among Swedish garrison troops in the Baltic during the early seventeenth century but not among contemporary Spanish forces in the Low Countries. On the other hand, sanitary and medical conditions in the military are not likely to have been at as high a standard as they

135

were during the principate. Armies under the stress of active campaigning may sometimes have been more prone to outbreaks of disease than the civilian population.[142] Given the frequent combat and the necessarily rugged conditions that Rome's constant warfare often compelled the soldiers to live under, one ought to expect that mortality, apart from deaths due directly to combat, would usually have exceeded somewhat that of the civilian population.

On these bases, then, a reasonable estimate might put the rate of mortality apart from death in battle between 25 and 75 percent higher than ordinary civilian mortality for men in the seventeen- to thirty-year-old age group. If we assume, as has been done throughout this study, that Coale-Demeny[2] Model 3 West Female approximates the age structure of the Roman and Italian population at this point, the ordinary rate of civilian mortality among men seventeen to thirty would be about 1.5 percent per year. Military mortality apart from combat would thus be between about 1.9 and 2.6 percent per year or a total of between 73,096 and 100,026 men from 200 through 168 B.C. on the assumption that an average of 116,580 citizens and allies were serving in the legions every year.[143] Coupled with mortality from combat, these figures yield a total of between 182,838 and 209,768 deaths for an overall rate of mortality among all recruits of between 4.75 and 5.45 percent per year in this period.[144] However, the net or "excess" mortality would be significantly lower because 1.5 percent of all conscripts, had they remained civilians, would have died annually of natural or other causes in the normal course of events. Subtracting them, therefore, yields an excess mortality attributable to warfare in this period of 130,271 to 157,147 deaths or 3.25 to 3.95 percent annually of all soldiers.[145]

Subsequently, the number of legions Rome fielded annually between 167 and 133 diminished somewhat from the levels of the previous thirty-three years, and northern Italy and Spain were generally peaceful for much of this period.[146] However, when serious fighting did erupt in the Spanish provinces, defeats were common. Unfortunately, the numbers of casualties Appian sometimes reports almost certainly include local auxiliary forces, making an estimate of Roman and Italian losses difficult.[147] The republic's final war against Carthage involved no major battles, only a lengthy positional struggle and siege, and although Andriscus destroyed a praetorian army in 148, the larger forces dispatched to combat him seem to have had little trouble defeating his army and later that of the Achaean League. Estimating the numbers of Romans and Italians who died as a re-

Mortality in War

sult of military service in these years on the basis of such limited and problematic evidence is impossible. Again, one can do no more than guess that the overall military mortality rate would not have been higher than that Rome's armies experienced during the first third of the century and might have been substantially lower. If Rome fielded on average 6.5 legions annually in these years, an overall mortality rate for all conscripts of 4.75 to 5.45 percent per year translates into a total of 128,070 to 146,943 deaths for the thirty-four-year period, while an excess mortality rate of 3.25 to 3.95 yields 87,627 to 106,500 deaths attributable to the effects of Roman warfare in the period beyond those that would ordinarily have occurred among men in this age-group.[148] Adding these figures to those for the period 200–168 brings the total losses at war for Rome and Italy from the beginning of the century down to the tribunate of Tiberius Gracchus to a total of 312,256 to 358,059 and the excess mortality to a total of 219,246 to 264,995. If military mortality between 167 and 133 averaged only 75 percent of that in the preceding period, total deaths among soldiers would have been 96,053 to 110,027 and those in excess of the civilian rate would have been 65,720 to 79,875, yielding a total number of deaths for the whole seventy-six-year period of 278,891 to 319,975 and an excess mortality of 195,991 to 237,022. Or put in other terms, if twelve years was the average term of service for all recruits, between 34 and 40 percent of all those Romans and Italians who left to fight Rome's wars might never have come back.[149]

These calculations, it must be stressed, cannot pretend to be anything more than an educated guess about the rate at which soldiers died during this period for the obvious reason that the results they yield are based on a series of assumptions about conditions and events for which our sources preserve little or no information. They provide no more than a starting point for how to think about the extent and impact of military mortality. In doing so, two critical points should be borne in mind in evaluating their usefulness. The first is the improbability that the actual number of Roman and Italian deaths was significantly smaller than that to which this exercise points. Few changes could be made in the assumptions on which the calculations are based that would result in a much lower number of deaths, whereas many changes would raise it considerably. For example, Table 3 is based on the premise that every "victorious legion" listed therein was at full strength when it entered combat, and this cannot have been true in all cases. But even allowing that "victorious legions" were 10 or 20 percent understrength would alter the number of deaths they suffered only marginally.[150] Some truly egregious losses also might be rejected on the grounds that a substantial portion of the deaths occurred among local *auxilia* or perhaps that Roman annalists exaggerated the number of casualties the republic suffered. Defeats in Spain in 194 and 190, in which casualties approached 50 percent, are two obvious cases.[151] But even so the overall rate would scarcely change.[152] And one might also eliminate some casualty figures in Table 2 as mere "conventions." Multiples of 300, a "proverbial" number in the view of some scholars, are the most attractive targets for such a step.[153] Again, however, doing so would not change the mortality rate greatly.[154] Even if all of these changes were adopted, the result would be to decrease the total number of deaths in this period by only a few thousands. Finally, the mortality rate assigned to those "victorious legions" in Table 3 whose losses are not reflected in Table 2 represents a critical variable, and some may feel that a figure midway between the mean and the median is too high in view of the fact that in half of all defeats where the number of deaths is known these were at 1.1 percent or less. But such small losses cannot have been typical in battles against enemies as formidable as the Gauls and Spanish even when the Romans won. As suggested, the heavy casualty figures sustained in battles against these opponents in the historical record probably stem from their rarity overall. They were remarkable and the annalists duly took note.

One might also argue that the estimate of combat deaths is too high on the grounds that many of the battles Livy records never in fact occurred, that they are the inventions of annalistic historians.[155] Consequently any reckoning of total casualties based on the aggregate of battles listed in Table 3 would be inherently flawed if a large number of them in fact did not take place. Such criticisms might seem to gain credibility from Livy's report that although Valerius Antias recounted a major battle fought by the consul of 199, P. Villius Tappulus, in Greece, the other Greek and Latin authors Livy consulted indicated that Villius did nothing worthy of note before his successor arrived.[156] And Appian's failure to record any fighting in Spain between Cato's command in 195 and the tenure of Q. Fulvius Flaccus in 180 might appear to belie the many battles there that Livy describes in those years.[157] Even Livy himself expresses skepticism at his sources' report that L. Postumius Albinus fought two battles in Spain in the summer of 179; in Livy's view Albinus must have reached his province too late in the year to accomplish anything.[158] But episodes such as these afford at best only a tenuous basis for claims of wholesale invention. Livy's skepticism about Postumius's achievements in 179 stems simply from his

Mortality in War

failure to recall that the proconsul had arrived in his province the year before.[159] The two triumphs and three ovations for Spanish victories between 195 and 180 recorded in the *fasti triumphales* demonstrate unequivocally that it is Appian's account of events there that is deficient for this period, not Livy's.[160] And it is noteworthy, too, that Livy himself signals the variation in his sources' accounts of Villius's achievements in Greece; he possessed more critical acuity than he is sometimes given credit for.[161] His failure to mention similar discrepancies for reports of other battles is significant, therefore. Further, the divergence that Livy found in accounts of Villius's activities in Greece was not between Polybius and the annalists but between Valerius Antias and "the other Greek *and Latin* authors" whose works he consulted.[162] The fabrication of Villius's "victory," in other words, was uniquely the work of Antias, not a product of the entire annalistic tradition.[163] It affords no grounds for postulating that similar invented battles were larded throughout the accounts of the Roman historians that Livy consulted.

But another and more substantial argument can be offered as well. One of the most striking features of Table 3 is the absence of significant fighting in so many years. The category "no combat" is far and away the table's most numerous, and the 153 legions represented here equal all reports for legions victorious, defeated, and engaged in minor actions combined. If annalistic historians were inventing fictitious battles to any significant extent in order to pad out their accounts of otherwise uneventful years, one would not expect them to have left so many provinces devoid of combat in any given year. But what need did annalists have for extensive fabrication when the era boasted plenty of real fighting to chronicle?[164] The statistical prevalence of "no combat" in Table 3, therefore, ought to rule out any a priori assumption that Livy's sources were fabricating "victories" for Rome wholesale. Although one cannot exclude the possibility that one or another battle that Livy or another source reports may not be genuine, a few cases of this sort would not be enough to alter significantly the conclusions to which Table 3 points.[165]

On the other hand, several considerations might suggest that the calculations offered here if anything substantially underestimate the numbers of Romans and Italians who died as a result of their military service. While it is at best merely possible that annalists exaggerated the numbers Rome lost in some battles, whether to glorify their victories or because they lumped *auxilia* into the figures along with citizens and *socii*, Roman generals are much more likely to have underreported their losses to the senate at least somewhat. Further, and more critically, the mortality rate assigned to "victorious legions" in Table 3 may well be too low. Assuming greater average losses in at least some cases would result in a higher overall rate of mortality. The most plausible argument for raising the average rate of death in combat for some legions is that three out of the four loss rates above 5 percent for victories occurred in Gaul. These totals however represent only six of the twenty legions that won battles there.[166] Quite possibly, therefore, the losses that the other fourteen legions sustained were significantly closer to the average of these six legions, which was 10.6 percent, rather than that of all legions in the period.[167] Putting the combat mortality for fourteen legions from Table 3 at this rate would pull the total number of deaths from Table 3 up substantially. That total would also rise dramatically on the assumption that the number of unreported combat deaths for "victorious legions" was much closer to the average of the losses in the victories reported in Table 2 rather than their median. That figure, 4.2 percent, might seem defensible in view of comparisons with the losses victorious hoplite armies sustained in classical Greece. Similarly, the figure for total losses due to minor battles and skirmishes could easily be increased simply by assuming a rate of death from these causes higher than the 1.325 percent used in the discussion here. The same is true for the numbers of deaths attributed to complications from wounds and sickness. The latter are based on fairly optimistic assumptions about the quality of medical care that the soldiers could provide one another and about the levels of sanitation in the camps and personal hygiene among the legionaries. Each of these claims is open to debate. Taken altogether, these considerations suggest that the results of the foregoing discussion ought to be taken as a minimum for mortality due to war, while the actual number of deaths attributable to that cause might if anything have been a good deal higher.

Scholars undoubtedly have long been aware in a general way of the terrible toll Rome's wars took upon its soldiers, but their full extent has for too long not been adequately gauged. To win the republic its empire in the second century, tens of thousands of young Roman and Italian men paid the ultimate price: they were killed in combat, died of complications to their wounds, or perished from the diseases prevalent in military camps. Death on this scale over so many years cannot but have profoundly affected the farms and families from which these men came.

Chapter 5

Military Mortality and Agrarian Crisis

If as many young men perished in the course of the Roman Republic's wars between 200 and 133 as the preceding chapter has argued, their deaths, coupled with the massive losses of life Rome and Italy suffered during the long struggle against Hannibal, cannot but have profoundly affected social and economic developments in the second century. But before any assessment can be offered of the consequences of these soldiers' deaths for the class of smallholders from which most of them were drawn as well as how they may have contributed to bringing about the crisis that Tiberius Gracchus addressed, it is critical to underscore where in families' life cycles their impact would have been felt. As argued earlier in this study, the conventional view errs in its assumption that the citizens Rome called to serve in its legions were often, if not usually, older, married fathers of families. Most conscripts were younger and as yet unmarried. Furthermore, older legionaries, those in their late twenties or early thirties and so of marriageable age, generally served as *triarii*. They numbered only between 600 and 740 in a legion in the middle republic, about 14 percent of the total, and the bulk of these probably came from the ranks of men who had not yet wed.[1] Equally important, the role of the *triarii* in combat tended to minimize their exposure to danger. As is well known, they composed a reserve force stationed in the rear of the line of battle upon whom the younger *hastati* and *principes* could fall back in cases of exceptional danger.[2] The latter bore the brunt of the fighting, and consequently their older comrades in the rear ought usually to have suffered the fewest casualties. If, for example, a legion of 5,200 infantry suffered 5 percent killed in a battle, that is 260 men, a proportional distribution of the casualties among the *velites, hastati, principes*, and *triarii* would mean that the latter accounted for only 36 or 37 of the dead. But because the *triarii* were much less exposed to the risk of injury than other categories of soldiers, the number of them killed ought to have been substantially lower.[3] And if few of the *triarii* were married, then the actual number of households that these deaths deprived of their head would have been very small indeed. Although we do not know that the age structures of the units that made up the allied *alae* were the same as those of the legions, there is no reason to expect that the *socii* preferred to conscript older, married men rather than younger, unmarried sons of families when they had the choice.[4] Allied communities therefore probably called up the same sorts of men that Rome did and in more or less the same proportions.

This conclusion does not mean, of course, that a son's death while away at war would have had no impact on his family, only that his survivors would have been more likely to feel its demographic and economic (as opposed to emotional) effects over the long run rather than in the short term. Because a son's departure for Rome's wars did not usually cripple his family's ability to support itself on its land, there is little reason to expect that when death removed him permanently from his family, its farm would suddenly cease to be viable. Over time, however, parents would grow old and either die or no longer have the strength to work the land. Sisters might marry and move away, and if no brother survived, the potential existed for a family's fields to lie fallow for want of someone to work them. Saller's computer simulations of Roman family structure suggest that among men aged twenty to thirty only between 55 and 59 percent might have had one or more living brothers and about a quarter would have been only children.[5] Possibly, then, in a very high proportion of cases, the death of a conscript would result in the loss of his farm's immediate heir.[6]

Yet that would not inevitably have meant a dearth of candidates to succeed to the estate. Testamentary disposition was always possible, although how commonly Roman smallholders made wills is unclear.[7] But for Roman citizens, at least, the law made elaborate provision for the succession to property in cases of intestacy, and as a result potential heirs abounded. Daughters who married *sine manu* remained in the *potestas* of their fathers; consequently they were *sui heredes* and so entitled to an equal share of his estate. By the early third century B.C., even sisters who had married *cum manu* were apparently recognized as heirs in the first instance along with their brothers. Were war to carry off an only son, therefore, a farm could pass to a sister, whose husband or members of his family might work it until her children inherited. In default of *sui heredes*, agnate relatives enjoyed first claim to an intestate estate followed, beginning early in the third century, by cognate kin. While these, as Saller has argued, might not have been numerous, how commonly will estates lacking a *suus heres* have escheated because an aunt, uncle, or cousin was nowhere to be found?[8] In one fashion or another, most farms whose only *suus heres* war had carried off would have found their way into the hands of another family eager for additional fields to till. Furthermore, the role that the *ager publicus* played in the subsistence economy simplified in many cases the problems with succession that such deaths would raise. As noted in a previous chapter, the bulk of the land that many smallholding families tilled was public.[9] Its occupation was fluid, therefore, and we ought to expect that anytime a family was no longer able to work its portion owing to death or old age, the fields would naturally pass to another family within the community that needed more land to cultivate.[10] Brunt calculates that some 120,000 Romans lost their lives during the war with Hannibal along with perhaps the same number of the republic's Italian allies.[11] If a quarter of these 240,000 men were only children, then as many as 60,000 farms might have lacked immediate heirs. By 200 B.C. a great deal of land throughout Italy had come onto the market, in many cases certainly the result of the effects of war deaths upon expected patterns of succession.[12] In the following sixty-seven years, the deaths of 312,000 to 358,000 or more young, unmarried Roman and Italian men in the republic's wars will have created similar disruptions in the transmission of many farms and opened up extensive tracts of *ager publicus* to new occupants.

While recognizing the terrible cost in lives lost to Roman and Italian families that the republic's acquisition of an empire entailed, therefore, we should not lose sight of the fact that for many of the young men who survived, these deaths would have opened up unparalleled opportunities. Some, perhaps many, would have found themselves, quite contrary to their expectations, able to marry when the demise of an heir allowed them to succeed to a farm on which to support a family. And in many cases they (as well as other men) may have wed on advantageous terms. Because young, unmarried men mostly carried the burden of mortality that Rome's conquests imposed, the number of brides seeking husbands far exceeded the supply, and so, as the preceding chapter argued, the prices that brides' families would have been willing to pay ought to have been high compared with the wealth that grooms would bring to a marriage. Men may have wed, too, at a slightly earlier age, and possibly their brides were on average younger as well.[13] Remarriage would have been much easier for older men whose wives had predeceased them. The deaths resulting from Roman warfare may also have caused many men's expectations about the amount and quality of land they would work to rise as overall demand for it declined. As population pressure eases, marginal land tends to fall out of cultivation since farmers concentrate their efforts on the most productive fields. Consequently, productivity rises because the same amount of work applied to the same quantity of better land yields larger harvests (or, conversely, less land of higher quality needs to be worked to produce the same yield). More land can be devoted to crops other than those required for basic subsistence, particularly more profitable ventures such as vines, arboriculture, and livestock. Land devoted to grazing benefits from the animals' manure, which in turn improves yields when it is again under crops. Diet tends to improve and become more varied.[14] Whatever market existed for casual, free labor in the countryside was likely to be very favorable to those in a position to supply it.[15]

To the extent that sexual abstinence, infant exposure, or sale of children into slavery represented means for families to cope with the danger of too many mouths to feed, these practices also will have declined when times were good.[16] Prosperity, under conditions of natural fertility, tends to raise the overall birthrate. One may compare on this point the hypothesis developed by Easterlin to account for the significantly higher birthrate among farming families in the western portions of the northern United States than in northeastern states during the nineteenth century.[17] Easterlin assumed that, on the one hand, fathers of families on farms both in the East and the West had as a primary goal establishing their sons and daughters on farms or in other positions in life at an economic level at least equal to their own when they first started out. On the other, he demonstrated that the abundance of land in the western frontier areas and in-migration allowed fathers of families there to expect much higher rates of return on their capital investments in agriculture than their eastern counterparts. Consequently, farmers in the West foresaw much less difficulty in accumulating the economic capital necessary to provide an appropriate start in life for a large number of children and so were willing to have more of them. In the East, on the contrary, fathers of farm families, unable to expect a comparable rate of increase in their wealth, took what steps they could to keep their families small.[18]

Other developments taking place in this period may have accelerated these trends, the most important of which was the growth of the city of Rome. Morley has recently emphasized that as Rome grew from a population of perhaps 200,000 around 200 to about 500,000 by the tribunate of Tiberius Gracchus, it annually absorbed new residents at a rate as great if not greater than that at which the republic's wars consumed the lives of its conscripts. Morley estimates that about 7,000 immigrants would have been necessary every year both to replace the excess of deaths over births among urban dwellers as well as to add 300,000 new residents to the capital's total by 133. Most of these would have come from the free, rural population of Italy and, more important, they would have been men and women in the prime of life, between the ages of perhaps fifteen and thirty.[19] In other words, the very same younger, unmarried men who composed the principal pool of recruits for the legions and who died in such large numbers in the republic's wars were in these same years fueling much of the capital's dramatic growth. Like those who died as a result of Rome's wars, immigrants to the capital significantly eased the burden on the land they left behind. In many cases the newcomers did not start a family or, if they did, they did not depend on access to a plot of land to cultivate in order to support it. Consequently, their departures ought to have diminished considerably the pressure on the existing supply of land for those families who stayed behind just as much as the deaths of conscripts did for those families whose sons returned home following a long stint with the legions. Also, a series of epidemics struck the rural population of Italy in the 180s and 170s B.C., and the many deaths these caused must have had much the same results for the survivors.[20] And the senate's extensive program of colonization during the first two decades of the century certainly also brought about a similar increase in the availability of land for those who remained after the emigrants left.[21] Indeed, one can suspect that the reason that colonization largely ceased after 181 was at least as much because the senate was running out of potential recipients as because the supply of land to allot was drying up.[22] Similar to some survivors of later European pandemics like the Black Death, many individual Roman and Italian smallholders in the second century may have faced short-term prospects that were anything but bleak.[23]

In broad social terms, however, prospects for this class may have been far less promising. An excess mortality on the order of that suggested in the previous chapter, it might be supposed, ought to have significantly constrained the social reproduction of the smallholder class. The deaths of so many potential husbands at war necessarily would have reduced the rate at which families of small farmers could produce new families to succeed them when their sons and daughters married. Consequently, the size of the smallholding class would have failed to increase and perhaps even declined. Certainly, this has been the dominant view of Italy's demographic development during the second century, a position championed by Beloch and most powerfully by Brunt.[24] Yet our principal evidence for population trends during this period, the census figures preserved in Livy and other historians, suggests precisely the reverse: over the period from 203 to 125/4, the number of Roman citizens counted by successive pairs of censors generally increased, a trend also reflected in the "corrected" census figures Brunt develops by adding in citizens, mainly men serving with the legions, whom the censors are likely to have missed.[25] He argues that between 203 and 168 B.C., the number of Roman citizens grew by 106,000, from about 240,000 to about 346,000 or at an annual rate of increase of nearly 1.1 percent. From 168 to 125/4, the population in his view increased from roughly 346,000 to 433,500 citizens or at an annual rate of about 0.5 percent.[26] Overall, therefore, Brunt's reconstruction of the population trend over the seventy-nine-year period from 203 to 124 posits a 0.75 percent annual rate of growth. But in fact the rate would have to have been substantially higher to allow for excess mortality due to war. From 203 through 168, where our best evidence for military casualties occurs, between about 35,414 and 63,966 citizens probably died as a result of Rome's wars in this period beyond ordinary civilian mortality.[27] Therefore, the actual increase in the citizen population in these years will have had to have been on the order of 141,414 to 169,966, or about 1.3 to 1.5 percent per year, in order to yield a net increase of 106,000 by 168.[28] Adding an estimate of the excess deaths of citizens resulting from Rome's military activity between 168 and 124 to the increase that Brunt calculates for this period yields a total increase for these forty-four years in the range of 104,924 to 148,848, or between about .6 and .8 percent per year.[29] Overall, adding the excess military mortality to the increase in the citizen population from Brunt's "corrected" census figures between 203 and 124 results in an average annual increase of between nearly .9 and 1.07 percent for the entire period.

A rate of increase on the order of 1.5 percent per year is staggeringly rapid for a preindustrial population, at least compared with the rates attested for those early modern Western European nations where population figures can be reliably reconstructed.[30] Even 1 percent per year is quite high. Brunt accounts for the latter figure by attributing much of the increase to manumission. He argues that the greatly increased numbers of slaves that Romans imported into Italy during the second century produced an equally large increase in the numbers of slaves who, upon being freed by their owners, became Roman citizens. It was these newly enfranchised freedmen rather than any significant increase in the overall numbers of smallholders in Brunt's view who swelled the census returns over the course of the second century.³¹ However, as Chapter 1 argued, the increase in the number of slaves entering Italy in this period was probably much smaller than Brunt assumes.³² Hence, the claim that a greatly increased population of slaves produced an equally large increase in the numbers of slaves who won their freedom and so increased the size of the citizen body lacks foundation. Nor are there obvious grounds for believing that the proportion of slaves who won their freedom increased significantly in this period. Manumission certainly may have made some contribution to the enlargement of the citizenry, but this phenomenon is unlikely to account for more than a small part of it. Yet even if one concedes that newly freed slaves accounted for as much as half of the increase over the course of these decades (which seems far too large), the remainder would suggest a robust rate of growth among the rural population, particularly in view of the number of deaths produced by epidemics in this period as well as the heightened mortality among immigrants to Rome.

The census returns would seem therefore, at least prima facie, to present something of a paradox. In spite of the high burden of mortality that war imposed on the Roman population, it nevertheless appears to have increased steadily and significantly over the course of the first two-thirds of the second century. Now it is at least within the realm of possibility that the census itself, and so the results obtained from it, was so flawed in its conception or execution that the figures preserved from it bear little or no relation to reality. But few if any scholars would go so far as to jettison this evidence altogether. As Brunt has argued, perhaps the main reason to believe that the census returns do in fact reflect the citizen population with some degree of accuracy is their general consistency over the whole period.³³ If that is so, then one must also believe not only that the citizen population grew overall from 203 to 125/4 but that the bulk of this increase occurred among the class of smallholders who formed the great majority of Romans.

To be sure, suggestions by scholars to this effect have not been lacking.³⁴ But the great difficulty with any hypothesis along these lines is to account for such growth. Under the environmental and mortality conditions generally obtaining in preindustrial societies before the demographic transition that introduced parity-specific family limitation, societies (at least in the West) tended to develop fertility regimes that result in an equilibrium between births and deaths.³⁵ Such regimes bring about moderate fertility among the population in order to keep it stable or, at most, growing only slowly and so more or less in balance with the economic and environmental resources available to it.[36] Consequently, one needs to explain how the Roman population (and one presumes the Italian, although evidence here is lacking) came to diverge so radically from demographic norms.

Possibly, much of this growth is illusory.[37] Population figures for early modern European populations include men, women, and children, whereas Roman censors recorded only free adult male citizens, and it cannot necessarily be assumed that the demographic trends among the latter reflect more or less those for the free Roman population as a whole in this period. If it is true that Rome fought the Second Punic War mainly with unmarried men under thirty, married men over that age probably kept reproducing at fairly normal rates throughout the war at the same time that men under thirty were dying in large numbers.[38] If so, the result by the war's end would have been a fairly high ratio of males under seventeen to adult male citizens compared, for example, with that prior to 218. Consequently, as the age-cohorts of boys born during the war reached adulthood during the first couple of decades in the second century and were enrolled as citizens by successive pairs of censors, the net increases in the total number of citizens that each census registered would have been fairly steep until the demographic consequences of the heavy losses among the seventeen to thirty year olds who fought the war began to make themselves felt. Because the war had killed so many of this generation, its male survivors will have contracted fewer marriages overall and consequently produced fewer children to come of age beginning in the late 180s, when sons born beginning around 200 would start to turn seventeen and be counted as adult citizens.

The full demographic complexities of this hypothesis cannot be worked out here, but it is suggestive that the census figure for 188 B.C. as corrected by Brunt reflects a nearly 1.3 percent annual rate of increase over the returns for 203. During the following two decades the rate of increase drops significantly, to slightly under .9 percent annually.[39] But the latter still suggests very strong population growth, and the trend from 203 to 168 overall, particularly in view of military mortality in the period and the other factors discussed previously, implies that rates of marriages and especially births among smallholders during the Hannibalic War and in the years following were at or above normal. And although Brunt denies that marriages and births could have been unaffected by the Hannibalic War to the extent the census and mortality figures suggest, his pessimistic assessment stands in need of reexamination, for it arises from a misunderstanding of the relationship of war, agriculture, and family life. In his view, conditions

Mortality and Crisis

during the war and the period following would have made it impossible for the population to increase to any great extent, and his analysis is worth quoting at length:

> None the less agricultural production may well have declined [during the Hannibalic War], even though Roman lands outside Campania were hardly affected by devastation, since *there must have been a serious diminution of the labour force* and since it was only from 211 that the rich Campanian land could again be worked for the benefit of Rome. Little food could be imported; the Italians, loyalists and rebels alike, had to depend mainly on what they could grow for themselves, and in 211 there was a grave, if temporary, shortage. The people certainly did not die *en masse* of famine, but undernourishment may have made them more susceptible to disease; serious epidemics are recorded in 208 and later in the war, associated with famine. . . . These conditions must also have been unfavorable to conceptions and births, and indeed the marriage rate probably declined; even though soldiers serving in Italy could return home on leave in the winter months (and also help on the farms), there were between 6 and 10 legions overseas for many years, some 24,000 to 40,000 citizens (less casualties), who had no opportunity for reunion with their wives or for consummating new marriages. . . . It is also likely that the women and children were also many fewer in 200 than in 218. Hence, the number of adult males may well not have risen fast after the end of the conflict.[40]

As argued in the preceding chapters, however, in most cases families disposed of enough labor to work their farms despite the conscription of sons for the legions. There is no reason to assume a priori "a serious diminution of the labour force" that the civilian population required to feed itself. While it may be true that little food could be imported, the rural population had never depended on imported food for its subsistence, and in any case it is not apparent how imports of grain from Sicily, Sardinia, or elsewhere would have been distributed to those in the countryside had they been in need. They all cannot have come to Rome. Indeed, the fact that the countryside was not depopulated during the war, as Brunt persuasively demonstrates elsewhere, argues strongly that people there were not starving.[41] Moreover, of the passages that Brunt adduces to substantiate the claim of "serious epidemics . . . associated with famine," all but one refer not to the civilian population but to armies.[42] Most of these have already been considered.[43] The epidemic of 212 or 211 struck not in Roman territory but among the legions besieging Syracuse; those mentioned at Livy 27.9.3 allegedly affected allied soldiers serving with the Roman forces. The passage at Livy 28.12.7–8 has nothing to do with civilians but refers rather to the difficulties Hannibal faced in feeding his army after his withdrawal to Bruttium, an area that would not have been able to feed an army the size of his even under the best of circumstances. The epidemic of 205 affected only the Roman and Punic forces in Bruttium, not the general population, and only in the case of the Carthaginians is there any association with shortage of food.[44] Only the pestilence of 208 touched the civilian population; its effects, while long lasting, were not particularly fatal, and Livy offers no reason to associate its outbreak with famine.[45] The "grave, if temporary, shortage" of food in 211 on the other hand had nothing to do with a lack of labor but with Hannibal's march on Rome, which resulted in a massive flight of the rural population into the city, creating scarcity and driving up prices, as Brunt himself acknowledges elsewhere.[46] In short, there is no evidence for the kind of pervasive, chronic shortage of food or prevalence of epidemics among the civilian population that Brunt is suggesting. He presumes that that famine was pervasive solely on the basis of his assumption that the men drafted into the legions provided the principal agricultural work force on Italy's small farms without whose presence those left behind would have been unable to work their land. As argued previously, this supposition is incorrect and hence so is the claim that undernourishment opened the way for serious outbreaks of disease in the *ager Romanus* during the Hannibalic War.[47]

More important, Brunt errs in his analysis of the effects the massive conscription beginning in 216 will have had on rates of conception. Men who had married by 218 largely stayed out of the fighting and so will have continued to father children at a normal rate. For this reason, the presence of 24,000 to 40,000 conscripts overseas will not have caused the rates of conception to decline dramatically. The war will have affected rates of nuptuality—apart from the large number of men who would never marry because they had been killed in the fighting—only in the cases of men in their late twenties or early thirties. The military crisis and the shortage of recruits for the legions during the war's middle years certainly forced many of them to continue their military service longer than usual and so prevented them from returning to civilian life to wed and become fathers.[48] However, one must also bear in mind that the senate demobilized six or seven legions in 210, another four in 207, and others from 205 onward. In the case of the legions dismissed in 210 at least, the *patres* ordered that none of their soldiers be reenrolled.[49] Although it is not clear whether these exemptions would have been preserved throughout the remainder of the war, without doubt the men least likely to have been called up again will have been the older ones, those of an age to marry and begin their families, both because recruiters, when given a choice, prefer younger soldiers to older ones, all other things being equal, and because the number of *triarii* required for the legions was much smaller than those for other age categories.

Prospects therefore for near normal rates of birth during the Hannibalic War, at least among the Romans and their remaining *socii*, would not necessarily have been as bleak as Brunt supposed. And as outlined earlier, the economic opportunities for families in the conditions of high mortality that Rome's wars produced would often have been quite favorable. However, to some extent any argument for a high rate of reproduction among smallholders in the late third and early second centuries depends on whether the number of potential husbands who died at war could have been replaced as spouses by surviving men who in other circumstances would never have married or by widowers who would not otherwise have remarried, so that the number of marriages overall remained more or less the same. For this to have happened, the Roman population would have to have contained a sizable pool of men who ordinarily would never have married, and its existence cannot be established on the basis of the surviving evidence. Certainly, Brunt claims that half the citizen population in 218 consisted of *proletarii* who in his view would have been unlikely to begin families; as argued elsewhere, however, this estimate is far too high.[50] Still, the existence of some sort of pool of substitute spouses cannot be ruled out; indeed, a rate of population increase on the order of what the census and military mortality figures suggest virtually requires a large supply of replacements for those potential husbands who died while on military service. Immigration, too, perhaps increased the pool of potential husbands. Latins had the right of *conubium* with Roman citizens.[51] Since high military mortality will have produced a surplus of prospective citizen brides over potential citizen grooms, in some cases the places of the latter may have been taken by Latin men. We know that the leaders of Latin colonies on more than one occasion expressed their concern to the *patres* that their territories were becoming depopulated due to migration of their citizens to Rome.[52] Marriage may have been the vehicle by which many of these men established themselves in Roman territory, where, availing themselves of the *ius migrationis*, they could become citizens whose offspring then enlarged subsequent census totals.[53]

Perhaps, too, in the wake of the terrible losses early in the Second Punic

War and subsequent deaths among soldiers, couples were willing to have more children, to the extent that they could control their fertility.[54] And continuing high mortality from the republic's wars in the decades that followed could have helped continue this trend, particularly as families sought to ensure that at least one male heir would survive to succeed to their property. Such a "postwar baby boom" is perhaps unlikely to have gotten under way much before Hannibal's departure from Italy in 203 brought a final sense of security from the possibility of a sudden military crisis and so would not have begun to affect the census figures until the later 180s, when the sons it produced reached adult status. Greater numbers of children per couple thus may help explain the continuing increase in the census figures after 188 despite a drop in the number of marriages contracted overall owing to the "hollow generation" of men that the war created. This decrease may also have been offset, at least temporarily, by an upsurge in marriages by men whom conscription for the duration of the Hannibalic War had forced to delay their nuptials.

Still, there are limits to the size of the increase that such a hypothesis can plausibly postulate. While the rate of natural increase in any human population can be very high in the absence of any restriction on fertility, in practice such limitations have always existed even before the advent of effective medical means of contraception. The mechanisms that control population growth, apart from environmental factors such as a high incidence of mortality due to disease, involve customs like the age of women at first marriage, the length of time infants are breast-fed and the intensity of their suckling, taboos governing when couples resume intercourse following the birth of a child, and the practice of infanticide, among others.[55] Next to nothing is known about most of these and similar aspects of Roman family life, but it cannot be doubted that some combination of these factors or ones like them had long affected the rate at which the population expanded.[56] Most of them seem unlikely to have altered rapidly in response to short-term developments. It is hard to believe, for example, that mothers throughout Roman territory (and presumably Italy as well) suddenly began nursing their infants for much shorter periods of time because of the disaster at Cannae or when Hannibal sailed back to Africa or even because more land was available for their husbands to work. But it is at least conceivable that a high ratio of prospective brides to potential grooms could have led to some decrease in women's ages at first marriage since families might have sought to arrange for their daughters to wed at the earliest opportunity for fear that another might not come along. Here, though, cultural expectations about when a young woman is ready to marry can also come into play. These might trump other, more pragmatic considerations and so limit the population increase that might be ascribed to this source.[57] But some reduction in the average age of women at first marriage cannot be altogether ruled out, so that overall the number of offspring couples produced may well have risen simply because women began having children earlier.[58]

The impact of these and similar changes, however, should probably be understood in light of another and more basic characteristic of the republic's demographic regime. Malthus long ago suggested that the republic's constant warfare prior to the Hannibalic War had already begun to keep the usual preventive checks on population growth—particularly delayed marriage and long intervals between births—from operating within Roman society. The mechanisms that in most societies before the demographic transition worked to keep the population increasing over the long term only very slowly if at all and well under the point at which it would have begun to outstrip its environmental carrying capacity simply were not as much in evidence at Rome as elsewhere. Instead, military mortality itself over the course of the late fourth and the third centuries down to 218, while not as heavy as that occurring during and after the Hannibalic War, represented a significant check on population growth.[59] As a consequence, other sorts of limits, such as a relatively high age for women at first marriage, did not need to play as great a role in limiting births, for an excess here would eventually be balanced by the high number of young men whom the republic's wars prevented from ever starting families.

Support for Malthus's hypothesis can be found in Rome's regular dispatch of colonies throughout this same era.[60] Colonization generally occurred on the initiative of the senate, and it is difficult to believe that the *patres* would have repeatedly embraced initiatives that depleted the republic's own reserves of legionary manpower, as the dispatch of citizens to the Latin colonies typically founded in this period would have done, if the citizen population had not been increasing.[61] And a steady demand among the Romans for land, which helped propel both the republic's war making and the colonial foundations and viritane distributions throughout Italy that were its consequence, is not likely to have been principally the result of expansion in the amount of territory occupied by large, slave-staffed estates.[62] Their growth was slow before the first century. Suggestive, too, is the popular support that Flaminius's tribunician bill to distribute the *ager Gallicus* attracted. The public's enthusiasm was sufficient not only to overcome senatorial opposition to the legislation but also to help the *novus homo* Flaminius to a pair of consulates and a censorship.[63] This agitation, coming at the end of a long period of regular colonization, demonstrates that a substantial number of citizens at that point still lacked land and, along with the senate's regular export of citizens to colonies, suggests that a steady growth in population characterized Roman society during the later fourth and third centuries if not earlier. But if the population had been steadily increasing throughout this period, then the regular opportunities to occupy new land that colonization offered as well as those created by the deaths of many young men at war must have worked to loosen the sorts of social and cultural constraints on population growth that typically operated in other pretransitional populations. Consequently, when the demographic catastrophe of the Hannibalic War's opening battles struck and then was followed by decades of very high military mortality, Roman society may already have had in place a demographic regime that encouraged a substantial excess of births over deaths. In other words, the alteration that may have occurred in the second century was not a sudden shift from a pattern of reproductive practices conducive to "no growth" to one that encouraged "high growth" but a move from strong to very rapid expansion by means of a further relaxation of cultural norms, such as women's age at first marriage, that were already quite pronatalist to begin with.

Surprisingly, therefore, the great many deaths of young Roman men between 218 and the last third or so of the second century are very likely to have made a significant contribution to the dramatic rise in population that took place following the defeat of Hannibal. By increasing the availability of land and its overall productivity and profitability, by enhancing opportunities for occasional paid labor and the bargaining position of those in a position to supply it, by improving diet, by fostering a reduction in the age when women married, and by increasing couples' willingness to have and raise more children, the era's high military mortality helped set in motion the cultural and social changes that brought about a rapid, dramatic growth of the Roman population. Possibly, too, at least some non-Roman peoples in the rest of Italy also benefited (if that is the right word) from similar developments. If the census figures can be believed, by the last third of the second century the Roman population had surpassed its prewar level by 25 percent or more. But in helping to bring about this result, war also set the stage for the turbulent political developments of the Gracchan era.

The hypothesis that high mortality among Roman (and Italian) soldiers coupled with emigration to colonies or Rome and outbreaks of epidemic disease helped to catalyze rapid population growth among the republic's citizens during the first two-thirds of the second century opens a very different perspective on social and economic background to the tribunate of Tiberius Gracchus. An increasing number of smallholders and the prevalence of partible inheritance among them offer an attractive alternative to conventional accounts of the origins of the agrarian crisis that Gracchus sought to solve.[64] Under this hypothesis, most of the men who supported the *lex agraria* had not been forced from their farms under the twin pressures of the conscription of too many of them to fight Rome's wars and competition from a growing number of large, slave-run estates. They were instead simply the inevitable result of too many people attempting to start out in life with too little land.

A reconstruction along these lines does not, of course, rule out the possibility of other, contributing causes. Indeed, the outpouring of public support that greeted the legislative programs of Tiberius Gracchus and the reformers who followed him indicates that the origins of the situation they confronted were complex and embraced a variety of factors.[65] Some, possibly many, of those who looked to Tiberius for succor may have formed part of the urban plebs.[66] The growth of the city of Rome in the decades since the end of the Hannibalic War may well have suggested to reformers that a resettlement program could transform some recent immigrants once again into smallholders and soldiers. Yet Diodorus and Appian note explicitly that the bulk of Gracchus's supporters came from the countryside.[67] Consequently, to conceive of his measure as exclusively an effort to rid Rome of an unproductive urban rabble fails to account for the avid support it evoked among his rural partisans. Possibly, too, a measure of truth lies behind Gaius Gracchus's claims about the inspiration for his brother's initiative, Tiberius's journey through Etruria in 137 on which he found a dearth of free inhabitants but many estates worked by imported slaves.[68] While the evidence for the growth of the "slave mode of production" clearly places its great efflorescence in the first century, the roots of this development certainly extended well back into the latter decades of the second.[69] That some small farmers were pushed off their land by larger neighbors therefore cannot be ruled out, even if the phenomenon may have loomed larger in the political diatribes of the period than it did in the agricultural changes actually taking place in the countryside.[70] And the slave wars that had broken out in Sicily shortly before may have led to fears of similar uprisings in Italy and so to a desire to increase the number of smallholders there to counter this threat.[71]

A principal aim of the *lex agraria* was also probably to enlarge the pool of recruits for the republic's armies.[72] Tiberius himself, if we can trust the reports of his speech preserved by Plutarch, claimed that men who fought for Rome wandered, homeless and unsettled, with their wives and children and had no family shrines, no ancestral tombs, and not a single piece of earth to call their own.[73] This statement cannot be literally true; Rome at this point was still insisting on a property qualification for those it conscripted.[74] And if the arguments presented in Chapter 3 are correct, the republic rarely called upon fathers of families to man its legions. Yet despite the speech's considerable hyperbole, it may well reflect an important element in the argument Tiberius and his supporters were making in support of the law. As the population multiplied and parents divided smaller and smaller inheritances among their children, the number of citizens whose wealth placed them among the *proletarii* may well have been increasing, too.[75] In this light, the possibility that around 141 the *patres* decided to reduce once again the minimum census required to be rated an *assiduus* is suggestive, although claims for such a reduction are problematic and, as in the case of the reduction of 212/11, any increase in the number of those claiming to be *proletarii* may have owed more to a general reluctance of potential recruits to serve in the Spanish wars than to actual poverty.[76]

The great weakness in any argument premised on overpopulation, however, lies in the difficulty of reconciling it with the evidence that the literary sources preserve. Appian states quite plainly that fears of a decline in population formed the background to Tiberius's initiative, not concerns over the consequences of too many people.[77] This evidence cannot simply be brushed aside, and a recent attempt to argue that Appian's language in the passages in question does not mean what it is usually taken to say fails to convince.[78] On the contrary, the numbers of citizens successive pairs of censors tallied between 164 and 136 show a nearly steady decrease.[79] Losses in some of the republic's combats in Spain had been appallingly high.[80] And the well-known speech of Q. Metellus Macedonicus during his censorship in 131, in which he urged Roman men to marry despite the irksomeness of the prospect, shows unmistakably that the republic's leadership had taken note of these developments and was duly concerned to increase the birthrate.[81] The speech and the census figures accord well with the testimony of Plutarch, who indicates that a tendency to avoid raising children—and, one presumes, to marry as well—lay behind the abortive land reform proposal of C. Laelius in 140 B.C.[82] And Appian's notice that the Italian population was shrinking likewise implies a reluctance on the part of many men to begin families.[83] The accounts of these writers clearly echo a genuine anxiety about a declining population and a possible shortage of manpower among Tiberius's contemporaries.

Still, the existence of a perception of population decline and its reality are two different things, and simply because the one was widespread among the *patres*, the other does not necessarily follow. Among French intellectuals of the eighteenth century, for example, the notion that the country was losing population was axiomatic, despite the fact that in truth the reverse was happening.[84] The Roman census depended in the last analysis on the voluntary compliance of citizens: if they did not register, there was little the government could do to compel them, despite draconian penalties for failure to do so.[85] And although conventional estimates of the actual number of citizens assume a 10 percent underregistration as a matter of course, the apparent jump in the census totals preserved by Livy, from 318,823 in 131/0 to 394,736 in the census of 125/4, strongly suggests that a rate of failure to register nearly double that was common throughout the preceding thirty-year period. In other words, while the Roman leadership may have believed that citizen numbers were falling, the basis for that belief was fallacious. The apparent decline in the number of citizens successive censors had counted since 146 was simply the product of an imperfect and unreliable mechanism for determining the size of the citizenry.[86] The failure of the citizens to come forward to register in censuses since the middle of the century is easily comprehensible in view of the widespread reluctance of conscripts to serve in Spain, particularly if the senate had, around 141, reduced the threshold for *assidui* by 60 percent and so obviated an escape from the draft on the plea of poverty.[87] Coupled with this phenomenon, some censors may have been far less energetic in their efforts to register citizens than others and so compounded the problem. If that is so, then Gracchus's program of reform was based on an error that had profound and fatal consequences for his measure's prospects. For if he and his supporters believed that the population was declining, then it must have stood to reason that land to distribute to those who lacked it ought to have been plentiful if only it could be pried from

Mortality and Crisis

the hands of those who illegally occupied excessively large tracts of the *ager publicus*. But in reality, any scheme to redistribute land in a context of a rising population faced the daunting prospect of having to acquire a large supply of an already scarce commodity to hand out.

To ascribe the agrarian crisis of the later second century simply to a growing population, partible inheritance, and a finite amount of farmland is however inadequate, for it fails to appreciate the complexity of the developments that led Rome to this impasse and, in particular, the full role that warfare played in bringing it about. For the impact of war on agricultural developments extended well beyond the deaths that military service inevitably produced among the citizens and *socii* who fought them. Simply by denying families the labor of their sons, conscription in and of itself held the potential to alter greatly the economic and social positions of Rome's and Italy's smallholding families. Consider, by way of illustration, a family that worked a fairly large farm, the labor requirements of which were at or near its maximum potential. Under the assumptions used in models developed in Chapter 3, no great amount of land would be required to reach this point.[88] If the mature family in the first example (a father aged fifty, a forty-year-old mother, two sons, twenty and fifteen, and a daughter, ten) worked a farm of sixty *iugera* (slightly more than fifteen hectares or thirty-seven acres) comprised of fifty *iugera* of grain and bean fields, two more of gardens, vines, and orchard, and another eight for pasture, the arable might require about 975 man-days of labor per year, the vegetable garden, vineyards, and orchards another 100, and the pasture 80 more. The total, 1,155 man-days, would put this hypothetical family very close to its full labor potential of 1,189 to 1,312 annual man-days of work. If one or both sons went off to war, the family would face a labor deficit. Such a shortage would not have meant starvation, of course. As Chapter 3 argued, this family could still have worked enough of its land to grow the food necessary to survive. Rather, conscription would have set up a very different sort of conflict, one that pitted the family's private economic and social status against the public benefit to Rome that military service represented. For a family with this much land could have expected much more from its farm than mere subsistence. At a 3:1 crop-to-seed ratio and a sowing rate of five *modi* per *iugerum*, fifty *iugera* would yield 3,325 kilograms of wheat, about double this hypothetical family's subsistence needs of 1,384 to 1,591 kilograms per year. And of course these figures are based on quite pessimistic assumptions about yield ratios and labor inputs. If yields per *iugerum* were greater, then the harvest would have been even larger.

The ability to work enough land to generate a substantial surplus would in turn have allowed this family's members to enjoy a markedly higher standard of living than mere survival would entail. The family might also have been in a position to dispose of some of its surplus in the form of gifts or loans to neighbors that could be called upon in times of need or that placed recipients unable to reciprocate in a socially inferior position. Obligations of this sort represented an important source of accumulated social capital that might have formed the basis for a degree of superiority over the family's neighbors. Or a surplus could have been saved, leading to the accumulation of economic capital that might have permitted the acquisition of more land and possibly even slaves with which to work it. Alternatively, its savings could have allowed the family to shift some production away from crops devoted to consumption and into more capital- and labor-intensive and so potentially more profitable forms of agriculture, such as viticulture. And to the extent that wealth correlated with family size during this era, as it did for example among serfs in nineteenth-century Russia and settlers in the western United States during the same era, a family with this much land might be expected to have been larger than average and hence to have possessed more potential manpower than most.[89] For a family to forgo all this by surrendering its young men to the army cannot have been easy. More important, the farm itself would have suffered because some of its fields would have been insufficiently cultivated or its orchards or vineyards neglected. One may compare the complaints voiced by prosperous French peasants during the Napoleonic era when the draft took their sons. Starvation was not widespread in the countryside during this period; rather, farmers simply did not have enough labor to exploit their land to its maximum potential when the military drew off so much rural manpower.[90]

To appreciate fully the handicap that conscription imposed on larger smallholdings, however, its effects over the long term need to be assessed, and here comparative material can be very illuminating. In the late nineteenth and early twentieth centuries, students of the Russian peasantry confidently predicted that it would ultimately split into two opposed classes, one composed of rich, capitalist entrepreneurs and the other made up of poor wage laborers. Yet in the event, it conspicuously failed to do so, resisting polarization and displaying a remarkable degree of stability over time.[91] In his classic study, *The Awkward Class*, Shanin postulated a theory to account for the peasantry's failure to develop as expected. He argued that behind this facade of apparent general stability individual families experienced a great deal of economic and social mobility over time. A typical family in the course of several decades would go through alternating periods of both increasing and declining prosperity but only rarely reach a point at one extreme or the other that either allowed it to escape the peasant class altogether or caused it to break up and cease to exist as an economic and social unit. He termed this phenomenon "cyclical mobility" and developed a complex model to account for it, one based on a variety of quantitative and empirical data and encompassing a number of different causes and kinds of movement.[92] At the model's heart lies Shanin's demonstration that what he termed "centrifugal trends"—that is, those developments that led to increases or decreases in families' wealth and social status—coexisted with and were to some extent counterbalanced by what he termed "centripetal trends." These tended to offset growing inequalities in wealth among peasant families, particularly the increasing wealth of some, and move all families back toward the middle. Among the latter trends he identified the partitioning of families that occurred when the members of a household, particularly one with greater than average wealth, agreed to split its assets among themselves to form new, independent families, but families much weaker in economic terms than the one they left.[93]

Certainly Shanin's model of cyclical mobility is not applicable in every respect to Roman and Italian small farming families of the second century B.C. However, the republic's long-standing practice of holding a census every five years in which, among other things, the censors reexamined every citizen's wealth and could alter his place within the census hierarchy if this had increased or decreased certainly implies the expectation among the *patres* that a good deal of movement up and down the economic scale would have taken place among the citizenry in the interval. And the economic deterioration that often resulted from family partitioning among Russian peasants suggests the possibility that when some Roman and Italian smallholding families gradually broke up as daughters married and sons established families of their own, the result, all other things being equal, would have been substantial reductions in their members' material well-being. Although one might suppose that the spouses whom sons and daughters married would have brought with them wealth roughly equivalent (or perhaps, in the case of daughters-in-law when husbands were scarce, somewhat greater) to their partner's portion of a family's estate and that this would have helped to counteract any drop in a new family's financial circumstances, such a matching of a dowry or inheritance would only have returned the new family to the parents' economic position when the children numbered fewer than three. If a family with three children made an equal division of its property among them and each wed a spouse who brought a comparable amount to the marriage, the resulting new families might each start out with about half the wealth of its birth family and ultimately inherit only two-thirds as much as the latter had had.[94] More children would naturally reduce the portions accordingly. And if any of these new families in its turn produced three children or more who survived to a marriageable age and failed to make substantial gains during its mature phase, the result, again if one assumes an equal division among the children and a spouse who brought similar wealth to the marriage, would have been new families that ultimately wound up possessing less than half the wealth of their grandparents. And to the extent that wealthier families had more children, the danger that their heirs' economic position would drop significantly when a family divided its resources among them was potentially that much greater. Saller's computer simulation suggests that family size tended to be small on average, about 1.6 surviving siblings for any "ordinary" thirty-year-old male.[95] If his model is accepted for the middle republic, it implies that families with three or four children surviving to adulthood, while not typical, would not have been uncommon either. However, Saller's simulation assumes a stationary population.[96] An even greater incidence of families with more than two adult children is a necessary corollary to the hypothesis advanced here that the population was increasing rapidly in this period.

Such a gradual decline certainly would not have occurred in every smallholding family or even a majority of them; many things could counteract it — the death of siblings or fewer than three children in the first place, a lucky inheritance, an advantageous marriage, or a run of particularly good harvests. However, one vital potential brake on any tendency toward steadily decreasing levels of wealth in families over several generations lay embedded in Roman families' sons' typically late age of marriage. Families could draw on their sons' full productive capacity during the long period between their late teens and the point at which they left to wed in their late twenties or early thirties. This abundance of labor relative to consumption gave families the potential to make substantial economic gains at this stage of their life cycles, particularly those in which the father survived. A family's accumulation of capital during this phase in turn meant that when the time came to begin dividing the farm's assets to allow sons to establish families of their own or a daughter to take her portion as dowry when she left to marry, the new households that resulted would be in a much stronger economic position.[97] They might expect to start out not simply with land but, as important, in possession of the full array of animals, tools, household goods, and other items essential to self-sufficiency. They would certainly have a stock of seed and food enough to tide them over until they made a crop, perhaps even for more than one year in case of a poor harvest. Furthermore, to the extent that a family had invested its surplus production in loans and gifts to neighbors or other forms of community generosity, its sons and daughters could expect to draw on that stock of social capital in their turn when the need arose.

But by sending sons off to war for most of the period between their coming of age and their marriages, the republic foreclosed this opportunity for families of small farmers to improve significantly their economic and social lot. In the terms of Shanin's model of cyclical mobility, conscription prevented the "centrifugal trend" of a family's increasing wealth during its mature years from offsetting the "centripetal trend" that the partitioning of households represented. Families' inability in many cases to put all their members' labor to work takes on added significance in the context of the forces outlined previously that acted to open up farmland during the late third and early second centuries: high military mortality, emigration to Rome, periodic epidemics, and colonization early in the century. Ironically, then, at a time when possibilities abounded to realize the full productive potential of many families' work force, they faced a shortage of workers. The farms that soldiers returned to upon their discharge would not usually have been ruined or loaded with debt. Rather, the danger was that those who remained behind on them would not have been able to augment or improve them much in their sons' absence. Military service in other words held the potential to force some of those families that had managed to accumulate somewhat greater wealth back down into the class of subsistence farmers as the partitioning of inheritances reduced the economic circumstances of succeeding generations without a corresponding opportunity to build them back up again during a family's mature years. And families with several children already at the subsistence level might not have been able to accumulate enough wealth to enable all of their sons and daughters to marry at all.

Colonies would have mitigated this downward trend, at least in some cases. For families whose assets were insufficient to establish all their children on farms of their own, these would have provided a means of doing so and in this way compensated them for the opportunities lost when they surrendered their sons to long service with the legions. And for many young men, going off to war and then to a colonial foundation after discharge may have been preferable—even with (or perhaps because of) the risks—to long years of toil on a small plot of land, slowly building up resources for the day when they might marry. But colonies would have provided little comfort to wealthier families for whom the conscription of a son for a decade or more threatened to reduce the succeeding generation to an economic level substantially below that of its parents. And by the century's middle years colonization had largely ceased.[98] Booty and donatives from successful warfare, too, might have compensated at least to some degree for wealth not accumulated during a son's time as a bachelor on the family farm. But victories also had become rare by the middle of the second century as Rome found itself bogged down in difficult wars in Spain. The response among many families that found themselves in this predicament, one suspects, was to replace the labor of sons with that of slaves on many larger smallholdings as well as on medium-sized farms where the labor force had previously comprised a mixture of servile and family members of the household. Evidence to substantiate this suggestion is, as one might expect, scarce, but two episodes from the Hannibalic War provide important testimony. In 214 and again in 210, the *patres*, facing a shortage of oarsmen for the republic's warships, ordered its wealthier citizens to supply slaves for that duty. Citizens whose property had been rated at between 50,000 and 100,000 *asses* in the prior census were to contribute one slave along with six months provisions; those above this level were to furnish more rowers.[99] The critical point to note is that 50,000 *asses* corresponds to the lower limit of the third census class.[100] In the senate's view, therefore, slave ownership was common enough among men at this rank for the *patres* to assume that each would have a slave to send to the fleet (or, perhaps, the wherewithal to purchase one).[101] We do not have any means of determining, of course, how large the farms were that citizens in the third census class might have owned, but it is difficult to imagine them as anything other than "medium-sized."[102] And regular slave ownership at this level of the republic's economic hierarchy at least suggests that the ownership of slaves might not have been wholly beyond the reach of many citizens in the fourth class. Indeed, the lack of evidence for the growth of vast, slave-run capitalist plantations during the first two-thirds of the century makes it seem very likely that many of the slaves Romans and Italians

Mortality and Crisis

were purchasing in this period were bought to replace the labor of sons that Rome took off to war.

The massive mobilization that the struggle with Hannibal required and the republic's continuing high levels of conscription in the ensuing decades therefore probably forced those smallholding families that could afford to buy slaves to do so and impelled middling families that had already acquired a few to buy more, enabling them to avoid the decline in productivity that the conscription of their sons would otherwise have entailed.[103] A trend along these lines, in turn, would have contributed substantially to a growing divide along the economic continuum.[104] Many families that owned a few *iugera* and worked enough common land to feed themselves and perhaps a bit more would never rise beyond this level because at the point where they ought to have been able to maximize their manpower and accumulate wealth, the republic drafted their sons. Above them would have been those families owing or exploiting considerably more land and who had been able to acquire slaves to help work it and who, faced with the prospect of losing their sons' labor for long stints with the legions, would have been eager to acquire more. Few families would have been able to remain in between, working substantial farms with only their own labor, because conscription and partition would eventually reduce succeeding generations to a subsistence level.

By 133 therefore Roman warfare in the years since Hannibal's invasion had not only contributed to a rapidly rising population but also produced a body of smallholders in the lowest census classes that was, overall, significantly poorer than their third-century counterparts and without much hope of improving their lot through their own efforts. And with the end of colonization in Italy after 181 the senate closed this safety valve for families unable to establish all their children on new farms.[105] Tiberius's rural supporters are likely therefore to have come both from the progeny of large families of smallholders and those whose fortunes the demands of winning an empire for Rome and partible inheritances had gradually eroded. Their numbers need not have been large, perhaps no more than 10 or 20 percent of the population, to have constituted a potent source of support for the measures Gracchus was proposing.[106] Moreover, a shortage of land may not have been their chief concern. Gracchus's success in winning their support suggests that they believed his claims that if some wealthy individuals who controlled too much *ager publicus* could be made to surrender it, this land would have sufficed to meet their needs while still leaving its former possessors with generous holdings to exploit. In the event, their and Gracchus's sanguine expectations were proved wrong. The fact that the commissioners seem to have moved fairly soon to recover public lands occupied by Rome's Italian allies seems to show that the amounts in the hands of Romans in excess of the limits Gracchus's agrarian law laid down were negligible or perhaps irrecoverable owing to their occupiers' political clout.

But in the run-up to the law's passage, what Tiberius's supporters may have felt to be their greatest need was the means to establish new, self-sufficient households on their allotments—the tools, livestock, seed grain, liquid capital, and the many other things necessary to set up an independent smallholding. If a family was unable to increase its wealth over the course of its life cycle, dividing that wealth would have not only crippled the viability of the parents' household but also made the resulting new ones unsustainable. As a result, some may have coped by modifying the typical neolocality of Roman families: sixteen members of the Aelii lived in a large, extended family in a small house supported by a single farm in the generation before Gracchus.[107] The case may have been memorable less because the arrangement was rare in this era than because it was unusual for members of a noble family to be reduced to such an expedient in order to marry. But the Aelii illustrate clearly how vulnerable large families were to the leveling effects of partible inheritances in the absence of the "centrifugal" effects of an overall increase in wealth. Other sons of smallholding families, particularly those at the poorest end of the economic spectrum, may have left the land altogether. Where a family's assets were insufficient to support more than itself, some siblings may have been forced to sell what property there was when their parents died, divide the proceeds, and seek to support wives and children through paid labor wherever they could find it—in the countryside, the towns, or Rome itself. These may be the homeless wanderers of whom Tiberius spoke in the speech Plutarch preserves.[108] And their search for work will have been made all the more difficult by the competition they faced from slaves.[109] While high military and other mortality in the late third and early second centuries may have led to an increased demand for seasonal and other occasional free laborers in the countryside and consequently raised wages for this type of work, employers there and also in Rome and Italy's towns with longer-term labor needs to fill may have preferred to employ servile workers owing to this same scarcity of freemen.[110] The result, ironically, would have been to depress wages for precisely the sorts of jobs that poor men would need to support a family, because the importation of slaves to fill them will have made full-time work scarce relative to the number free Romans and Italians seeking it. And the situation will only have gotten worse as the population began to rise rapidly again.

The attraction of Gracchus's proposal and other land laws that followed, therefore, may have lain as much in the fact that they sought to create working farms by providing the equipment and livestock necessary to make a smallholding self-sufficient as because they made allotments available in the first place. Tiberius appropriated the Attalid bequest specifically to provide these necessities following the passage of his law, when he encountered senatorial resistance to its implementation in the form of a refusal to grant more than a token sum for the operation of the land commission.[111] But even before this point, Tiberius's legislation—and, more important, its supporters—must have envisioned that those who received land under its terms would be subsidized at public expense in some way. When the senate resolved in 180 to deport the Ligurian Apuani from their homes to public land in Samnium, it appropriated 150,000 *denarii* to cover the costs of buying what was necessary for their new homes.[112] Although explicit testimony is lacking for Roman or Latin colonies in the second century, it is difficult to believe that the *patres* would have shown such solicitude when they voted to relocate the Apuani and not regularly taken similar steps to ensure the success of new foundations of citizens and allies.[113] The natural expectation among the citizenry would have been therefore that the Roman fisc—the *aerarium*—would provide what many lacked, the resources necessary to set up an independent small farm that would enable its possessor to marry and raise a family.

Once the Attalid bequest became available, however, Tiberius apparently sought to go further, proposing to supply cash grants in lieu of allotments to those eligible for them.[114] Livy's epitomator indicates that the reason Tiberius took this step was that the supply of *ager publicus* upon which to place colonists proved to be inadequate. Because little land was apparently distributed before his death, this cannot be correct, but Gracchus may have anticipated that the land commissioners would at some point run out of land to allot.[115] However, Tiberius's solution hardly amounted to pandering to the greed of the mob.[116] It was an effort to achieve the same aims as the land distributions only in a different, and novel, manner. For if the problem that the sons and daughters of some small farming families believed they faced was not primarily a lack of land but insufficient wealth to begin families of their own, the distribution of the Attalid bequest directly in cash ought to have solved it as surely as turning the money into equipment and livestock for colonists. Individual settlement in the *ager Gallicus* of northern Italy represented an attractive possibility for migrants from elsewhere in Italy, provided they had the resources to establish themselves.[117] Tiberius may have avoided a proposal to provide would-be settlers with them initially because such a measure was unprecedented as well as because of the potentially enormous costs involved, but such objections may have seemed less compelling after Attalus left his kingdom with its wealth to Rome and the senate had signaled that it would take the equally unprecedented step of refusing to fund the settlement of the law's beneficiaries.

In conclusion, we may return to the question raised at the start of this investigation, that of the aristocracy's responsibility for the ruin of Rome's and Italy's smallholders. To answer it, this study has explored how warfare, small-scale agriculture, and patterns of family formation could interact in such a way as to make continuous war and a high rate of military mobilization sustainable for a very long time. The "fit" among these three elements enabled Rome to recruit the extraordinarily high proportion of its population necessary to emerge victorious from its long struggle against Hannibal and then go on to become the ruler of a Mediterranean-wide empire in the space of fifty years without causing the smallholding economy to collapse, as the consensus view has long maintained. In so doing, the study has exonerated the republic's political leaders from culpability in that nonevent. Their corporate failure to limit the competitive drives of their individual members for the glory and wealth derived from conquest did not overburden smallholders with endless military service. They did not pursue their drive for empire heedless of the harm they were causing—even though the *patres* were farmers, too, and they if anyone ought to have understood what the consequences would be for smallholders. Rather, war represented a means for commoners as well as members of the senatorial class to acquire wealth (among other things). Colonial allotments and loot represented tangible benefits to families of small farmers, after all.[118] The warfare that won them was simply a different way of employing the surplus labor characteristic of subsistence agriculture and a strategy for simultaneously diversifying the risks inherent in Mediterranean farming. Aristocratic competition to win *gloria* and *laus* along with the riches that military victory bestowed can in this light be viewed as complementary to the interests of smallholders, not inimical.

Instead, the harm being done was largely invisible. On one level, the whole system was ultimately going to prove as unsustainable as any Ponzi scheme. It could only be kept going as long as those tangible benefits were there to be acquired. Once the colonies in northern Italy ceased, and once Rome found itself embroiled in hard, unprofitable fighting in Spain and elsewhere, few if any gains offset the costs, and things began to come unglued. But on a more fundamental level, the system required a higher birthrate than Rome's and its Italian allies' environmental resources could sustain in order to counter the effects of military mortality and the regular colonization that was essential to holding the territorial gains Rome made in Italy. War and colonization could be said in this (very limited) sense to have "caused" the high birthrate, because neither could be carried on for any length of time without it.[119] But when the enormous military mortality of the late third and early decades of the second century altered dramatically the demographic calculus that governed the interplay of population growth and environmental resources heavily in favor of the former, it thereby loosened significantly whatever preventive checks had previously controlled the birthrate. Women wed earlier; men who ordinarily might never have wed found spouses; and couples were willing to have more children because they supposed that the economic good times from which they themselves were benefiting would continue when their children grew up and began families of their own.

But Rome's wars grew less bloody overall despite some very costly losses in Spain, and more young men survived their time with the legions or never served at all. Yet the preventive checks were not reapplied and tightened; the population continued to grow and began to reach the point, at least in some places, where those in need of land exceeded its supply. Where land was available, its occupation by smallholders was in many cases impeded by the downward economic pressures exerted by conscription and cyclical mobility. The republic's wars of conquest harmed Italy's small farms not in the short run, by removing husbands and fathers and so causing a crisis of subsistence for those left behind, but over the long haul—by removing young, unmarried men and so preventing their labor from countering the centrifugal tendencies inherent in partible inheritance and neolocal family formation. Smallholding families grew poorer when they divided their property to enable sons and daughters to begin new families because the wealth from Roman warfare failed to compensate for the gains forgone when farms lost their sons' labor to the military and colonization ceased. At the same time, the initial stages of plantation agriculture just getting under way in some parts of Italy also may have begun to compete with smallholders for access to the *ager publicus* on which the latter depended.

What determined how the resulting political cataclysm played out, however, was the failure of the republic's political leaders to perceive that population growth was a fundamental cause of the poverty they saw around them. Census returns showed a declining number of citizens, wars in Spain had been costly of lives, and slave labor was increasing in the countryside. They drew the wrong conclusions, and accordingly in 133 Tiberius Gracchus and his advisers sought to apply the wrong remedy. Their error led not simply to a constitutional crisis and political tragedy, as serious as those consequences were, but introduced into the political arena an issue—the redistribution of the *ager publicus* in Italy—that would inflame sentiments on both sides at the same time that it could not but fail to solve the basic problem. For there was just no longer enough land to go around. Too many people needed farms, and subsequently the great increase in plantation agriculture during the first century, along with the enormous concentrations of property in the hands of Sulla's aristocratic partisans in the wake of his victory and confiscations, only exacerbated the problem. In the end, overseas colonization coupled with more rounds of heavy military (and civilian) mortality opened the way to a solution, but only at the cost of terrible suffering, the collapse of the Roman Republic, and the establishment of a monarchy on its ruins.

Appendix 1   The Number of Roman Slaves in 168 B.C.

Dumont has argued that the slave population in 168 was between 151,320 and 756,000 based on his conclusion that freedmen in that year composed between 10 and 50 percent of Rome's citizen population. Dumont claims that the censors of that year, Ti. Sempronius Gracchus and C. Claudius Pulcher, restricted enrollment of most freedmen to the four urban tribes because they had become by that date numerous enough to be perceived as a potential force in politics. Consequently, he reasons, a measure to restrict the voting power of *libertini* must mean that they had come to make up at least 10 and perhaps as much as 50 percent of the citizenry.[1] Yet in order to support this reconstruction, Dumont must insist that Gracchus and Claudius had been the first censors in some time to restrict freedmen to the urban tribes and that their action represents a sudden awareness of the increased size of the freedman population following the Hannibalic War. Yet Livy recorded that the censors of 230 had restricted the enrollment of freedmen to the four urban tribes, and most scholars have taken Livy's account of the matter at 45.15.1–2 to summarize what had been standard practice since that date.[2] The passage opens after a lacuna and states (following Briscoe's Teubner edition here and below; compare Weissenborn and Mueller's edition): *in quattuor urbanas tribus discripti erant libertini, praeter eos quibus filius quinquenni maior ex se natus esset—eos ubi proxumo lustro censi essent censeri iusserunt,—et eos qui praedium praediaue rustica pluris sestertium triginta milium haberent,—<. . .> censendi ius factum est.* But Dumont insists this passage can be read to say that the censors by whom the freedmen had been distributed into the urban tribes were identical with those who ordered freedmen with sons older than five to be enrolled where they had been (that is, in a rural tribe) in the prior census, and that in both cases the censors were Gracchus and Claudius.[3] Others have taken the passage to say that the *libertini* had been distributed by *previous censors* into the urban tribes except those whom Gracchus and Claudius had excepted from this and enrolled where they had been enrolled in the previous census.

While Dumont's interpretation of this passage is admittedly possible,

Appendix 1

although strained, it founders within the larger context of Livy's narrative at 45.15.1–6. When the passage opens, a controversy between Claudius and Gracchus has arisen over the enrollment of the freedmen and, in particular, over some measure Gracchus wishes to take in regard to them. Livy's description of what that measure was has unfortunately been lost in the lacuna that precedes the passage. The measure can only be reconstructed on the basis of Claudius's argument against it of which 15.1–2 forms a part and which 15.3–4 continues. The text here is difficult: *hoc cum ita seruatum esset, negabat Claudius suffragii lationem iniussu populi censorem cuiquam homini, nedum ordini uniuerso adimere posse. neque enim si tribu mouere posset, quod sit nihil aliud quam mutare iubere tribum, ideo omnibus quinque et triginta tribubus emouere posse, id est ciuitatem libertatemque eripere, non ubi censeatur finire, sed censu excludere.* Treggiari renders it, "Since this (sc. right) was kept, Claudius denied that the censor could, without a mandate from the people, deprive any individual, let alone a whole order, of the franchise. For his capacity to move a man from a particular tribe (which only meant that he ordered him to change his tribe) did not imply that he could remove him completely from all thirty-five tribes, which was equivalent to depriving him of citizenship and liberty, and was not to define where a man should appear on the census-list but to exclude him from it."[4] The supposition that Gracchus's proposal was to deny freedmen enrollment in *any* of the thirty-five tribes makes good sense out of the conflict between the censors, for their controversy was eventually settled by a compromise by which freedmen were to be henceforth enrolled in *one* of the four urban tribes chosen by lot (although it is not clear which if any of the exemptions from enrollment in an urban tribe described at Livy 45.15.1–2 were preserved). The compromise strikes a balance between Gracchus's harsh position and Claudius's desire to keep things as they had been with freedmen enrolled in the *four* urban tribes.

But if this is so, then Dumont's interpretation of the passage must be discarded. For Livy at 45.15.1–2 is clearly describing an enrollment of freedmen in the four urban tribes that had been carried out prior to that stage of the debate, either by previous pairs of censors on the standard view or on Dumont's reading of the passage by Gracchus and Claudius themselves. But the latter reading makes little sense: why, if Gracchus wished to deny freedmen enrollment altogether, had he already enrolled them in the four urban tribes? Admittedly, Cicero's brief remark in the *de Oratore* complicates matters, for there he describes Gracchus as having been responsible for transferring freedmen into the urban tribes.[5] But it cannot justify making nonsense out of the historian's narrative. If Cicero's evidence has any value for understanding the events of this censorship, it probably reflects a tightening up of the exemptions from enrollment of freedmen in an urban tribe that Livy describes at 45.15.1–2.

But if no sudden perception arose in 168 of a political danger posed by the votes of freedmen enrolled in rural tribes, then no grounds exist here for imagining that their presence in the citizen population—and, by implication, the number of slaves the Romans owned—had increased dramatically in the interval. Gracchus's attempt to deny them enrollment in any tribe, and the censors' eventual compromise by placing them in only one of the four urban tribes, are to be viewed in symbolic terms rather than as a response to the perception of a political threat. As for Dumont's assertion that freedmen could have made up even as much as 10 percent of the citizen body, its implications for estimates of the contemporary slave population are very difficult to accept.[6]

Appendix 2 The Accuracy of the Roman Calendar before 218 B.C.

Perfect accord between the solar year and the pre-Julian Roman calendar never existed owing to the flaws inherent in the latter.[1] Every twelve-month Roman year ended about ten days before seasonal time returned to the corresponding solar date and so required the insertion of an intercalary month every two years or so in order to bring the two cycles into approximate harmony once again. Thus any given Roman date might have been about two to four weeks in advance of the corresponding solar date because we cannot in most cases know when the pontifices would have ordered intercalation. However, discrepancies between seasonal and Roman dates greater than this should not be assumed to have been usual prior to 218.

THE SECOND PUNIC WAR

Most scholars have held that the Roman calendar was generally in accord with seasonal time in the early years of the Second Punic War.[2] Derow, however, calculates that the kalands of January 218 fell on a seasonal date of November 2, 219.[3] Hence, in his view the calendar in 218 was running nearly two months ahead of the seasons. However, the basis for his reconstruction of the calendar during the Hannibalic War rests largely on claims for a synchronism between the date of the battle of Cirta in 203, which Ovid recorded as taking place on June 22, and a solar date of about May 23 or perhaps ten days earlier, and this correlation is anything but secure.[4] Derow relies here upon Soltau's claims that the events reported by Polybius at the beginning of book 14, namely the very beginning of spring and Scipio's launching of his fleet, must fall between March 5 and March 15, 203, because spring always begins in coastal Algeria early in March and that the opening of the sailing season at the same time is "generally known."[5] From this date, a computation of the days that elapsed until the battle at Cirta brings Soltau and Derow to a seasonal date of May 23. However, both of the points on which this dating rests are open to doubt. Spring in coastal Algeria seems to begin sometime in April, not

March.⁶ In addition, mid-March represented the extreme outer limit of the sailing season in antiquity; April and May were much more normal months for the beginning of naval activity.⁷ Hence Polybius's language cannot be pressed to yield a date more precise than sometime in late March or April.⁸ Furthermore, the seventy days that Soltau and Derow allow for the events that transpired between the beginning of spring and the battle may be too brief.⁹ If that is so, then if Scipio launched his ships in early April rather than early March, the battle at Cirta may have fallen on a seasonal date very close to Ovid's June 22. On the other hand, Derow is certainly correct to argue, with the majority of scholars, that the few chronological indications that our literary sources preserve elsewhere suggest that the Roman calendar was at most no more than a month behind the seasons in 217–215.¹⁰

This position, however, has been challenged by Desy who argues that Polybius's reference to Hannibal's army harvesting grain around Luceria following its victory at Lake Trasimene in 217 dates that battle to around April 15.¹¹ Because Ovid gives June 22 as the Roman date for the battle, Desy concludes that the Roman calendar was running some sixty-eight days in advance of the seasons.¹² To accept this correspondence, however, means disregarding Polybius's synchronism between the battle and Philip V's receipt of the news of it at the time of the Nemean games in July.¹³ Since it is unlikely that the report of so momentous an event as Rome's defeat took two months or more to reach Philip, we must either believe, with Desy, that Polybius has deliberately falsified the chronology here or that Polybius has misunderstood a reference in his sources to Hannibal's soldiers collecting grain from farmhouses and granaries that already had been harvested and stored away and taken this to mean that the soldiers were harvesting the grain themselves.¹⁴ The former seems far less likely than the latter.

A more decisive objection to claims that the Roman calendar was running seriously in advance of the seasons in this period arises when the foundation date of Rome's colony at Placentia in 218 is synchronized with the chronology of Hannibal's march from New Carthage to northern Italy. The detailed accounts that Polybius and Livy provide of this event support these authors' general statements that the journey took five months to complete.¹⁵ Polybius also puts Hannibal's army on the summit of the pass leading across the Alps around the time of the setting of the Pleiades, which scholars usually hold signals a seasonal date around the third week in September.¹⁶ Proctor, however, has argued forcefully for a date

Appendix 2

in the first week of November, although he concedes that Polybius's language is imprecise and might be stretched to imply a date as early as October 16.[17] These fixed points enable the other events of the march to be dated with some precision. The most important for present purposes is Hannibal's departure from New Carthage, which falls between mid-June in Proctor's chronology and the end of April according to Walbank and others.[18] Polybius reports that when the Boii in Gaul learned that Hannibal would shortly arrive in Italy, they attacked the colonies at Placentia and Cremona, which had only just been founded.[19] Asconius reports the foundation date of the former as the last day of May.[20] If the calendar were running as far in advance of the seasons as Dusy claims and the end of May fell on a seasonal date around late March, synchronizing the revolt with Hannibal's departure makes nonsense out of Polybius's account of the attack on the colonies. On Desy's chronology, either the Boii struck at a seasonal date around early June and the colonists had already been in Gaul for more than two months when the assault came, or if the assault occurred soon after the colonists arrived at the sites of their settlements, Hannibal's departure from New Carthage was still at least four and perhaps as many as ten weeks off. Only on the assumption that the Roman calendar and the solar year were in fairly close accord can a coherent chronology of events in 218 be developed.

THE FIRST PUNIC WAR

Most scholars have held that the calendar accurately reflected seasonal time during the First Punic War.[21] Recently Morgan has argued that it was in accord with the seasons after circa 258 but a month or two in advance of them prior to that date.[22] His defense of the alignment of the calendar after 259 is very persuasive against those who argue for serious displacement. However, the evidence does not support his claim that the calendar was running behind the seasons early in the war. Morgan argues that the events that fell between the entry of the consuls of 262 into office on May 1 and the battle at Panormus, in early or mid-June, would have required at least eight weeks to transact. But in the first place it is not clear, *pace* Morgan, that two weeks would have passed between the levying of four new legions and the start of the soldiers' journey to Rhegium. Despite Morgan's belief that levying four legions "could certainly have been done inside a fortnight," the episode he appeals to in support of this contention was quite exceptional.[23] In that case, the praetor who enrolled four legions in eleven days did not begin the process from scratch but took over conduct of the levy from the consuls.²⁴ The latter had complained to the senate of their difficulties in finding enough recruits and in turn had been criticized for being too free in awarding exemptions. At that point the levy had been entrusted to the praetors.²⁵ In addition, the censors of that year announced that they would crack down on unwarranted exemptions from the draft and circulated this proclamation widely. As a result, a large number of potential draftees presented themselves at Rome.²⁶ The levy, in other words, was already well along when the praetor took over, and his task was made easier by an exceptional turnout of men when he came to complete the job. Far more time than two weeks would typically have been required: in 77, Pompey boasted to the senate of having levied an army in only forty days.²⁷ It is most unlikely that consuls entering office on the first of May in this period could have enrolled new armies and been ready to march in time to threaten the ripening crops of their enemies. Far more likely, new legions were enrolled when necessary by the consuls of the previous year toward the end of their term of office in order to have a military force ready to take the field at the beginning of summer. Hence, any new forces levied for the consuls of 262 were probably ready to begin their march to Rhegium when they entered office, if they had not already been dispatched there.

But the assumption that the legions these consuls commanded were newly levied is itself open to question. Although Polybius informs his readers that the Romans mobilize four legions each year, his statement is not to be taken literally for the period in which he wrote, as Afzelius's study of the legionary disposition during the second century clearly demonstrates.²⁸ We have no reason to assume a priori therefore on the basis of Polybius's statements in the passages that Morgan cites that armies during the First Punic War were invariably demobilized and reenlisted from year to year. As Chapter 2 has demonstrated, legions regularly wintered in the field during the third century. Consequently, nothing compels us to assume that one of the consuls of 263, M'. Otacilius Crassus, brought his legions with him when he returned to Rome. They may have wintered in Rhegium or somewhere on Sicily.²⁹ Nor can we be certain that the troops of his colleague, M'. Valerius Laevinus, did not go into winter quarters after his triumph, where they would have been available to his successor.³⁰ Consequently, the four weeks that Morgan allots for the levying and deployment to Rhegium of the armies the consuls of 262 would command can be considerably reduced. These forces are likely to have been ready to march on May 1.

Appendix 2

Morgan's other arguments are equally unpersuasive. Despite Morgan's claim to the contrary, Polybius's choice of words clearly indicates that the consuls' decision to begin the siege of Agrigentum, which followed shortly upon their arrival before the city, coincided with "the height of the harvest."[31] Similarly, temperatures on the southern coast of Sicily in December are not severe. Consequently, Zonaras's statement that the consuls returned to Messinia following the siege "on account of winter" need not mean that this event must have taken place in that month rather than in mid-January, as is generally assumed. Average minimum daily temperatures on the coast of Sicily do not fall below 40 degrees Fahrenheit and frost is almost unheard of.[32] The claim that the military tribune C. Caecilius must have served under one of the consuls of 261 and not Scipio Asina, the consul of 260, also fails to convince. Although it is true that Zonaras reports that Scipio's military tribunes were seized with him, it is highly unlikely that Scipio left no one in command when he and his officers met the Carthaginian commander for the parlay that led to their capture.[33] There is no reason to think that Caecilius had not been given that assignment by Scipio. And while it is true that the consuls of 258 found the Carthaginians still in their winter quarters when they arrived in Sicily, all that need mean is that the Carthaginians were not prepared to take the offensive in that year, as their refusal to meet the Romans in battle when they arrived indicates.[34] Finally, little can be built on the discrepancy between Zonaras's statement that Duilius, consul in 260, returned to Rome "summer now being past" and the date of Duilius's triumph in the intercalary month in 259. Because Morgan assumes that the campaigning season ended in mid-October, he argues the gap between this and a triumph at a seasonal date of late February cannot be explained. Hence, the Roman calendar must have been two months in advance of the seasons, so that Duilius's triumph can be placed at a seasonal date around mid-December, a more plausible interval in Morgan's view. But Zonaras's account is highly compressed; certainly Duilius did return to Rome after the summer had ended, but Zonaras's words cannot be pressed to mean that he did so immediately. In fact, a good deal of fighting occurred during Duilius's year in office. There is no reason to assume that all of it took place before October 15 or, more important, that the need to secure the gains Rome had made would not have kept Duilius in Sicily after his victory at Mylae through the fall and into the early winter months.[35] Nor would a lengthy delay between a signal victory and the triumph celebrating it be unprecedented in view of the five months that elapsed between the battle

Appendix 2

at Sentium and Fabius Rullianus's triumph in September 295.[36] Consequently, the date of his triumph offers no support for the claim that the Roman calendar was as much as two months ahead of seasonal time at this point.

THE CALENDAR PRIOR TO 264

The accuracy of the Roman calendar prior to the First Punic War is much harder to ascertain in the absence of Livy's narrative for most of this period. However, recently Brind'Amour has revived the arguments of Holzapfel that in 295–293 the civil calendar was about two months in advance of the seasons.[37] Brind'Amour and Holzapfel note that very shortly after his election the consul of 295, Q. Fabius Maximus Rullianus, sought to toughen his army, which had encamped near Perusia during the winter, by forcing it to abandon its camp and undertake daily marches.[38] In addition Livy remarks that *fiebant autem itinera quanta fieri sinebat hiemps hauddum exacta* and then notes that in the early spring Fabius returned to Rome to consult the senate about the war.[39] These passages lead Brind'Amour and Holzapfel to conclude that the seasonal date of Fabius's entry into the consulship was early February and consequently the calendar was about two months in advance of the solar year. In Brind'Amour's case, however, his reconstruction is predicated on the assumption that consuls in this period entered office around March 1, and the date on which magistrates began their term of office in the late fourth and early third centuries is anything but secure.[40] For the period 320–294, Holzapfel and Soltau argue that the consuls entered office on December 1, which would fit precisely Livy's description of Fabius's winter marches in 295.[41] Hence, this passage offers no indication that the calendar was running behind the seasons at this point. On the contrary, it might suggest, on the assumption that the civil year began on December 1, that the two were in close accord at this point. Only in 293, Holzapfel and Soltau continue, do consuls begin their term on May 1. Holzapfel's assumption that the calendar was two months ahead of the season in 295, on the other hand, stems from his acceptance of Fränkel's argument that Livy's mention of the legions' suffering on account of cold and snow in Samnium in 293 shortly before they returned to Rome coupled with the *fasti*'s dating of these consuls' triumphs to February 13 and 17 must indicate that the latter were celebrated at a seasonal date of early or mid-December.[42] But Fränkel's argument is predicated on the supposition that heavy snow cover cannot have been lacking in the area where the legions were encamped before Febru-

Appendix 2

ary. Rather, he assumes the snow Livy mentions must have fallen much earlier. But this supposition is fallacious, the product, as Proctor argues, of northern writers' tendency "to foreshorten the passage of the seasons as conceived by ancient Mediterranean writers."

Armies in Mediterranean regions regularly operate into January.[43] The legions in late 293 were in the central Appennine highlands, at Aufidena and Saepinum, which lie above 3,000 feet.[44] To judge by the average daily minimum temperatures for L'Aquila, at 2,400 feet, which only dip below the freezing point in January, there is no reason to suppose that Livy's reference to the cold and snow must indicate a date in December rather than one in mid- to late January.[45] The order for the consuls to withdraw their forces came only *after* the soldiers could no longer endure the cold or live out of doors—hence, in the depth of winter.[46] If the consuls required a few weeks time to march their legions back to Rome, address the senate, and prepare for their celebrations, then their triumphs in mid-February make perfect sense. The evidence of Livy coupled with that of the *fasti* suggests that the calendar was in accord with the seasons even in the early third century.

Prior to this date the evidence is generally unreliable, but a fragment from Ennius's *Annales* may represent an important exception. The passage (163 Vahlens = 153 Skutch, preserved by Cicero *Rep.* 1.25) mentions an eclipse of the sun that occurred on June 5 about 350 years after the foundation of the city: *Nonis Iunis soli luna obstitit et nox.* Scholars have generally identified this eclipse with one known to have taken place at a seasonal date of June 21, 400, although some hold that the event in question is one that occurred a few years later, on June 12, 391.[47] Either date, however, would indicate that the Roman calendar was only two to four weeks behind solar time in the early fourth century, well within the variation one would expect given the inherent shortcomings of the pre-Julian calendar.

Appendix 3   Tenancy

De Ligt has recently challenged De Neeve's finding that tenancy was not widespread in Italy before the first century B.C. De Ligt argues that provisions in the Twelve Tables concerning *pignoris capio* and *venditio filii* could have provided the legal framework for agricultural tenancy in the early and middle republic and that "this finding, together with the fact that tenancy of one kind or another has been important in virtually all preindustrial societies, makes it at least highly probable that cultivating tenants were widespread long before the final decades of the second century."[1] His case however is not persuasive, for while he may be correct that the provisions of the Twelve Tables that he highlights *could have* served to establish a legal basis for agricultural tenancy, his argument that they *in fact did so* rests on nothing more than the assertion that tenancy's importance in other societies means it must have been so too in the early and middle republic. But everything we know about Rome in this era suggests that it was not like other preindustrial states: "Typically, then, preindustrial states were monarchies . . . republics [were] abhorred as positively unnatural in so far as they were known at all: the republicanism of Mediterranean antiquity is distinctly unusual."[2] Most important, plebeians at Rome were exceptional in their organizational abilities and their political power, as their repeated and successful efforts through *secessio* first to limit and ultimately to end *nexum* demonstrate. Although the power of the *populus* in the middle republic may not have been as extensive as Millar has maintained, the evidence he presents unquestionably demonstrates the people's active participation in the government of the *res publica*, a far cry from the normal relationship between the citizenry and their rulers in most preindustrial societies.[3]

Consequently, there is no reason to assume a priori that citizens in the middle republic would have allowed themselves to be forced into the position of having to lease land from the rich in order to survive, as was the bulk of their preindustrial counterparts. Their frequent agitation for colonial foundations and viritane distributions and their regular support for

Appendix 3

the wars that won the land necessary for these indicates that they were able to pursue a long-term political and diplomatic agenda that enabled them to avoid economic dependency.[4] Equally important, the argument from silence on which De Neeve bases his case is considerably stronger than De Ligt appears to realize. Its strength lies not simply in the failure of the legal texts explicitly to reflect an institution that allegedly was central to the economic structure of republican society but in the absence of any echo of it anywhere in our evidence for this period. When well-to-do Romans for example complained of the economic harm they were suffering because of the republic's enormous demands for manpower during the Hannibalic War, they stressed that their fields lacked cultivators because their *slaves* were being taken off to war, not their tenants.[5] Yet the conscription of the latter ought to have hurt them as much as the former if tenancy had been widespread at this stage of the republic's economic development.[6] Indeed, Appian remarks in describing agrarian conditions prior to Tiberius Gracchus's reforms that the rich preferred to cultivate their estates with slaves rather than freemen because of the latter's liability to be conscripted into the legions, a liability that certainly existed before as well as after the Hannibalic War.[7] Evidence of this sort is difficult to reconcile with the notion that agricultural tenancy was common during the middle republic.

Appendix 4   The Minimum Age for Military Service

Although most scholars accept at face value the evidence preserved in the ancient sources that Roman men became eligible for conscription at seventeen, Mommsen, followed by Brunt, supposes that military service in fact began two years later, at nineteen.[1] While Mommsen is certainly aware of the evidence that military service began at seventeen, he subscribes to the curious notion that the Romans reckoned ages inclusively, so that one was not considered seventeen until the final day of one's eighteenth year of life rather than the first, so that a Roman was not considered seventeen until he had reached what moderns would think of as his eighteenth birthday. Mommsen further believes, based on Cicero *pro Cael.* 11, that all citizens spent a year in a kind of boot camp, the *tirocinium*, so that military service did not really begin until the following year, when a man would be, by contemporary reckoning, nineteen. But the latter, if it had any general application during the middle republic, surely applied only to members of the upper class, not all citizens. We hear nothing of any such practice elsewhere. More tellingly, Cicero says nothing at *pro Cael.* 11 to indicate that this year of training began at (on modern reckoning) seventeen; it may well have begun at sixteen (or even younger) in anticipation of the beginning of military service the following year.

More tellingly, the evidence that Mommsen cites in support of the notion that a Roman would view as only seventeen years old someone who in contemporary terms would be considered eighteen is not at all convincing. Rather, it demonstrates exactly the opposite, that Roman ideas about how age was reckoned accord quite closely to contemporary ones. Consider *Dig.* 50.16.134, which states that someone is considered a "one-year-old" not when he is born but after he has lived 365 days—not 730, as Mommsen's claim would imply. *Dig.* 40.1.1 establishes that someone could legally manumit a slave following the last day of the owner's twentieth year, since the law forbids this privilege to those younger than twenty. But this statement does not mean that the Romans imagined that someone was not twenty until he had reached what a modern would term his twenty-first

Appendix 4

birthday but rather his twentieth since this would be the point at which he had completed his twentieth year, in the same way that someone on his first birthday can be thought of as having completed his first year.[2] The case envisioned is clearly one in which someone turns nineteen and claims no longer to be younger than twenty since he was now in his twentieth year of life. The point the jurist is making here, in other words, is the same as that being made in *Dig.* 50.16.134. No reason exists, therefore, to reject the sources' testimony that military service for Romans began at seventeen.

Appendix 5   The Proportion of *Assidui* in the Roman Population

The optimistic assessment of the demands that republican warfare placed on the fathers of young families is open to the objection that not all Roman men were *assidui*, that is, citizens whose wealth sufficed to qualify them for service with the legions. Brunt, the only scholar to address the problem in any detail, has argued that, in 218, about half of all Roman citizens fell below this threshold. They were *proletarii* and so served only as rowers in Rome's fleets.[1] And although the amount of property needed to be rated an *assiduus* was lowered in 212 or 211, the new minimum amounted to only slightly less than 75 percent of the former sum.[2] So, while somewhat more *proletarii* probably qualified as *assidui* after this date, a very large proportion of the adult male population, on Brunt's reconstruction, was still too poor to serve in the infantry. Consequently, if Brunt's estimate of the size of the Roman proletariat in this period is correct, then obviously the republic would have had to draft a substantially greater proportion of *assidui* over thirty than what has been suggested previously in order to meet its need for *triarii*.[3]

However, Brunt's claims should be rejected.[4] The desperate shortage of *proletarii*, whose sole military obligation was to serve as rowers, to man the republic's warships during the Second Punic War clearly demonstrates that as many as half of all Roman citizens cannot have been available for naval service, given Brunt's own estimate that, in 214, when the senate was forced to recruit slaves for this duty, no more than 20,000 *proletarii* were serving in that capacity.[5] That figure suggests that *proletarii* composed only about 10 percent of the citizenry.[6] Nor is it easy to discover any plausible reason why the senate in the period prior to the Hannibalic War would have set the minimum for assiduate status so high that half of all citizens avoided legionary service. The poor, after all, were the primary beneficiaries of the republic's conquests in these years, obtaining land in colonies or through individual grants. Why should the senate have exempted them from a share in the burdens of fighting the wars that made these allotments

Appendix 5

possible? Neither political nor ideological explanations convince. And it is difficult to argue that half the population was too poor to afford arms, for the amount of land that could qualify a citizen as an *assiduus* could be quite small, as little as two *iugera* in some cases.[7] By this point, too, the republic was probably making weapons available for conscripts who lacked them either by grant or through purchase. Indeed, when the historian Q. Fabius Pictor, in recounting the Gallic crisis of 225, wished to offer his readers an idea of the republic's total military potential, he presented a figure representing Rome's entire adult male citizen population, and this must mean, in his view at least, that the city could have mobilized nearly all of it to meet the Gallic threat if necessary.[8] Certainly, some proportion of this number represented *proletarii*, and such men were important to Rome's naval power. But the Gauls were not coming by sea. *Proletarii* could also be swept into the legions in a *tumultus*, an emergency levy, but that would be of little use in a crisis unless such men were armed, and it surpasses belief that if the city had arms available for more than 50 percent of the male population on an emergency basis, it would have denied these men their use under ordinary circumstances to fight Rome's wars. Pictor's citation of this figure in this context only makes sense on the assumption that he wanted his readers to understand that the overwhelming majority of Roman citizens, on the order of 90 percent or more, were available if necessary to take the field against the Gauls. There is no basis, therefore, for the claim that fewer than half of the citizenry in the late third century could qualify as *assidui*. Instead, the general picture outlined previously of the degree to which Rome avoided burdening men of marriageable age with military obligations ought to be accepted.

Confirmation comes from a brief passage in Livy where the historian reports that the censors of 214 discovered among the *iuniores* no more than 2,000 men who over the preceding four years had not served in Rome's military forces and who did not have a legitimate exemption or illness as an excuse.[9] Brunt has estimated the republic's citizen population on the eve of the Hannibalic War at about 285,000, leaving aside the Campanians who went into revolt in 216.[10] Of these, about 70 percent, or 200,000, would have been between the ages of seventeen and forty-six.[11] Brunt calculates that of these, about 100,000 were under arms or had been killed by 214. Hence, the censors' finding implies that nearly all of the remaining *iuniores*, about 100,000 men, had obtained a formal excuse from military service or had been prevented because of ill health.[12] Illness or infirmity is unlikely to have affected more than a minority of cases, and, as we have seen, these men cannot be *proletarii*, as Brunt claims. Hence, *vacationes* must account for the bulk of these exemptions.[13] Scholars usually follow Mommsen and assume that only those in a few narrowly defined categories could obtain *vacationes* from military service: men over the age of forty-six or who had served the maximum number of campaigns required, members of maritime colonies, and a handful of others.[14] But clearly those who fell into these categories will not have constituted 50 percent of the male citizen population between seventeen and forty-six in 214. The vast majority of them, rather, must have been men who had obtained exemptions from the consuls or their representatives on the grounds that their labor was required for the operation of their farms. Consuls won no popularity by forcing unwilling recruits to serve, and because Rome's usual deployment of four or even six legions each year in the period prior to the Second Punic War did not require drafting more than a minority of the eligible men, the custom arose of liberally granting *vacationes* to those who sought them.[15] Their principal recipients will have been married men with dependent families.

This practice only led to problems following Hannibal's invasion of Italy, when the staggering loss of life in the catastrophes of 218–216 and Rome's massive effort to avoid defeat in their wake drained the pool of *assidui* of those men whose labor was not essential on their farms. Yet even then the city seems to have been reluctant to force men off their farms who were unwilling to go. In 216 the crisis following Cannae forced the dictator Iunius to levy slaves, debtors, condemned criminals, and boys recently turned seventeen and even some under that age in order to man new legions. The reason was not, as Livy claims, a lack of free men—on any realistic appraisal of the population, there were plenty of free men—but because these draftees were men whose labor was dispensable or irrelevant on the republic's small farms.[16] This policy apparently changed only in 214, when the city mustered five or six new legions after having been able to field only a single new one in the preceding year. Brunt claims that a lowering of the minimum census required for military service accounts for this sudden increase in available manpower, but as noted in Chapter 1, a much better candidate for the date of the reduction is 212 or 211, when the senate reformed the city's coinage.[17] Hence, the additional manpower to make up the new legions of 214 must have come from some other source, and the most likely candidate is a decision by the senate to

Appendix 5

reduce significantly the number of exemptions the consuls granted to prospective recruits. Yet even then the city's continuing difficulties in finding soldiers during the remainder of the conflict indicates that it had to permit a substantial number of smallholders to remain on their farms, despite the demands of the war.[18]

Appendix 6 The Duration of Military Service in the Second Century B.C.

Steinwender, whom most scholars follow, long ago argued that six years represented the normal term of military service in the third and second centuries, but the texts he cites to support this contention are not persuasive.[1] So in 210 the senate ordered seven legions discharged, but not because the soldiers' tours of duty had come to an end, as Steinwender assumes. These legions had won critical victories at Capua and Syracuse in the preceding year or served in Greece where Laevinus had successfully entangled Philip in a war with Aetolia. The senate's action rewarded the men for their achievements as well as signaled the city's confidence that it had gained the upper hand in the war.[2] More important, we do not know that all these men had only begun their military service when their legions had been formed. Many may have been veteran campaigners. And although the *patres* exempted these soldiers from the draft for that year, nothing was said of levies thereafter.[3] Hence, their discharge may not have marked the end of their military service. Finally, note that as many as three of these seven legions had been mobilized four years earlier, not six.[4] Likewise in 184 and 180, the question of discharging the legions in Spain arose only in the context of important victories there in each of the preceding years.[5] Their commanders sought to return with their armies to enhance their claims to triumphs as well as to add luster to those celebrations and reward their troops.[6] The senate, however, responded to the pleas of their successors not to remove veteran armies from these provinces and sought a compromise: legionaries who had distinguished themselves in battle as well as, in the dispute that took place in 180, those soldiers who had been serving in Spain six years or longer were to be replaced by fresh troops.[7] The senate may well have considered service in Spain hardship duty and six years there enough to ask of any recruit.[8] Certainly, the *patres* had ample reason for concern. In 206 the mutiny of 8,000 Roman and Italian soldiers serving in Spain arose in large part because many of these men had been there since 218 or 217 and now sought discharge.[9] In light of such a precedent, the senate would naturally have sought to limit con-

Appendix 6

tinuous duty in Spain. Livy notes explicitly its worry in 180 that the soldiers in Spain would desert or mutiny if they were not brought home.[10] Once again, however, nothing suggests that those discharged after six years in Spain had begun their period of military service when they went out to those provinces or that they would not, after their return to Italy, be called upon to serve again.[11]

On the other hand, Polybius's statement that the maximum term of service in his day was normally sixteen years but twenty in case of military emergencies strongly suggests that periods of service approaching sixteen years were common: why else specify a higher limit in cases of emergencies?[12] If average terms of service were on the order of six years, the city could have demanded up to a decade of additional service from its *assidui* before breaching the legal maximum of sixteen years. Significant as well in this regard is the ten-year term of service that Polybius specifies for a cavalryman along with his statement that no one could run for public office until he had completed his ten years with the legions. Those who sought political careers—the sons of senators as well as other members of the republic's wealthy equestrian class—served in the cavalry. Hence, their military service was timed to end when a new phase in their public lives ought to have been getting under way for them. Much the same would have been true for ordinary Romans if twelve- or even fourteen-year terms of service were common: their military obligations would end roughly at the point where marriage ought to have embarked them upon the next stage of their lives. If so, then the senate, by stipulating sixteen-year terms of service as the maximum under ordinary circumstances, had not simply plucked some arbitrary figure from the air but set the limit at a point where, if service had been continuous or nearly so, a man would have been nearing the outside range for starting a family.[13] More years of service would have imposed a serious hardship on his personal life as well as harmed the state by diminishing his ability to produce future citizens and their wives.

Appendix 7   The Number of Citizen Deaths as a Result
of Military Service between 203 and 168 B.C.

Chapter 4 estimated the number of excess deaths above normal mortality attributable to military service in this period at about 130,271 to 157,147 men.[1] How many of these were Roman citizens cannot be determined precisely, but developing an estimate might begin by supposing that citizens died of disease more or less in proportion to the percentage of an army that they represented. Citizens normally made up about 40 percent of the soldiers in a Roman army during the period 200 to 168.[2] Consequently, citizen deaths from illness are likely to have totaled between about 29,238 and 40,010.[3] Mortality due to combat is more difficult to estimate. One might suppose that here, too, citizens died more or less in proportion to their presence within the city's military forces, which would yield about 43,897 deaths from combat and so a total military mortality among citizens of between about 73,135 and 83,907.[4] However, the two figures for deaths in combat that Livy preserves where he distinguishes between citizen and other casualties suggest that this estimate may be too high.[5] And it is not difficult to imagine that legionaries often would suffer fewer combat fatalities than other allies, whether owing to better training and equipment or because Roman generals were, for obvious reasons, eager to spare the lives of fellow citizens whenever possible. At Pydna in 168, for example, a contingent of *socii*, the Paeligni, sustained the bulk of Rome's casualties in that battle.[6] Estimating the citizen deaths at only 25 percent of all deaths would bring the figure down substantially, to 27,436 for the period, and yield a total mortality in the range of 56,674 to 67,446, or an annual rate of 3.6 to 4.3 percent.[7] Assuming, therefore, that combat mortality among citizens typically might fall somewhere between 25 and 40 percent of all Roman and Italian deaths results in an estimate of the total military mortality between 56,674 and 83,907 deaths. Excess mortality would fall between 33,591 and 59,824.[8]

Two casualty figures preserved in our sources for battles late in the Hannibalic War need to be added to these figures. One occurred in Gaul in 203 in which Rome lost 2,300 men; the other is Zama, where 1,500 Romans

Appendix 7

and allies fell in 202.[9] The sources do not preserve any figure for the number of Romans who fell in the third major battle fought in this period, the Battle of the Great Plains in Africa in 203.[10] If we assume, as has been done in the case of Roman victories between 200 and 168 for which no casualty figures are reported, that about 2.65 percent of the soldiers died, then Scipio lost about 466 men in this encounter.[11] The total number of deaths in combat is thus 4,266. Wounds would add another 427 deaths. The ratio of citizens to Italian allies in the armies of this period was probably about 1:1.[12] One might therefore put Roman deaths in these battles at between about 25 and 50 percent of the total, or 1,174 to 2,347. In addition, Rome fielded an average of about 17 legions per year between 203 and 201. If we assume, with Brunt, that these were generally understrength and averaged about 3,200 citizens each, an additional 3,101 to 4,243 citizen deaths from disease per year would boost the total of Roman deaths to between 4,275 and 6,590.[13] Excess deaths above ordinary mortality might therefore fall somewhere between 1,823 and 4,142, yielding a total number of excess Roman deaths attributable to warfare of 35,414 to 63,966 between the census of 203 and that of 168.[14]

Notes

Chapter 1

1. Rosenstein, *Imperatores Victi*, 1–8; Rosenstein, "Competition and Crisis"; Rosenstein, "Sorting Out the Lot."
2. Harris, *War and Imperialism*, 9–41, is fundamental here. Of course other important benefits accruing to Rome and individual soldiers also impelled Rome down the road to empire, and many other factors were involved in any specific decision to go to war: see, in general, Gruen, *The Hellenistic World*; Richardson, *Hispaniae*; Eckstein, *Senate and General*; and for an interesting recent attempt to apply "neorealist" theories of interstate relations to Roman imperialism, Eckstein, "Brigands, Emperors, and Anarchy," 868–79.
3. Here and throughout this work, the terms "smallholder," "small farmer," and the like are preferred to "peasant" in describing ordinary Romans and Italians who drew their livelihood from small farms, that is, farms of perhaps forty *iugera* (= ten hectares or twenty-five acres) or fewer, that were worked primarily with the labor of the families they supported, who had limited involvement with the market. This preference derives in part from the considerable difficulties with "peasant" as a category for social and economic analysis, on which see, recently, Horden and Purcell, *The Corrupting Sea*, 271–78. But in large measure it also stems from the fact that nearly every definition of "peasant" assumes that those so designated were the victims of an unequal power relationship that subordinated them to the domination of a group that exploited them by extracting their economic surplus in the form of taxes and/or rents in kind, labor, or money; see, for example, Shanin, "Nature and Logic of the Peasant Economy," 76–77; Wolf, *Peasants*, 3–4, 11; Foster, "What Is a Peasant?," 8–9; Scott, *The Peasantries of Europe*, 1–19. The precise nature of the exploitation might vary considerably, however; see Dalton, "How Exactly Are Peasants Exploited?," 553–61. For antiquity, see also de Ste. Croix, *Class Struggle*, 205–8. The degree to which Roman small farmers were politically powerless and economically exploited remains controversial; see Chapter 3, n. 72.

4. For example, Holmes, *The Roman Republic*, 11–13; Gabba, "Origins of the Professional Army," 11; Gabba, "The Roman Professional Army," 39; Yeo, "The Development of the Roman Plantation," 323; Steiner, "The Fortunate Farmer," 58; Smith, *Service in the Post-Marian Roman Army*, 8; Earl, *Tiberius Gracchus*, 34; Toynbee, *Hannibal's Legacy*, 2:9, 73–74, 80, 166; Astin, *Scipio Aemilianus*, 163; Gabba, Review of Toynbee, 157; Brunt, *Italian Manpower*, 10, 68, 77, 155, 399, 401, 403–4, 426; Brunt, *Social Conflicts*, 16–17; Badian, "Tiberius Gracchus," 683; Bernstein, *Tiberius Sempronius Gracchus*, 74–77; Stockton, *The Gracchi*, 7–9; Hopkins, *Conquerors and Slaves*, 30; Meier, *Caesar*, 29; Cornell, "The End of Roman Imperial Expansion," 156–57; Nicolet, *CAH*^{2}, 9:619. Compare the summary of the consensus view in De Neeve, *Peasants in Peril*, 8–9.

5. For example, Brunt, *Italian Manpower*, 399: "In early days campaigns might often have lasted no more than six months. The hardship of serving only a little more often than every alternate year was plainly much lighter when the farmer had to absent himself from his land for such a relatively short period"; compare 400, 640–41. Note, too, Adcock, *The Roman Art of War*, 19; Hopkins, *Conquerors and Slaves*, 29; Nicolet, *World of the Citizen*, 97; Harris, *War and Imperialism*, 45; Keppie, *The Making of the Roman Army*, 51–53.

6. Brunt, "The Army and the Land," 75 (similarly Brunt, *Fall of the Roman Republic*, 73, 256). Compare Brunt, *Social Conflicts*, 17; Brunt, *Italian Manpower*, 642–43; and, in the same vein, Astin, *Scipio Aemilianus*, 163 n. 3; Momigliano, *Alien Wisdom*, 45; Hopkins, *Conquerors and Slaves*, 29–30; Evans, *War, Women and Children*, 107; David, *The Roman Conquest of Italy*, 80–81; Cornell, "Hannibal's Legacy," 110–11; Nicolet, *CAH*^{2}, 9:619, and the works cited in the following note. On the veracity of the story, however, see De Neeve, *Colonus*, 43–44.

7. Bernstein, *Tiberius Sempronius Gracchus*, 73–101; Stockton, *The Gracchi*, 6–22; Astin, *Cato the Censor*, 241–42.

8. De Neeve, *Peasants in Peril*.

9. Brunt, *Italian Manpower*, 155. Compare Brunt, *Fall of the Roman Republic*, 73, and the analysis of Hopkins, *Conquerors and Slaves*, 1–98.

10. Sources: Plut. *TG* 8.1–3; App. *BC* 1.7; Sall. *Iug.* 41.7–8; compare Florus 2.2. Early doubts: Earl, *Tiberius Gracchus*, 23–30; Badian, "Tiberius Gracchus," 670–73, 682–90.

11. Frederiksen, "The Contribution of Archaeology," 330–57; Evans, "*Plebs Rustica*," 19–47; Evans, *War, Women and Children*, 108–13; Dyson, *Community and Society in Roman Italy*, 23–55; Potter, *The Changing Landscape*, 125; Potter, *Roman Italy*, 115–16. A good survey of recent scholarship on the archaeology of

Roman Italy in Morley, *Metropolis and Hinterland*, 95–103, 129–35, 146–58. For doubts about this interpretation of the archaeological evidence, however, see below, n. 73.

12. For example, the *ager Faliscus* in southern Etruria: Camilli et al., "Recongizioni nell'Ager Faliscus," 399; the *ager Capenatus*: Camilli and Vitali Rosati, "Nuove Ricerche nell'Agro Capenate," 407; northern Campania: Arthur, *Romans in Northern Campania*, 64. Western Lucania, near Buxentum: Gualtieri and de Polignac, "A Rural Landscape in Western Lucania," 200; Apulia: Barker, Lloyd, and Webley, "A Classical Landscape," 42–43, 48–49; Barker, *A Mediterranean Valley*, 192–97, 200–201. Other areas discussed in Dyson, *Community and Society*, 28–33, with further references.

13. See especially Frederiksen, "Cambiamenti delle strutture agrarie," 268–72, along with the other contributions to volume 1 of Giardina and Schiavone, *Società romana e produzione schiavistica*, summarized by Rathbone, "The Slave Mode of Production in Italy," 162, 167–68; Spurr, "Slavery and the Economy in Roman Italy," 125–27. Compare Curti, Dench, and Patterson, "The Archaeology of Central and Southern Italy," 177.

14. Northern Campania: Arthur, *Romans in Northern Campania*, 73; compare Curti, Dench, and Patterson, "The Archaeology of Central and Southern Italy," 176–77, for further discussion and references to additional scholarship. On the date for the introduction of Dressel 1: Tchernia, *Le Vin de l'Italie romaine*, 44, but compare Tchernia, "Italian Wine in Gaul at the End of the Republic," 87; Panella, "La Distribuzione," 2:67, 70–71, 74. On this type of amphora at Cosa: Will, "The Roman Amphoras," 173–77, 182–83.

15. Purcell, "Wine and Wealth," 5–9.

16. First noted by Liverani, "L'Ager veientanus," 36–48; see also, for example, Quilici and Quilici Gigli, *Ficulea*, 481–86; Quilici and Quilici Gigli, *Crustumerium*, 294–95; Quilici and Quilici Gigli, *Fidenae* 404–13. A good synthesis and overview in Morley, *Metropolis and Hinterland*, 95–103. Doubts on desolation: Potter, "Towns and Territories," 199–200. See now Sallares, *Malaria and Rome*, 234–61, especially 248–53, who argues that endemic malaria was responsible for the abandonment of much of Latium by its free population and its eventual replacement by chattel slaves.

17. De Neeve, *Peasants in Peril*, 7–22; Morley, *Metropolis and Hinterland*, 86–90.

18. Livy 21.63.3–4.

19. Plut. *Cato Mai.* 3.2, although Astin, *Cato the Censor*, 9, suspects some "imaginative embroidery" in Plutarch's account of Valerius's admiration of Cato's industrious and frugal way of life.

20. Slave volunteers: Livy 22.57.11; oarsmen: Livy 24.11.5–9; 26.35.1–36.12, with

Notes to Page 8

Brunt, *Italian Manpower*, 67, for the number of rowers. Agriculture: Brunt, *Italian Manpower*, 67 and 273–75; Brunt, *Social Conflicts*, 18–19; Finley, *Ancient Slavery*, 84, although see Erdkamp, *Hunger and the Sword*, 84–102, for a discussion of overseas sources of food for the legions during this and other wars.

21. See Chapter 3 at n. 79.
22. Finley, *Ancient Slavery*, 83–86.
23. See especially Habinek, *Politics of Latin Literature*, 45–49, on the ideological subtext of the *De Agricultura*, and compare on Cato's ideological stance David, *La République romaine*, 83–89; Gruen, *Culture and National Identity*, 52–83; and Dench, *From Barbarians to New Men*, 83–85. Frederiksen, "Cambiamenti delle strutture agrarie," 271–72, argues that the *De Agricultura* dates from late in the author's life and so mainly reflects midcentury developments. However, there is no consensus on when the work was written: Astin, *Cato the Censor*, 190–91.
24. Morley, *Metropolis and Hinterland*, 126.
25. Gabba, "Urbanizzazione e rinnovamenti," 73–112; Gabba, "Considerazaioni politiche," 315–26.
26. Tchernia, *Le Vin de l'Italie romaine*, 98–100, and compare 77–94.
27. Livy 31.13.2–9.
28. The evidence for this largely depends on the fines that the aediles of 196 and 193 levied on grazers: Livy 33.42.10, 35.10.11; Gabba, "Considerazioni sulla decadenza," 275–76; Pasquinucci, "La Transumanza nell'Italia romana," 92–95. Relevant here, too, is the question of the date of the passage of the *lex de modo agrorum* and of whether modifications directed specifically at limiting the number of animals that could be pastured on public land were added in the early second century, on which see now the discussion of Oakley, *Commentary on Livy*, 1:654–59, with further bibliography, to which add Rathbone, Review of Flach, 28–30.
29. Labor: White, *Roman Farming*, 399; Brunt, *Italian Manpower*, 284; Spurr, "Slavery and the Economy in Roman Italy," 124. Cato the Elder on grazing: Cic. *de Off.* 2.89; Col. 6 praef. 4; Pliny *NH* 18.29–30. Frayn, *Sheep-Rearing*, 60, notes a sixteenth-century estimate that 1 sheep required 720 square meters of the best grazing land or 938 of poor. One *iugerum* of the best land thus could support about 3.5 sheep, a fairly high ratio compared with Tibiletti's calculations, "Il Possesso dell'*Ager Publicus*," 11, where he gives modern figures of 5–6 sheep per hectare on the best land, or 1.5 at most per *iugerum*. Varro *RR* 1.11.10–11 gives a ratio of 1 herdsman for every 80 or 100 sheep and indicates that the ratio may be higher in larger flocks. Thus a shepherd

tending 100 sheep would occupy about 28.5 *iugera* of prime land; a shepherd on poor land or with a larger flock would occupy more. The agricultural writers give manning ratios for arable of 1 man for every 25 *iugera*; for an olive yard, 1:22; for *arbustum*, 1:16; for vineyards, between 1:10 and 1:7: sources and discussion in Duncan-Jones, *The Economy of the Roman Empire*, 327–28.

30. How much displacement smallholders suffered in southern Italy, where large-scale stock rearing developed, as a consequence of this type of ranching is not clear. The rapid development of large-scale transhumant stock rearing in the period following the Hannibalic War is far less certain than often supposed by scholars: see Dench, *From Barbarians to New Men*, 116–25. What principally had driven farmers from their land here was the widespread devastation visited upon the region by years of combat between Rome and Hannibal and, after the latter's departure, the confiscations of territory that Rome visited upon some allies as punishment for their disloyalty during that struggle: Earl, *Tiberius Gracchus*, 24–25. On the devastation of the region, see Cornell, "Hannibal's Legacy," reviving Toynbee's claims in *Hannibal's Legacy*, 2:10–35, and rebutting Brunt's arguments against these in *Italian Manpower*, 269–77; and see now Erdkamp, *Hunger and the Sword*, 270–96.
31. On the size of donatives to enlisted men in this period: Brunt, *Italian Manpower*, 394–95.
32. See now Churchill, "*Ex qua quod vellent facerent*," challenging Shatzman's argument in "The Roman General's Authority over Booty" that Roman generals were free to appropriate booty for their personal enrichment without restriction.
33. Sources and discussion in Gruen, "The 'Fall' of the Scipios," 70–74; Gruen, *Studies in Greek Culture*, 133–37.
34. Polyb. 18.35.4–6, 31.22.1–4, and compare Livy *Per.* 46; Diod. 31.26.1.
35. Frederiksen, "Cambiamenti delle strutture agrarie," 267–70, and compare Shatzman, *Senatorial Wealth*, 12–46.
36. And note that Polybius, writing after 167, indicates that while many citizens were involved in the various contracts let by the censors at Rome, these involved mainly construction and tax collection, not the disposal of booty and captives: 6.17.1–4.
37. The *volones*, who were slaves serving as legionaries, were freed: Livy 24.14.3–16.10; whether the slaves conscripted as rowers for the fleet and who survived were returned to their owners at the conclusion of the war is unclear. On the lack of money and slaves in the civilian economy during the war: Livy 26.35.4–6; 28.11.9.

38. Livy 41.21.6; Ziolkowski, "The Plundering of Epirus," 76–80.
39. Dumont's claim, *Servus*, 57–71, that the Romans possessed between 151,320 and 756,000 slaves in 168 is not persuasive: see Scheidel, "The Slave Population of Roman Italy," 132–33; and also see Appendix 1.
40. Brunt, *Italian Manpower*, 121–24.
41. Ibid., 124.
42. Lo Cascio "Recruitment and the Size of the Roman Population," 111 n. 2. Compare Scheidel, "The Slave Population of Roman Italy," 134.
43. As Brunt himself concedes (*Italian Manpower*, 125), "This hypothesis [of 3 million slaves] permits us to believe that the population of Italy had grown by 50 percent since 225 B.C., though the increase was largely in the servile element."
44. Scheidel, "The Slave Population of Roman Italy," 134–35, further suggests that Brunt's choice of a figure for the slave population may have been influenced by the proportion of slaves in the American South prior to the Civil War, but even so, this "correspondence" proves nothing because, without an independent basis on which to estimate the slave population of Italy, there is no correspondence.
45. Brunt, *Italian Manpower*, 67, although at 121 he seems to imply that the total number of slaves in Italy would have been about 600,000, meaning perhaps to include in the latter figure slaves in the hands of non-Romans.
46. Evidence on the numbers enslaved in Toynbee, *Hannibal's Legacy*, 2:171–73; Ziolkowski, "The Plundering of Epirus," 74–75.
47. Scheidel, "The Slave Population of Roman Italy," 135–38. On slave fertility, see Harris, "Demography, Geography," 64–72, and compare Brunt, *Italian Manpower*, 145–46.
48. At an annual rate of decrease of 1 percent, a total of 6 million slaves would have had to have been imported over 200 years to reach 3 million by the time of Augustus; if the decrease was 2 percent, 9.5 million would have been required: Scheidel, "The Slave Population of Roman Italy," table 1.
49. On the size of the slave population in Sicily at the time of the First Slave War: Bradley, *Slavery and Rebellion*, 64; slavery in the Hellenistic world: de Ste. Croix, *Class Struggle*, 506–9, with further references. Carthage: Yeo, "The Development of the Roman Plantation," 329, with further references.
50. Scheidel, "The Slave Population of Roman Italy," 139–42, and compare Scheidel, "Progress and Problems in Roman Demography," 56, who summarizes and critiques an argument presented by Zelener, "The Slavish Gene," an unpublished paper delivered at the Second Finley Conference. Note also now Sallares's theory (*Malaria and Rome*, 252–56) that endemic malaria in Latium

and elsewhere would have killed a high proportion of the agricultural slaves working on large estates.
51. Piazza et al., "A Genetic History of Italy," 203–13.
52. Beloch, *Die Bevölkerung*, 435–43; Brunt, *Italian Manpower*, especially 91–130.
53. Brunt, *Italian Manpower*, 121, and compare the passage quoted above at n. 9 and in general 121–30.
54. Lo Cascio, "The Size of the Roman Population," 23–40. Earlier criticism of Beloch's interpretation of the census figures in Frank, "Roman Census Statistics."
55. Frank, "Roman Census Statistics"; Jones, *Ancient Economic History*, 3–7.
56. Scheidel, *Measuring Sex, Age and Death*, 167–68, and Morley, *Metropolis and Hinterland*, 46–50, are both critical of Lo Cascio's arguments. The latter responds to Morley in "Recruitment and the Size of the Roman Population," 113–19. Lo Cascio will presumably set out his case in full in his forthcoming monograph.
57. Morley, *Metropolis and Hinterland*, 33–54, especially 46.
58. The two scholars differ somewhat in their reconstructions: Gabba, *Republican Rome*, 2–12; Brunt, *Italian Manpower*, 74–83, 402–8, though see Brunt, *Fall of the Roman Republic*, 253 n. 58, for second thoughts.
59. Rich, "The Supposed Roman Manpower Shortage," 316–21. So, too, Shochat, *Recruitment and the Programme of Tiberius Gracchus*, 46–65.
60. Rich, "The Supposed Roman Manpower Shortage," 287–331. Rathbone, "The Census Qualifications," 121–52, argues for a reduction from 1,100 liberal *asses* (that is, *asses* coined at a standard weight of about 10 ounces of bronze each) to 4,000 sestantal *asses* (coined at a 2 ounce standard weight) in 212/11 and then to 375 *sestertii* (a coin worth 4 *asses*) around 140 B.C. without, however, holding that they imply anything about the availability of recruits or the condition of Rome's small farmers: see especially 147–48; compare also Rich, "The Supposed Roman Manpower Shortage," 314. Lo Cascio, "Ancora sui Censi Minimi," 289, 299–300, also argues for a reduction from 1,100 liberal to 4,000 sestantal *asses*, but links this to a need to expand the pool of potential draftees in the middle years of the Hannibalic War. Lo Cascio further disputes the very existence of the figure of 1,500 *asses* (= 375 *sestertii*) in the manuscript of Cicero's *De Republica*, "Ancora sui Censi Minimi," 287–88, although compare Rathbone, "The Census Qualifications," 140 n. 12.
61. Rathbone, "The Development of Agriculture," 12–15; Rathbone, "The Slave Mode of Production in Italy," 162; Spurr, "Slavery and the Economy in Roman Italy," 126; Dyson, *Community and Society*, 35; Scheidel, *Grundpacht und Lohnarbeit*, 159–200.

62. Laurence, *The Roads of Roman Italy*, 101–5; Erdkamp, *Hunger and the Sword*, 62–66, apropos of military transport but quite relevant to the problems of civilian transport, and compare Horden and Purcell, *The Corrupting Sea*, 155, 377. Note, too, that much of the grain that fed Rome itself probably arrived by land transport: Brunt, *Italian Manpower*, 286; De Neeve, Review of Rickman, 445–46, criticizing the traditional view in Rickman, *The Corn Supply of Ancient Rome*, 36–38; Erdkamp, *Hunger and the Sword*, 92 n. 24. In general, see now ibid., 85–94, and Erdkamp, "Feeding Rome," 53–70, on the development of the system for supplying grain to the capital during the second century.
63. Dyson, *Community and Society*, 41–44. On the importance of symbolic capital in the Republic's political economy, see David, *La République romaine*, 22–23, 33–35.
64. Laurence, *The Roads of Roman Italy*, 95–108, for a recent discussion of the economics of transporting crops to market.
65. Morley, *Metropolis and Hinterland*, 63–68, and compare 80–81; De Neeve, *Peasants in Peril*, 25–29; Dyson, *Community and Society*, 33–36.
66. Pessimism in De Neeve, *Peasants in Peril*, 30–35, followed by Morley, *Metropolis and Hinterland*, 81.
67. On the productivity of peasant agriculture, see in general Vergopoulos, "Capitalism and Peasant Productivity."
68. Ibid., 451–52.
69. See above at n. 47 on self-replacement of slave populations.
70. Morley, *Metropolis and Hinterland*, 122–29.
71. Vergopoulos, "Capitalism and Peasant Productivity," 453.
72. On storage, see Evans, "*Plebs Rustica*," 144–47; Gallant, *Risk and Survival*, 94–98; Forbes and Foxhall, "Ethnoarchaeology and Storage," 69–86; Horden and Purcell, *The Corrupting Sea*, 204–5.
73. Brunt, *Fall of the Roman Republic*, 247; Frayn, *Subsistence Farming*, 22. Note, too, Badian's point, "Tiberius Gracchus," 673, that if *coloni* in the Gracchan settlement program remained liable to conscription, then *coloni* on private estates may well have been, too.
74. See Chapter 5 at n. 34.
75. For example, Evans, "Resistance at Home," 123; De Neeve, *Peasants in Peril*, 40.
76. To my knowledge, only Crawford, *The Roman Republic*^{2}, 103 n. 2, has expressed doubts.
77. Note, by way of example, Brunt, *Italian Manpower*, 77: "But prolonged absence on campaigns both in the Hannibalic and in later wars . . . must have

ruined many small farmers.... The poorest ... were the most vulnerable to economic distress precisely because they had the smallest resources." Compare ibid., 108 ("The protracted absence of the husbandman from a small farm ... inevitably meant that it was likely to be inefficiently cultivated, that money had to be borrowed, and that in the end there was no alternative to sale") and 640–41. Note, too, from a Marxist perspective de Ste. Croix, *Class Struggle*, 207–8: "For all those below my 'propertied class,' conscription, diverting them from the activities by which they earned their daily bread, could be a real menace, and those who were furthest from belonging to the propertied class would presumably suffer the most."

78. Clark and Haswell, *The Economics of Subsistence Agriculture*, 93–130. For Roman Italy: White, *Roman Farming*, 345; Finley, *The Ancient Economy*, 105–7; Hopkins, *Conquerors and Slaves*, 24; Crawford, *The Roman Republic*²2, 102–3; Evans, "Plebs Rustica," 137.

79. Note especially the remarks of Humphreys, "Diskussion," 549–50, to whose insight there the argument presented here owes much. Compare her similar point regarding Homeric warfare in *Anthropology and the Greeks*, 165, and also Hopkins, *Conquerors and Slaves*, 24–25.

Chapter 2

1. On agriculture in Italy, see especially Spurr, *Arable Cultivation*, an excellent study to which the discussion throughout is much indebted. See also White, *Roman Farming*; Frayn, *Subsistence Farming*.

2. Pliny *NH* 18.174–75, 242, Col. 2.4.11, and in general: Spurr, *Arable Cultivation*, 23–40. Compare also Davis, *Land and Family in Pisticci*, app. VI, for a synoptic chart of the distribution of the labor needed to cultivate wheat and other crops in southern Italy over the agricultural year.

3. Spurr, *Arable Cultivation*, 43–44; compare White, *Roman Farming*, 180. On "risk management" among ancient farmers, see especially Garnsey, *Famine and Food Supply*, 49–53, and Gallant, *Risk and Survival*, 36–38.

4. Spurr, *Arable Cultivation*, 66–67; for a detailed table of Italian harvest dates, see Azzi, "Il clima del grano in Italia," 544–45, and compare A. L. Broughton, "The *Menologia Rustica*." Note the autumn grain harvest mentioned at Sall. *Hist.* 3.98M (= 3.66 McGushin).

5. However, things could be quite different in the mountains: see McNeill, *The Mountains of the Mediterranean World*, 108–14, for summer as the season of greatest agricultural activity in early modern Lucania.

6. Gomme, *Commentary on Thucydides*, 1:11–12; Grundy, *Thucydides and the History of His Age*, 1:244–57; Vidal-Naquet, "La Tradition de l'hoplite athénien," 166; Anderson, *Military Theory and Practice*, 2–3; Garlan, *Recherches de poliorcétique*, 22–26; Hanson, *Warfare and Agriculture*^{2}, 42–46, 122–26, and compare Hanson, *The Western Way of War*, 33–35; Ober, *Fortress Attica*, 33–34; Foxhall, "Farming and Fighting in Ancient Greece," 138–42.

7. Soldiers could, of course, purchase food, but most hoplites paid their own expenses on campaigns, and subsistence farmers were apparently quite reluctant to part with what little ready cash they had in such cases: Thuc. 1.146.5; Dem. *Olyn.* 1.27.

8. In the spring of 425 the Peloponnesian force invading Attica arrived to find the grain not yet ripened, and the shortage of food they experienced as a result forced them to curtail their operations after only fifteen days: Thuc. 4.6.1–2. As Pritchett, *The Greek State at War*, 1:38–39, notes, this is the best evidence that an invading army depended "to a large extent upon plundering for its sustenance." See now Erdkamp, *Hunger and the Sword*, 122–40, for a fairly pessimistic assessment of an ancient army's ability to "live off the land." See, too, Forbes and Foxhall, "Ethnoarchaeology and Storage," 82, for a modern instance of peasants' hiding their foodstocks during unsettled times.

9. Hanson, *Warfare and Agriculture*^{2}, 49–55, although burning standing grain might not be easy: Foxhall, "Farming and Fighting in Ancient Greece," 140. However, because the ears of overripe grain shatter easily if left standing, simply preventing defenders from harvesting wheat or barley until they reached that stage would result in the loss of much if not all of the crop. See below, n. 140.

10. Thuc. 3.15.2. See also Grundy, *Thucydides and the History of His Age*, 1:257; Gomme, *Commentary on Thucydides*, 1:11–12.

11. Why defenders chose to offer battle rather than simply wait in safety behind their fortifications for attackers to leave is much discussed in the literature. Threats to food supply: Grundy, *Thucydides and the History of His Age*, 1:247–48, 257; Gomme, *Commentary on Thucydides*, 1:11; political tensions because of the uneven distribution of the damage inflicted by the invading army: Osborn, *Classical Landscape with Figures*, 154; Foxhall, "Farming and Fighting in Ancient Greece," 142–43, and compare Garlan, *Recherches de poliorcétique*, 24–25; agonal mentality and revulsion: Ober, *Fortress Attica*, 33–35; Hanson, *The Western Way of War*, 4–6. There is no reason, however, why all of these motives might not have been in play when defenders made the decision to offer battle.

12. Hanson, *The Western Way of War*, 31–32.

13. January campaigning: Proctor, *Hannibal's March*, 17. Rains and equipment: Grundy, *Thucydides and the History of His Age*, 1:249–50.
14. On the date of the introduction of pay, see n. 16. On the Roman phalanx, see Oakley, *Commentary on Livy*, 2:452–55.
15. On the topography of Samnium, for example, see Salmon, *Samnium*, 14–23.
16. Pay for Roman legionaries and rations for the allied forces: Polyb. 6.39.12–15. The date of the introduction of payment for service is a notorious crux. Some follow the ancient sources (Livy 4.59.11–60.8; Diod. 14.16.5) and date the practice to the last years of the fifth century in connection with Rome's siege of Veii, but others believe this is far too early, at least on a regular basis, and suggest a point a century later, during Rome's wars with the Samnites. Late fifth century: see, for example, Crawford, *Coinage and Money*, 22–24; Cornell, *Beginnings of Rome*, 187, 313; Oakley, *Commentary on Livy*, 1:630–32; for the late fourth century, see Raaflaub, "Born to Be Wolves?," 290.
17. On Rome's military commissariat and the constraints on an army feeding itself by foraging, see now the excellent study by Erdkamp, *Hunger and the Sword*, 46–140. He seems to place the beginnings of this system in 264 at the onset of the First Punic War, but there is every reason to assume that its foundations were laid much earlier when the republic's military commitments in Italy forced it to maintain armies far from Rome.
18. On changes in the conduct of war in fourth-century Greece, see Ober, *Fortress Attica*, 32–50; Garlan, *Recherches de poliorcétique*, 20–86. On these developments at Rome, see further below at n. 132. On the difficulty of agricultural devastation: Hanson, *Warfare and Agriculture*^{2}, 42–76.
19. On the manipular army in combat: Adcock, *The Roman Art of War*, 10–13; Meyer, *Kleine Schriften*, 2:197, 210; Rosenstein, *Imperatores Victi*, 95–96; and, now, Sabin, "The Face of Roman Battle," 1–17. See also Goldsworthy, *The Roman Army at War*, 171–282, for a detailed description of infantry combat in the cohort army of the late republic and empire, when tactical operations were not fundamentally dissimilar to those of the earlier manipular formation. The date at which maniples were introduced is, like much else, controversial. Estimates range from 406 to the early third century: see, recently, Oakley, *Commentary on Livy*, 2:455–57, and Cornell, *Beginnings of Rome*, 354 (who suggests 311 B.C., although at 187–88 he opts for 406 B.C.) for discussion and references to earlier works.
20. On training programs for troops in the middle republic, note the senate's insistence on training the soldiers who would face Hannibal in 216: Polyb. 3.106.4–5, and compare 108.6–7; Scipio's efforts at New Carthage in 210: Polyb. 10.20.1–4; Livy 26.51.4–5, and at Syracuse in 205, where he spent an

entire year readying his soldiers for the campaign in Africa: 29.22.2. Note, too, Cato the Elder in Spain in 195: Astin, *Cato the Censor*, 36.

21. Livy 8.16.6; Oakley, *Commentary on Livy*, 2:584; Livy 8.17.1-2, 17.8. However, difficulties arise since for much of the fourth century consuls apparently entered office on the Kalands of Quinctilis (that is, the first of July): Livy 8.20.3, although see Oakley, *Commentary on Livy*, 2:612-14, for discussion of the problems raised by this passage. Consequently, one cannot rule out the possibility that the armies these consuls took command of after July 1 had only been levied in the spring or early summer of these years. Yet, even so, this would mean that many of these recruits had been kept away from the harvests on their farms.

22. For 327, see Dion. Hal. 15.10.2 and compare Livy 8.22.9-10; Oakley, *Commentary on Livy*, 2:649; for 316-314, see Livy 9.21.1, 22.1, 24.1; for 313, see Livy 9.28.1, although Salmon, *Samnium*, 237-38, dismisses Livy's report there of a siege of Bovianum as "preposterous." Salmon believes that these forces instead carried out the genocide of the Aurunici. But if so, this atrocity certainly is likely to have required the legions to spend the winter accomplishing it. For 310, see Livy 9.33.1-2.

23. For 297/6, see Livy 10.16.1-2, to judge from the prorogation of the consuls of 297 and the proconsul Decius's operations in 296. For 296/5, see 10.25.4-11, note especially 25.11 (*hiemps hauddum exacta*), and compare Salmon, *Samnium*, 262, 265. For 294/3, see Livy 10.39.1.

24. Livy 10.46.1-16. Note the dates of the triumphs, January 13 and February 13, 293/2: Degrassi, *Inscriptiones Italiae*, 13.1:72, 544. De Sanctis, *Storia*², 2:342-44; Salmon, *Samnium*, 270-73; and Harris, *Rome in Etruria*, 76-77, accept Livy's basic account of both consuls operating in Samnium and then proceeding to Etruria and Campania; *contra*, however, Beloch, *Römische Geschichte*, 447-48, who argues that only one consul went south while the other operated in Etruria. But Beloch offers no compelling reason why Livy's narrative should be so extensively modified, and his reconstruction ought to be rejected.

25. Dion. Hal. 17/18.4-6; Dio frag. 36.30-31; Zon. 8.1, other sources in T. Broughton, *MRR*, 1:182-83; Degrassi, *Inscriptiones Italiae*, 13.1:72-73, 544: the year is clear on the stone. See Salmon, *Samnium*, 274-75, for discussion and a defense of the victory in 291 against the skepticism of Beloch, *Römische Geschichte*, 449.

26. On the time required to assemble an army, note Pompey's boast of having levied the army he led against Sertorius within forty days: Sall. *Hist.* 2.98.4M (= 2.82.4 McGushin), and compare Proctor, *Hannibal's March*, 56. If Pompey

made this achievement a point of pride, it is likely that other commanders usually took considerably longer.

27. As early as 343 Livy reports that Roman garrisons wintered in a number of Campanian cities: 7.38.4–39.6; Dion. Hal. 15.2.3, and compare 15.3.3–15; Salmon, *Samnium*, 196. See Oakley, *Commentary on Livy*, 2:365–66, for further examples and discussion.
28. Zon. 8.2; Lévêque, *Pyrrhos*, 310, and compare 282; Garoufalis, *Pyrrhus*, 313 n. 37; Franke, *CAH*², 7.2:467.
29. Front. *Str.* 4.1.24.
30. App. *Sam.* 10.3; Zon. 8.4; Degrassi, *Inscriptiones Italiae*, 13.1:72, 545.
31. Degrassi, *Inscriptiones Italiae*, 13.1:74, 547. On the correspondence of these calendar dates to seasonal time, see Appendix 2.
32. Beloch, *Römische Geschichte*, 89 (compare De Sanctis, *Storia*², 2:403 n. 101), however, argues that the four triumphs that the *fasti* record for Pera and Pictor result from dittography and that in fact each consul celebrated only one: one consul fought in Umbria, the other in Calabria. Beloch bases his claim on the length of time it would have taken an army to march from Rome to Calabria and the improbabilities of a winter campaign, of two consular armies being sent successively against two such small peoples as the Sassini and the Sallantini, and of a single magistrate triumphing twice in one year. He also believes that a consul who crossed the *pomerium* to triumph probably lost the *auspicia*, and he questions whether auspices renewed for a second campaign would have permitted a consul to triumph. But none of these objections is conclusive. It is true that the Sallantini and the Messapians do not seem to have been very numerous. Polyb. 2.24.11 gives a total of 50,000 potential soldiers for the Apulians and Messapians together, and the latter, among whom the Sallantini seem to be included (Walbank, *Commentary on Polybius*, 2:201), will not have composed more than half this number. However, the *fasti* indicate that the Romans had dispatched two consular armies against them in the preceding year: Degrassi, *Inscriptiones Italiae*, 13.1:547, accepted, it seems, by Beloch, *Römische Geschichte*, 89 and 473. It is not therefore a priori improbable that they would have done so again. Beloch estimates that the armies' marching time to Calabria and back (over 500 kilometers in his estimate) would have been two months, leaving only two months for the campaign in the South. But the journey might have been accomplished in less time: as is well known, in 218 Sempronius's legions marched from Sicily to Gaul, over 1,300 kilometers, in forty days: Polyb. 3.68.12–14, and see Proctor, *Hannibal's March*, 67–70, for discussion. Yet in the absence of Livy's narrative, we simply do not know whether two months' time would have sufficed to

Notes to Page 33

pacify the region. Given the victories there the year before, one would tend to think so. Nor can we assume that the senate did not deem the crisis in Calabria urgent enough to warrant sending both armies there late in the year. Possibly, rebelliousness in the region led the *patres* to make a prompt show of force to quash incipient unrest. The same thinking may have lain behind the senate's decision to dispatch both consuls to Umbria earlier, and anyway, no other war was in the offing to occupy one of them. Certainly, as the campaign in 293 indicates, nothing precluded winter operations for the legions. As for the validity of renewed auspices for a triumph and two triumphs in a single year, there was precedent: M'. Curius Dentatus in 290, which Beloch is compelled to reject as well, along with the tradition that Dentatus fought in both Samnium and Sabine territory: *Römische Geschichte*, 428–29. Again however there is no good reason to reject the sources' account; Dentatus's double triumph is accepted as genuine by De Sanctis, *Storia*², 2:345–46, and Salmon, *Samnium*, 276. The double triumphs in 266 therefore should be retained.

33. De Sanctis, *Storia*², 3.1:254; followed by Walbank, *Commentary on Polybius*, 1:70, and Lazenby, *First Punic War*, 58. Morgan, "Calendars and Chronology," 94–95, argues that the siege ended in December, but his arguments are not persuasive: see Appendix 2.

34. Zon. 8.10–11; Morgan, "Calendars and Chronology," 96.

35. Polyb. 1.24.8; Zon. 8.11; compare Lazenby, *First Punic War*, 75, who calls Aquilius's stay in Sicily over the winter "an unusual departure, since previous consuls seem to have returned to Italy before the winter." On May 1 as the first day of the consular year in this period, Mommsen, *Römische Chronologie*, 101–2; Mommsen, *Römische Staatsrecht*, 1:599; De Sanctis, *Storia*², 2:370–72 n. 21; Varese, *Il calendario romano*, 2–4; Soltau, *Römische Chronologie*, 302–3. Brind'Amour, *Le Calendrier romain*, 175–76, arguing for March 1, is unpersuasive. See Appendix 2 for further discussion.

36. Polyb. 1.25.8. The two legions Cn. Cornelius Blasio had commanded probably wintered over; Caiatinus's legions presumably returned with him to Rome for his triumph in January and were replaced either when he left or soon thereafter by new legions: Lazenby, *First Punic War*, 86.

37. Walbank, *Commentary on Polybius*, 1:91, for a discussion of the date of Regulus's disaster. On the size of his forces: Polybius 1.29.9. On the following year: De Sanctis, *Storia*², 3.1:160; Lazenby, *First Punic War*, 116.

38. Polyb. 1.40.1; see Walbank, *Commentary on Polybius*, 1:102, for the dating; compare Lazenby, *First Punic War*, 119–20. If Metellus's colleague left in the spring rather than the fall, then he, too, probably kept his legions with him in Sicily.

39. For the events and discussion, see Lazenby, *First Punic War*, 123–59.
40. Recently, for example, Drummond, *CAH*², 7.2:19; Cornell, *CAH*², 7.2:289–90. Even the hypercritical Beloch remarks: "Gegen die Tagdaten aus dieser Zeit [i.e., 300–264] ist nichts einzuwenden." *Römische Geschichte*, 90–91, and compare 88–89. He explicitly rejects only seven triumphs between 300 and 241, in most cases groundlessly: see above, n. 32, for one example. But even if one were to accept his doubts, eliminating these triumphs (one each in February, March, April, Sextilis, and December and two in October) from the calculations that follow would not significantly alter the proportions of triumphs celebrated between January and April. The *Annales Maximi* are often cited as the source for the dates that the *fasti triumphales* record, based on Servius *Comm.* 1.373: so, recently, Forsythe, *The Historian, L. Calpurnius Piso*, 62–63. *Contra*, however, Frier, *Libri Annales*, 90–91, with references to earlier discussions. This is not to maintain that the reliability of the *fasti* for periods earlier than circa 300 B.C. is generally accepted: see Mora, *Fasti e schemi chronologici*, 13–14, for discussion of current opinion on this question.
41. The variation is due to uncertainties about whether M'. Curius Dentatus's triumph in 275 fell in January or February and the month of M. Atilius Regulus's triumph in 267. I have followed Degrassi, *Inscriptiones Italiae*, 13.1:547, in dating the latter to the Kalands of February, the same day that his consular colleague L. Iulius Libo celebrated his. The former I have placed in February since his colleague, L. Cornelius Lentulus Caudinus, triumphed on the first of March, and one should presume that Dentatus's celebration took place not long before. But even if it be placed in January, it would not affect the argument being offered here. Two triumphs that I have placed in February (Gurges' in 276 and Caninia's in 273) took place on the Quirinalia. Warrior, "Notes on Intercalation," argues forcefully that such dates in fact indicate that the events in question occurred in the intercalary month of these years. Yet accepting her argument would not affect the point being made here.
42. On the popularity of February, March, and the intercalary month for triumphs, see also Sumner, "The Chronology of the Outbreak," 10. Although one might suspect that triumphs celebrated in these months were timed to coincide with consular elections, as was certainly the case in 207 (see below at n. 93), and on occasion subsequently, the consular year began on the first of May prior to 222, not the first of March (above, n. 35). One would therefore have expected a far higher number of third-century triumphs to have taken place in April if their timing had been politically motivated.
43. The argument here assumes that soldiers regularly returned with their commanders to march in triumphs in the third century. Richardson's comment,

however, in "The Triumph, the Praetors and the Senate," 54–55, that "[t]he argument which caused Marcellus' [proconsul in 211 who sought a triumph from the senate for his capture of Syracuse] cause to founder, that he had not brought back his army from the province, has the air of a technicality introduced for the purpose" might suggest that the *deportatio exercitum* was not a regular feature of previous triumphs. But this seems unlikely. Had Marcellus been able to cite precedent for a magistrate triumphing without bringing his army home, it is hard to see how the failure of Marcellus's army to return to Rome with him could have derailed his bid for a triumph. More plausibly, the question had up to that point never arisen because magistrates had always returned with their armies to celebrate triumphs, and the question that caused the long debate in 211 was precisely whether a triumph could be awarded when this was not the case: Livy 26.21.1–4; Plut. *Marc.* 22.1, and compare Richardson, "The Triumph, the Praetors and the Senate," 62. To be sure, the senate early in the second century reversed its position and allowed a few magistrates to triumph without their armies (on which see Richardson, "The Triumph, the Praetors and the Senate," 52–54, 61–62), but these cases do not invalidate the argument advanced here. Whether soldiers were furloughed in the fall with orders to reassemble for triumphs is discussed later.

44. So the Roman calendar date of July 11, 190, corresponded to a seasonal date of March 14 while Roman September 3, 167, in fact fell on a seasonal date of June 21: see Derow, "The Roman Calendar, 190–168 B.C.," 345–46, for discussion.
45. See Appendix 2 for full discussion.
46. For example, in 262: Polyb. 1.17.9 with De Sanctis, *Storia*², 3.1:246–47; Walbank, *Commentary on Polybius*, 1:70; for 255–254, see Polyb. 1.36.10 with De Sanctis, *Storia*², 3.1:252–54; Walbank, *Commentary on Polybius*, 1:91, on Polyb. 1.32.8. See Appendix 2 for detailed discussion.
47. Above, n. 24. Compare De Sanctis, *Storia*², 2:371–72 n. 21, at 372. Again, see Appendix 2 for full discussion.
48. The pontifices may have judged it expedient for generals fighting in Spain, Greece, or the Near East during the early second century to enter office at a seasonal date of December or January in order to be able to reach their provinces with the whole of the campaigning season ahead of them: so Derow, "The Roman Calendar, 190–168 B.C.," 348 n. 12. On the other hand, political jealousy may have been involved as well, in that intercalation extended the terms of office of some magistrates, which may have been opposed by rivals. Recently, Warrior, "Intercalation," 131–32, has revived a suggestion made by Michels, *Calendar of the Roman Republic*, 170, that intercalation was often omit-

ted "because of the difficulty of communicating the pontifices' decision to magistrates serving in the provinces" (131) and the resulting confusion over when civil and religious events were supposed to take place. Her argument has much to commend it.

49. De Sanctis, below, n. 52; Gabba, *Republican Rome*, 39; Toynbee, *Hannibal's Legacy*, 2:73; Brunt, *Italian Manpower*, 68, 399.
50. Note, for example, that P. Cornelius Scipio Nasica discharged his men after a victory in 191 with orders to reassemble in Rome on the day of his triumph: Livy 36.39.4. Unfortunately, the *fasti* do not preserve the date, on which see below, n. 103. And compare Pompey's similar instructions when he dismissed his troops in 62 upon landing at Brundisium after his victories in the East: Vell. 2.40.3; Plut. *Pomp.* 43.2; Dio 37.20.6; App. *Mith.* 116.
51. Sowing dates: above at n. 1.
52. De Sanctis, *Storia*², 3.2:245–46 and n. 109; 278–79 and n. 147.
53. Livy 23.48.2: *[Fabius Maximus] M. Claudio proconsuli imperavit ut retendo Nolae necessario ad tuendam urbem praesidio ceteros milites dimiteret Romam ne oneri sociis et sumptui rei publicae essent.* De Sanctis must, of course, take *Romam* in the passage to mean "to their homes": compare Wissenborn and Müller's comment on the passage in their edition of Livy.
54. Livy 23.48.1–2.
55. On early October as the usual sowing time: Pliny 18.224; other times: Spurr, *Arable Cultivation*, 42–43. On factors affecting the growth of wheat, see Nuttonson, *Wheat-Climate Relationships*, 3–38, although it should be born in mind that since we do not know the particular strains of wheat grown in Campania in antiquity, Nuttonson's material is useful only for establishing general parameters; individual types of wheat vary tremendously in their growth characteristics and often have adapted to the particular constraints imposed by their environment. On average temperatures in Campania: Great Britain, Naval Intelligence Division, *Italy*, 1:528; on the length of days at 41° latitude, see Nuttonson, *Wheat-Climate Relationships*, 26, fig. 4. Pliny *NH* 18.111, it is true, claims that Campania was in crop year-round, being sown twice with emmer and then with millet, and this might suggest a winter-sown crop fully ripe and ready to harvest at the end of winter with a crop of spring wheat then sown and ready to reap in June or July, after which came the millet over the summer. But Pliny here almost certainly exaggerates: Jongman, *Economy and Society of Pompeii*, 104, and compare Spurr, *Arable Cultivation*, 118 n. 3.
56. Livy 23.21.1–6, 48.4–12.
57. Livy 24.11.7–9; on the financial crisis at this point: Nicolet, "A Rome pendant la seconde guerre punique"; Crawford, "War and Finance"; Crawford,

Roman Republican Coinage, 1:29–35; Crawford, Coinage and Money, 52–62, all with citations to earlier works, and compare Loomis, "The Introduction of the Denarius." See also below, n. 157.

58. Marchetti, Histoire économique, 57.
59. Livy 24.13.8; compare Erdkamp, Hunger and the Sword, 174.
60. De Sanctis, Storia², 3.2:246 n. 109; Kromayer, Antike Schlachtfelder, 3.1:402 n. 1.
61. Livy 24.9.9: Absens Marcellus consul creatus cum ad exercitum esset.
62. Livy 24.7.10–12; Lazenby, Hannibal's War, 97.
63. After Flaminius's disaster, there would have been no question but that Marcellus would return to Rome to assume his duties as consul: see Rosenstein, Imperatores Victi, 77–78.
64. Fabius's election of censors: Livy 24.10.2, 11.6; Marcellus's march from Cales to Nola: 24.13.9–10.
65. Polyb. 9.6.6, Paton's translation: οἱ γὰρ περὶ τὸν Γνάιον καὶ Πόπλιον τοῦ μὲν ἑνὸς στρατοπέδου πρότερον πεποιημένοι τὴν καταγραφὴν ἐνόρκους ἔχον τοὺς στρατιώτας εἰς ἐκείνην τὴν ἡμέραν ἥξειν ἐν τοῖς ὅπλοις εἰς τὴν Ῥώμην, τοῦ δ' ἑτέρου τότε τὰς καταγραφὰς ἐποιοῦντο καὶ δοκιμασίας. ἐξ οὗ συνέβη πλῆθος ἀνδρῶν αὐτομάτως ἀθροισθῆναι πρὸς τὸν δέοντα καιρὸν εἰς τὴν Ῥώμην.
66. Walbank, Commentary on Polybius, 2:126, with additional bibliography and discussion. Urban legions for 211: Livy 26.28.4.
67. Brunt, Italian Manpower, 628, but compare Gelzer, Kleine Schriften, 3:241.
68. Polyb. 6.19.5–21.3, 21.6.
69. Polyb. 6.21.7–26.1.
70. Polyb. 6.26.2–4, and compare Cincius in Gell. NA 16.4.3–4.
71. Livy 25.3.7, and compare 26.8.5. The details of Polyb. 9.6.6 tell against the hypothesis that by στρατοπέδον Polybius (or his source) means "legion," for a key result of the procedure Polybius describes in book 6 for levying armies is the similar quality of the soldiers in each legion. This equity resulted from the practice of having the tribunes of each legion rotate the order in which they picked one man from each group presented to them, and to achieve this required the simultaneous enrollment of the legions: Polyb. 6.20.1–6. Translating στρατοπέδον at 9.6.6 as "legion," however, would mean that in this case the legions were being levied sequentially: the first legion's enrollment being completed, the consuls were now enrolling the second.
72. Marchetti, Histoire économique, 68–69; Livy 25.5.5–9.
73. Polyb. 3.40.2, 41.2–3, 61.8–9; Livy 21.17.1, 5–6.
74. On the date, see Polyb. 3.72.3 with Walbank, Commentary on Polybius, 1:406;

Proctor, *Hannibal's March*, 70–74; on winter quarters: Livy 21.56.9, and compare Polyb. 3.74.7–8, 75.3. Sempronius's soldiers were dismissed at Rhegium and placed under oath to reassemble at Ariminum on a specific day. It is unlikely, however, that they could have found time to return home, plow, plant, and report back to their units within the specified time. The journey overland from Rhegium to Ariminum took forty days of continuous marching: Polyb. 3.61.10, 68.14, with Walbank, *Commentary on Polybius*, 1:396; Proctor, *Hannibal's March*, 67–70.

75. Hannibal's intention to invade Italy will have been clear by late August: Polyb. 3.49.1–4 with Walbank, *Commentary on Polybius*, 1:377; on the date of Hannibal's arrival in Italy, see Polyb. 3.54.1 with Walbank, *Commentary on Polybius*, 1:390, arguing for early October; however, Proctor, *Hannibal's March*, 35–45, puts the arrival in November.

76. Polyb. 3.100.1–106.4; Livy 22.18.5–10, 23.9–24.14, 27.1–30.6, 31.7, 32.1–3; and see Walbank, *Commentary on Polybius*, 1:435, for the chronology.

77. Livy 23.17.2–3, 19.1–12. The fact that as the time for the consular elections approached the various commanders in Campania, upon summons to Rome, left legates in command of their legions likewise indicates that these troops had remained in camp throughout the winter: Livy 23.24.1–5. The date will have been in February or early March because the new magistrates were due to enter office shortly after the elections: Livy 23.24.5. Note too that L. Postumius Albinus, praetor in 216 and consul-elect for the following year, set out very early in the campaigning season of 215, before the beginning of the consular year, on his disastrous attack upon the Gauls: Livy 23.24.6–13, and compare 30.18, and see Walbank, *Commentary on Polybius*, 1:448–49, on the date. His early start suggests that he, too, had kept his legions together over the winter, rather than suffer the delay of reassembling them at the opening of the following spring.

78. Livy 24.3.16–17.

79. Livy 25.13.8–9; on the date: Livy 25.12.1–2.

80. So De Sanctis, *Storia*², 3.2:278–79 and n. 147; on the assignment of armies for 212: Livy 25.3.3.

81. Supplementa: Livy 25.3.4, 5.5; on the distance from the *castra Claudiana* to Beneventum, see Livy 25.15.1: *duos consules ad Beneventum esse, diei iter a Capua*.

82. Livy 25.14.12.

83. Livy 25.22.8–13, and compare 26.12.5. Walbank, *Commentary on Polybius*, 2:7–8, assumes that Capua was surrounded in the autumn of 212 and therefore argues on the basis of Livy's synchronism at 25.23.1 between the circumvallation of Capua and the fall of Syracuse that the latter is to be dated to the late

summer of 212. However, Livy is likely to be referring here to the beginning of the siegeworks' construction, not their completion. Compare De Sanctis, *Storia*², 3.2:283–84.

84. Massiveness: Livy 25.22.8; App. *Hann.* 37; counterattacks: Livy 25.22.8–9, 26.4.1–10; App. *Hann.* 37; food supply: Livy 25.20.1–3, 22.5.
85. Livy 27.4.1–2; Plut. *Marc.* 24.6.
86. Dates of consular elections: for 217, see Polyb. 3.30.7 with Walbank, *Commentary on Polybius*, 1:404; Livy 21.57.4, and compare 21.53.6; and Proctor, *Hannibal's March*, 70–74, for full discussion. For 216, note the fact that an *interregnum* occurred when a *vitium* compelled a dictator appointed to conduct elections to resign fourteen days after he took office, indicating that the appointment was made at most less than two weeks before the end of consular 217 and therefore that the senate's initial request to the consuls that one of them return to conduct the elections must have been made not long before that, that is, in late February: Livy 22.33.11–12. On the elections for 215, see above, n. 77. Elections for 214: Livy 24.7.10; elections for 211: Livy 25.41.8. Note too that elections for 209 seem to have taken place about the time of Valerius's raid on Africa, which De Sanctis, *Storia*², 3.2:448, dates to the beginning of spring: Livy 27.5.8–6.2. For the senate's fears that consular 209 would begin without magistrates in office: Livy 27.4.1–4. Datable consular *comitia* in the second century likewise fall late in the official year: Warrior, "Intercalation," 133.
87. Livy 27.2.1–12; Plut. *Marc.* 24.1–6.
88. Livy 27.12.7.
89. Livy 27.12.7–17; Plut. *Marc.* 25.4; a second and supposedly victorious battle on the following day is invention: Münzer, *RE*, 3:2752–53; exaggeration: T. Broughton, *MRR*, 1:289 n. 5; or another defeat: Livy 27.21.3.
90. Livy 27.20.10, 21.4; Plut. *Marc.* 26.4.
91. Plut. *Marc.* 27.2.
92. Livy 27.21.6–8; Plut. *Marc.* 28.1.
93. Livy 28.9.1–10.5. On the date of the battle at the Metaurus, Ovid *Fasti* 6.770, although the interpretation of this passage is controversial. See Scullard, *Scipio Africanus in the Second Punic War*, 324–25, for discussion.
94. Livy 28.9.1.
95. See above, n. 86.
96. The dedication of the temple to Jupiter Victor vowed by Fabius during the battle on April 13 probably took place on the anniversary of the victory: Ovid *Fasti* 4.621; Livy 10.29.14; Holzapfel, *Römische Chronologie*, 98; Soltau, *Römische Chronologie*, 303; on the date of the triumph, see Degrassi, *Inscriptiones Italiae*,

Notes to Pages 42–44

13.1:543; on the accuracy of the Roman calendar in this period, see Appendix 2.
97. Livy 28.9.3.
98. Livy 29.10.1–3.
99. Livy 29.38.5.
100. Scullard, *Scipio Africanus in the Second Punic War*, 326–27.
101. Livy 30.24.1–4.
102. Livy 43.11.9–10, 14.7–10. Livy dates the return of the embassy to February; in view of the fact that three years later the calendar was about two and a half months in advance of the seasons, the embassy must have reached Rome sometime in the late fall: above, n. 44. Derow, "The Roman Calendar, 190–168 B.C.," 348–49, calculates the date as October, but note the criticism of his assumption that intercalation occurred regularly every other year after 191: Warrior, "Intercalation," 137–43, and compare the leaves granted to soldiers in Greece in 196: Livy 33.29.2.
103. Livy 36.39.3–4, and for the events: 36.37.6–38.5. The calendar was running about four months in advance of the seasons at this point. Although the date of Nasica's triumph does not survive on the *fasti*, the ovation preceding it fell *xv Kal. Ian* (Degrassi, *Inscriptiones Italiae*, 13.1:553) and took place at about the time Nasica was winning his victory: Livy 36.38.7–39.2. The ovation probably occurred at a seasonal date of early August: Derow, "The Roman Calendar, 190–168 B.C.," 348–49, and thus Nasica might have celebrated his subsequent triumph in late August or September.
104. Livy 36.38.7–39.3.
105. Livy 36.39.6–10.
106. Livy 39.20.9.
107. Derow, "The Roman Calendar, 190–168 B.C.," 348–49, calculates that the Kalends of March 182 fell on a seasonal date of November 12, 183; however, if Warrior, "Intercalation," 137–43, is correct that intercalation occurred in several successive years following 190, the effect would be to move the seasonal date of the Kalends of March later, into late November or early December. Marcellus's dismissal of his legions seems to have been due to the lack of any enemy for them to fight and, perhaps, the senate's fear that otherwise the consul would find a way to start a war against the Histrians: Livy 39.54.1–13, 55.4–5, 56.3–4.
108. This is based on the fact that the dismissal occurred during the consulship of A. Postumius Albinus and Q. Fulvius Flaccus *priusquam hi consules venirent ad exercitum qui Pisas indictus erat* (40.41.7; see Briscoe's Teubner edition). Postumius was in the North in nominal command, but he had apparently gone

to Placentia before going to the army at Pisa; his colleague was still at Rome: Livy 40.41.7, 9–10. Despite Livy's remark that the consuls were levying new legions at the beginning of spring (40.37.8), one can make the best sense out of the episode by placing it in the winter months. The Ides of March, 180, when the consuls would have taken office, fell at a seasonal date of about December 1, 181. Thus the tribune will have dismissed the legion sometime between December and March intending, since there was no prospect of fighting until late spring or early summer, to win a measure of popularity with the soldiers: compare the motives for the lavish granting of leaves to troops from the legions in Greece in 170/69 (Livy 43.11.10, 14.7). Of course, one cannot rule out the possibility that in fact the dismissal took place in April or May, after the beginning of spring, when Livy places the levy, but then the tribune's motive becomes very difficult to fathom.

109. Livy 40.41.9–11.
110. Livy 43.9.1–3.
111. Livy 43.9.3.
112. Livy 45.12.10–12.
113. Livy 40.17.7–18.1, 19.8. The Kalends of March 182 will have fallen on a seasonal date in mid-November 183 or later: above, n. 107.
114. Livy 41.5.12, 10.1, 11.2. On the events generally: Livy 41.1.1–6.3, 10.1–11.1, and compare 7.4–10.
115. Livy 42.9.2.
116. Dismissals explicitly mentioned: Livy 31.8.5, 32.1.3–6, 32.8.3, 32.9.5, 40.36.6–7, 41.11.2. Livy frequently notes the senate's instructions to enroll new legions without recording orders at the same time to dismiss old ones. In view of the cases just noted, however, one is justified in presuming that such orders were in fact given. See Afzelius, *Römische Kriegsmacht*, 34–46, for discussion.
117. These dates derive from the two explicit chronological reference points we have for this period, the notices of eclipses in Livy 34.7.7 and 44.37.8. They do not, however, imply, as Derow, "The Roman Calendar, 190–168 B.C.," 346–50, suggests, that intercalation occurred in the intervening years at regular intervals, allowing us to calculate the seasonal date of any Ides of March between these two dates: see Warrior, "Intercalation," 137–43.
118. Polyb. 6.26.1–3, and compare 21.6. See also, for example, Livy 31.11.1, and compare 8.5; 36.3.13; 37.4.1; 41.10.11; 41.17.8, and compare 14.10; 42.48.5, and compare 31.2. Nicolet, *World of the Citizen*, 102.
119. For example, Livy 34.8.5, and compare 33.43.3–5; 34.56.3–4; 37.4.1; 41.5.6–7; 45.12.10–12.

120. Polyb. 3.61.10, 68.10, with Walbank, *Commentary on Polybius*, 1:396; Proctor, *Hannibal's March*, 67–70.
121. Livy 34.56.3–9.
122. Livy 45.12.10–12. Afzelius, *Römische Kriegsmacht*, 46, assumes that Licinius's legions were those that his predecessor in Gaul had commanded, while command of the urban legions of 169 went to the praetor Anicius. However, this disposition is very difficult to reconcile with the passage of Livy cited here. If Licinius's legions were those that had served in Gaul in the previous year, one would have expected to find them either in that province or, if they had been dismissed to their homes for the winter, scattered across the peninsula. Licinius remained in Rome until after the victory at Pydna in early June, when the news reached the city (Livy 45.1.11), and it was there that he made his flawed proclamation of a day to assemble, as reference of the question to the augurs shows. There is no reason why legions in Gaul should have marched there while he awaited the outcome of events in Greece. Better sense can be made on the assumption that Anicius received the legions that had been in Gaul in 169. Their journey from there to the Roman camp on the Genusus River could have been accomplished as quickly via the Po and Adriatic as from Rome, if not more so. The urban legions of 169 remained at Rome along with those Licinius was levying for 168. After Pydna, the senate ordered the consul to dismiss most of the forces that had been held in reserve up till then (Livy 45.2.1–2, 2.9–11), and these probably included one or the other of these two armies, the other of which Licinius intended to take with him to Gaul. Noteworthy, too, is the consul Sempronius's order for his soldiers to assemble in 215 at Sinuessa. He commanded the legions of *volones*, slave volunteers enrolled the previous summer amid the crisis ushered in by Cannae, and if there ever were soldiers who did not need to go home to tend their farms during the winter of 216/15, these were the ones: Livy 23.35.5.
123. Farms made up of a number of small, scattered parcels were certainly the norm in Greece: Foxhall, "Farming and Fighting in Ancient Greece," 136–37, with references to earlier discussions, although Halstead, "Traditional and Ancient Rural Economy," 82, properly cautions against retrojecting land use and landholding patterns from later periods uncritically back to antiquity. That a similar pattern was typical among smallholders in Italy is, admittedly, conjectural, but it is difficult to see how farms could have been kept compact in the face of the practices that would have worked to disperse fields. Note, for the imperial era, the fragmented holdings of the landowners recorded in the inscriptions from Ligures Baebiani and Velia (*CIL* 9.1455 and 11.1147), on which see Pachtère, *La Table hypothécaire de Veleia*, and De Neeve, *Colonus*,

224–30. These individuals, however, were wealthy, and the relevance of their preferences in landownership to small farmers in the middle republic is uncertain.
124. Spurr, *Arable Cultivation*, 20, but compare 57–58 and Cato *Agr.* 155 on the importance of keeping drainage ditches clear during the autumn and winter rains—more work for the farmer in the fall.
125. Spurr, *Arable Cultivation*, 21.
126. Pliny *NH* 18.204; Spurr, *Arable Cultivation*, 43.
127. Garnsey, *Famine and Food Supply*, 49–53; Gallant, *Risk and Survival*, 36–38. The practice was so common as to be axiomatic even among the owners of large estates: Pliny *Ep.* 1.20.16.
128. Pliny *NH* 18.71, 202, 204, with Spurr, *Arable Cultivation*, 43. Although emmer wheat was the mainstay of the Roman diet down to the end of the republic, there is no reason to suppose that Italian farmers grew this to the exclusion of all other types of grain. Archaeological evidence attests the presence in Latium of einkorn wheat as well as barley; other data suggest that both hard, durum wheat and soft bread-type wheats were grown in Italy in the Roman period. For archaeological data: Ampolo, "Le condizioni materiali," 15–16; barley and wheat varieties: Spurr, *Arable Cultivation*, 10–17.
129. Note, too, that some Italian farmers may have prepared fields for planting with mattocks: Pliny *NH* 18.178; Horace *Odes* 3.6.38–39, which of course would have increased the time required dramatically. How widespread the practice was is unclear, however; for opinions, see Chapter 3, n. 52.
130. For the legitimate pretexts a soldier could use to excuse failure to report for duty, see Cincius in Gellius *NA* 16.4.3–4; compare Polyb. 6.26.4.
131. Livy 43.14.7, and compare 11.9–10, 14.5–8.
132. On Hellenistic armies preventing the enemy from sowing its crops, see Ober, *Fortress Attica*, 40; Hanson, *Warfare and Agriculture*, 46 and n. 33.
133. The infamous *legio Campana* garrisoning Rhegium during the Pyrrhic War is the only exception to my knowledge, but there may have been others. Compare the irregulars at Rhegium in 210–209: Livy 26.40.16–18, 27.12.4–6; and the 8,000 volunteers who accompanied M. Centenius Paenula on his way to disaster in 212: Livy 25.19.8–17.
134. See above, n. 16.
135. In general, see Erdkamp, *Hunger and the Sword*, 123–30, on foraging and pillaging, which represented in most cases two sides of the same coin. As he notes, the light troops and cavalry were used to *defend* foragers because of their speed in meeting enemy threats. Erdkamp is, however, too pessimistic

on the effectiveness of Roman and allied cavalry in such combats; see now McCall, *Cavalry of the Roman Republic*, 13–99.

136. Polyb. 6.37.10–12, 1.17.9–11, and compare 10.16.2–17.1, although on the latter passage, see Ziolkowski, "*Urbs direpta*," for important cautions; note, too, the oath at 6.33.1, which will have had particular relevance to a force covering pillagers as well as soldiers fighting in line of battle, and compare Spence, "Perikles and the Defense of Attika," 100 n. 60: "[P]resumably one of the main problems involved in detailing troops to protect foragers or ravagers was that they were probably unwilling to stand by and watch others have the pick of the available food and/or booty."

137. Livy 25.13.1–2, although some grain remained in the fields even then for the Romans to destroy: Livy 25.25.28. Note also Fabius's ravaging of the Capuans' territory in 215: Livy 23.46.9–11; Cornell, "Hannibal's Legacy," 106–7.

138. Certainly, Fabius in 215 withdrew from Campania to allow the Capuans to do their sowing, but this indicates precisely that preventing this was an important strategic option that Roman commanders could elect to pursue or not as circumstances dictated. In this case, Fabius seems to have intended to use the young wheat as fodder for his animals: Livy 23.48.1–2. Compare Agesilaus's decision *not* to interfere with the Akarnanians' planting in 389: Xen. *Hell.* 4.6.13–14 and the modern discussions cited above, n. 132.

139. Note, for example, the capture of Tarentum in the winter of 213/12: Polyb. 8.34.13; Livy 25.11.20. On the date, see De Sanctis, *Storia*², 3.2:322–24.

140. If grain is left standing until it reaches the dead-ripe stage, the heads shatter easily when the stalks are harvested, which causes much of the seed to fall to the ground and so greatly reduces the yield: Percival, *The Wheat Plant*, 141. Further, summer storms as well as other agents can cause heads to shatter even at the ripe stage. And once ripe, wet grain is susceptible to sprouting in the ear and to attack by fungi: Kent, *Technology of Cereals*, 11.

141. Compare Chayanov, *The Theory of Peasant Economy*, 74–76 and figs. 2-1 and 2-2.

142. See above, n. 4.

143. Polyb. 1.17.9: see above at nn. 33 and 44 for discussions of the calendar date.

144. See Eckstein, "Two Notes"; Proctor, *Hannibal's March*, 52–53. Note that another legion was already in Gaul at the time of the revolt and this legion may also have been drawn from the army Scipio levied in 218; see Eckstein, "Two Notes," 270 n. 41, for discussion.

145. Ovid *Fasti* 6.767–68. The calendar at this point was not far out of accord with the seasons: De Sanctis, *Storia*², 3.2:115–17; Walbank, *Commentary on Polybius*,

1:412–13, for discussion. Desy, "Il grano," dates the battle to mid-April, but his argument is not persuasive: see Appendix 2.

146. Polyb. 3.107.1–7; De Sanctis, *Storia*², 3.2:131–32; Walbank, *Commentary on Polybius*, 1:438–39, for discussion of the date.

147. Livy 31.2.7–8; although some of these soldiers were part of a hastily levied force of allies, four cohorts came from the consul's army: Livy 31.2.5–6.

148. See above at n. 4.

149. Sall. *BJ* 41.7: *populus militia atque inopia urgebatur*; Plut. *TG* 8.3: ἐξωσθέντες οἱ πένητες οὔτε ταῖς στρατείαις ἔτι προθύμους παρεῖχον ἑαυτούς; App. *BC* 1.7: τοὺς δ' Ἰταλιώτας... τρυχομένους πενίᾳ τε καὶ ἐσφοραῖς καὶ στρατείαις. The passage from Appian is highly problematic in its reference to Italians, implying as it does that only *socii* rather than Roman citizens were affected in this way and formed the object of Gracchus's reforms. See now Mouritsen, *Italian Unification*, 15–22, for analysis and references to earlier discussions. Note also the second-century historian Cassius Hemina, frag. 17 (Peter, *HRR* 1:103): *Quicumque propter plebitatem agro publico eiecti sunt*. However, there is no indication of the passage's context; Peter (ibid.) argues that it formed part of Cassius's account of the Struggle of the Orders, when conflict over the *ager publicus* undoubtedly took place, and this may well be where the passage belongs. If so, then it tells us nothing about conditions in the second century.

150. Events of the Spanish wars in this period: Simon, *Roms Kriege*; Astin, *Scipio Aemilianus*, 35–47, 137–55. On the unwillingness of recruits to serve, see, for example, Polyb. 35.4.1–7; App. *Iber.* 49; Livy *Per.* 45; further sources and discussion in Astin, *Scipio Aemilianus*, 167–70; Shochat, *Recruitment and the Programme of Tiberius Gracchus*, 58–60; Rich, "The Supposed Roman Manpower Shortage," 317–18. Lack of booty: Pliny *NH* 33.141; App. *Iber.* 54, although compare Brunt, *Italian Manpower*, 396 n. 2. On the loss of life, see below, Chapter 4. On army morale, note Scipio's efforts to restore discipline at Numantia: sources in Astin, *Scipio Aemilianus*, 136.

151. Brunt, *Italian Manpower*, 639–44; Cornell, "Hannibal's Legacy," 111. Somewhat different is Gabba's claim, "Motivazioni economiche," 136–38, that Dion. Hal. 8.73.5 reflects a recognition within optimate circles around the time of Tiberius Gracchus that small farming and war were incompatible. But Gabba's interpretation goes beyond the evidence. The passage presents Ap. Claudius as foreseeing that recipients of small parcels would be unable to cultivate them because of their poverty and unable to lease them to anyone except a powerful neighbor. However, Appius does not connect any inability to cultivate a small parcel of land with the obligation to perform military service, as Gabba believes. Military service is mentioned only in connection with

Notes to Pages 53–54

the provision of pay and equipment by the public treasury, rather than at the private expense of the soldiers, out of the revenues to be derived if the land in question is rented out in large units rather than distributed in small parcels.

152. Sulla's veterans might seem an obvious counterexample, but of course their indebtedness did not result from their military service. They fell into debt only after their discharge and settlement by the dictator on farms: Cic. *Cat.* 2.20; Sall. *Cat.* 16.4, 28.4; App. *BC* 2.2; discussion in Brunt, *Italian Manpower*, 310–11, who assumes many were again called up for military service, although there is no mention of this in the sources, and Gruen, *Last Generation*, 424. Of course, they never paid *tributum*.

153. Livy 2.23.5–6, and compare 23.1, Dion. Hal. 6.26.1–3; 24.6–8, and compare Dion. Hal. 5.10.1–10; 6.14.3; 6.29.1, although Oakley, *Commentary on Livy*, 1:518, regards this story as a doublet of that at 2.23.1–8; 6.31.4; 7.27.4, 38.7; compare Dion. Hal. 15.3.6. References to *tributum* are, of course, anachronistic before the late fifth century at the earliest, when the introduction of payment for military service brought about the creation of the tax that funded it; see above, n. 16.

154. Suspension: Cic. *Off.* 2.76; Val. Max. 4.3.8; Pliny *NH* 33.56; Plut. *Aem.* 38.1; reimposition: Cic. *Fam.* 417 (12.30).4; *ad Brut.* 24 (26).5; Dio 46.31.3–4. This makes it difficult to credit Appian's reference to the ἐσφοραῖς at *BC* 1.7 as among the burdens of the poor unless he is in fact referring only to the taxes paid by Rome's Italian allies to fund the pay for their contingents—so, for example, Gabba, *Appiani, Bellorum Civilum Liber Primus*, 18.

155. See in general Nicolet, *World of the Citizen*, 153–60; Nicolet, *Tributum*, 35–37, who, however, doubts that the normal tax rate was as high as one per thousand. Legionary pay is generally reckoned at three *asses* per day on the basis of Polyb. 6.39.12; see Crawford, *Coinage and Money*, 146–47, for a recent discussion of this interpretation of the passage and other hypotheses. Since *tributum* was levied on all *assidui*, each adult male citizen in a family was presumably liable for its payment, but this fact does not obviate the point that the amounts in question would not have been particularly onerous for families in the lowest census classes.

156. Livy 5.10.5, with Nicolet, *Tributum*, 29, 33. Brunt's statement, "the annalists suggest that the burden of conscription was augmented by the simultaneous levy of taxes," is misleading, therefore; see *Italian Manpower*, 641.

157. Sources for 215: Livy 23.31.1, and compare 38.12–13, 48.4; for 214: 24.11.7–9, and compare 24.18.10–15; 210: 26.35.1–36.12. It is noteworthy that in these accounts, the financial burden is placed principally on the higher cen-

sus classes. In the special levy for 210, for example, citizens in the fourth and fifth classes paid nothing. See also above, n. 57.

158. It is also possible that exactions in era of the civil wars did provide suggestive analogies for historians writing in the first century; compare Brunt, *Italian Manpower*, 641.

159. Livy 5.10.9, especially *incultaque omnia*.

160. Discussion recently in Cornell, *Beginnings of Rome*, 311–12.

161. Above, n. 16, although there is less consensus on whether payment for military service was regularly and continuously offered after this date.

162. So, for example, Brunt, *Italian Manpower*, 108, quoted above in Chapter 1, n. 77.

163. See Thuc. 6.43.1 on the number of soldiers; Thuc. 6.9.1–14.1, 20.1–23.4, for the debate. Possibly slaves would take on the principal burden of work on the farms while their owners were away on military service, as Jameson, "Agricultural Slavery," 140, suggests. However, the view that slaves were common on the farms of ordinary Athenians has been strongly challenged by Wood, *Peasant-Citizen and Slave*, 42–80. For a recent survey of the problem, see Osborne, "Economics and Politics of Slavery," 32–33, in reference to slavery on small farms.

164. Errington, *A History of Macedon*, 239–41. Although Alexander did send recently married men back home during his first winter in Asia Minor, this was to visit their wives, not see to their farms: Arrian 1.24.1. And the practice seems to have ceased thereafter.

165. Although Brunt would not endorse such claims: "Those [soldiers] who had previously been small farmers would often have nothing to return to; even two or three years in the legions might have been enough to ruin them"; see "The Army and the Land," 75 (= *Fall of the Roman Republic*, 256). Brunt here is speaking of conscripts in the second and first centuries, but in terms of the impact their absence would have had on the operation of their farms, there is no reason to suppose that matters would have been any different for conscripts before the Hannibalic War. Compare Cornell, "Hannibal's Legacy," 109.

166. On second-century practice, see Afzelius, *Römische Kriegsmacht*, 34–46, and especially the table at 47, reproduced in Toynbee, *Hannibal's Legacy*, 2:652.

167. Livy 39.38.4–12.

168. Note especially the veterans of Scipio's African campaigns, who were forced into the army being assembled against Macedon in 201 despite explicit assurances to the contrary: Livy 32.3.4–7, and compare 31.8.6; other examples: 31.8.8; 32.9.1; 34.56.9.

169. Brunt, *Italian Manpower*, 64–66, 417–20. See also Appendix 5.
170. See above, n. 77.
171. Livy 24.18.7–8. On what follows, see Rosenstein, "Marriage and Manpower," and Appendix 5.
172. Livy 24.11.5–9, 26.35.1–36.12; compare Libourel, "Galley Slaves." Shochat, *Recruitment and the Programme of Tiberius Gracchus*, 66, similarly argues that the senate's decision to compel citizens resident in maritime colonies to serve as rowers in 190 indicates a shortage of *proletarii* at that date: Livy 36.3.4–6. Estimates of the number of rowers serving: De Sanctis, *Storia*², 3.2:313; Brunt, *Italian Manpower*, 65.
173. On the reasons for these *vacationes*, see Rosenstein, "Marriage and Manpower," 179–86.
174. On what follows, see Chapter 1 at n. 13 and the works cited there; 1,100 libral *asses* would weigh 11,000 ounces; 4,000 sextantal *asses* would weigh 8,000 ounces.
175. On the notional value of an individual's property, see Rathbone, "The Census Qualifications," 132.
176. See Brunt, *Italian Manpower*, 15–16; Nicolet, *World of the Citizen*, 67–73, for the procedures.
177. Rathbone, "The Census Qualifications," 144–45; see also Crawford, "The Early Roman Economy," 206.
178. See Chapter 3, n. 68, for the sizes of third-century land distributions, which certainly were intended to place those who received them in the ranks of the *assidui*, and compare Rathbone, "The Census Qualifications," 145–46.
179. The exemption (*vacatio*) granted to participants in maritime colonies is known from the senate's decision in 207 B.C. to revoke this privilege: Livy 27.38.2–3. On the size of colonists' plots: Livy 8.21.11; Salmon, *Roman Colonization*, 72, and compare 98 and see further Rosenstein, "Marriage and Manpower," 190–91.
180. Livy 24.11.7, and see further Chapter 5 at n. 99.
181. Compare the models of four- and six-hectare subsistence farms in ancient Greece developed by Gallant in *Risk and Survival*, 78–110, especially 89; the labor shortages that develop are much greater for the larger farm than the smaller.
182. See Chapter 3 at n. 194.
183. Moore, *Conscription and Conflict*, 68, on the exemption for owners of twenty or more slaves, and 52–108, on exemptions generally.
184. Polyb. 2.21.7–8; other sources in T. Broughton, *MRR*, 1:225.
185. Compare Toynbee, *Hannibal's Legacy*, 2:178–210, especially 209–10 on the

Notes to Pages 59–60

situation following the end of the Hannibalic War. On the cessation of colonization after 181, see Salmon, *Roman Colonization*, 108–13; see also Salmon, "The *Coloniae Maritimae*," 10–13, who argues that Vell. 1.15.3 wrongly dates the colony at Auximum to 157; the correct date for its foundation is in his view 128. However, see now North, "Deconstructing Stone Theaters," for an important challenge to Salmon's rejection of Velleius's notice, in this same passage, of the demolition of the stone theater constructed by the censor Cassius, on which much of Salmon's argument for the dating of Auximum rests.

186. Harris, *War and Imperialism*, 41–53, especially 46–47, is fundamental on the citizens' willingness to participate in Rome's wars, countering earlier authors, for example, Toynbee, *Hannibal's Legacy*, 2:76; and Brunt, *Italian Manpower*, 391–415, who had maintained that exactly the reverse was true in the first half of the second century. See also Evans, "Resistance at Home."

187. On preindustrial policing, see in general Crone, *Pre-industrial Societies*, 35–57, especially 52–54. Brunt, *Italian Manpower*, 391–92, is too optimistic on the enforceability of legal penalties for evading the draft in the face of widespread failure to comply and the absence of an extensive police force. On policing at Rome during the middle republic, see Nippel, "Policing Rome"; Nippel, *Public Order in Ancient Rome*, 1–84.

188. On the military considerations behind the foundation of colonies in this period, see Salmon, *Roman Colonization*, 55–81.

189. Colonies: Carseoli (298), Venusia (291), Hadria (289), Paestum and Cosa (273), Ariminum and Beneventum (268), Firmum (264) and Aesernia (263), Brundisium (244), Spoletium (241), and Placentia and Cremona (218); viritane distributions: 299, circa 290, 241, and 232, and see below, n. 196.

190. Compare Brunt's descriptions of the consequences of a farmer's absence for his family, cited above, Chapter 1 at nn. 6 and 165.

191. Poly 1.11.2, Paton's translation: τετρυμένοι μὲν ὑπὸ τῶν προγεγονότων πολέμων καὶ προσδεόμενοι παντοδαπῆς ἐπανορθώσεως. The interpretation of this passage is controversial. Calderone, "Di un antico problema"; Calderone et al., "Polibio 1,11,1 sq."; and Eckstein, "Polybius on the Rôle of the Senate," have revived the view that Polybius is describing here the passing of a *senatus consultum* by the Roman senate to aid the Mamertines and makes no reference to an assembly of the people until 1.11.3. Hoyos, "Polybius' Roman οἱ πολλοί," however, convincingly defends the more usual view, that the passage in question describes the senate's inability to reach a decision and the subsequent passage of a bill to send aid by a *comitia* of the citizens. See also Scullard, *CAH*², 7.2:542–43, whose brief discussion also upholds the latter reading of the passage and provides references to earlier bibliography.

192. Livy 31.6.3–8.1. Harris suggests that approval of the motion may have come only after a promise to the voters that none of Scipio's African veterans would be required to serve in the new war against his will: Livy 31.8.6, and compare 14.2; and *War and Imperialism*, 218, although the promise was apparently not kept: Livy 32.3.3–7.

193. For example, in 169, when the Third Macedonian War was going badly (Livy 43.14.2–6), or in 151 when reports about the hard conditions of warfare in Spain kept recruits from coming forward (Polyb. 34.4.1–14). See further Chapter 1, n. 59.

194. This is based on an average of 4,500 men to a legion, four legions levied every year, for 80 years = 1,440,000 man-years. If each man drafted served a total of 6 years (1,440,000/6), this total will have required 240,000 men. See Chapter 3 at n. 145 for further discussion. Of course, if soldiers served a shorter term of service in this era, then more men will have served.

195. For Latin colonies 298–263, see conveniently Cornell, *Beginnings of Rome*, 381, table 9, who estimates their total at 40,500 adult male colonists, although Hopkins, *Conquerors and Slaves*, 21, is cautious about accepting the figures our sources preserve for such settlements at face value. Four Latin colonies were founded subsequently: Brundisium, Spoletium, Placentia, and Cremona. Afzelius, *Die römische Eroberung*, 133, estimates the first two at 4,000 men each; Polybius (3.40.4) gives 6,000 each for the latter pair. In addition, the seven citizen colonies Rome founded between 298 and 218 can be reckoned at 300 men each: Salmon, *Roman Colonization*, 67–80.

196. Cornell, *Beginnings of Rome*, 380, estimates the recipients of viritane distributions connected with the creation of six new rural tribes in 332, 318, and 299, as well as those carried out circa 290 following the conquests of M'. Curius Dentatus at 20,000–30,000. Strictly speaking, only this last falls within the period 298–218, but opportunities for allotments afforded by the creation of the tribes Aniensis and Terentia in 299 may have been available for a few years thereafter. Assuming therefore that each of the seven distributions involved approximately equal numbers, the total number of new settlers accommodated under the last three will have been between 9,000 and 13,500. Thereafter, similar resettlement schemes were carried out in 241 with the creation of the final two rural tribes and subsequently in 232 as a consequence of Flaminius's legislation distributing land in the *ager Gallicus*. Again, the numbers involved are largely a matter of guesswork, but 9,000–13,500 in all would seem about the right order of magnitude for a total of 18,000–27,000. See also Taylor, *Voting Districts*, 53–68.

197. One might be tempted to argue that, because the formation of large slave-

run agricultural estates only began to appear after the Hannibalic War, smallholders would be unlikely to lose their land since there would be few wealthy neighbors to buy or seize their land while they were away at war. However, it is clear that a slave-based agricultural economy was developing as early as the late fourth century, when chattel slavery began to replace debt bondage as the principal source of dependent labor on the estates of the rich: Cornell, *Beginnings of Rome*, 332–33, 367, and see further Chapter 3 below. And there can be no doubt that Rome's conquests in the late fourth and third centuries enriched its ruling class: Harris, *War and Imperialism*, 58–60, 65–67. Hence there is no warrant to assume that the economic forces that, in the consensus view, were helping to remove ruined farmers from their land in the second and first centuries would not have operated in the third as well.

Chapter 3

1. Rosivach, "Manning the Athenian Fleet," however, disputes the usual view that the majority of the *thētes* who rowed Athens's warships did not work on the land.
2. On mercenary oarsmen in Athens, note especially Thuc. 1.121.3; on mercenaries in general, see Parke, *Greek Mercenary Soldiers*; G. T. Griffith, *Mercenaries of the Hellenistic World*; Millar, "The Practical and Economic Background."
3. Note that they received not only pay but a food allowance: Pritchett, *The Greek State at War*, 1:3–4, with further references.
4. Note especially for this period Cato *Agr.*, praef. 1–4; Pliny *NH* 7.139–40, especially: *pecuniam magnam bono modo invenire*. This is not to say that aristocrats could not be tempted by the profits to be made from commerce, for otherwise there would have been no need for the *Lex Claudia* of circa 218: Livy 21.68.3–4, and compare Plaut. *Merc.* 78; Cic. *Verr.* 2.5.45; even the upright Cato himself was not immune: Plut. *Cato Mai.* 21.5–8. Attitudes were complex and not without contradiction, as D'Arms, *Commerce and Social Standing*, demonstrates for senators of the late republic and empire.
5. On the aristocracy's military ethos, Harris, *War and Imperialism*, 10–41, is fundamental; on its origins: Hölkeskamp, "Conquest, Competition and Consensus," 20–31; Raaflaub, "Born to Be Wolves?," 290–92.
6. The date at which the republic began to provide food to allied troops campaigning with the legions is not known.
7. See Chapter 2 at nn. 4 and 123 and, in general, Spurr, *Arable Cultivation*. Here

Notes to Pages 66–67

and throughout, Italy and Italian agriculture include the farms of Rome's citizens as well as those of the republic's *socii*.

8. *Energy and Protein Requirements: Report of a Joint FAO/WHO/UNU Expert Consultation*.

9. These and the following recommendations come from table 15. To calculate the energy required to sustain various activities, the authors of the report multiply the basal metabolic rate (a measure of the energy required per unit of body weight to sustain life at complete rest), or BMR, by a factor representing the amount of energy expended over and above the BMR itself. Since a person's activities throughout the day may range from sleep, rated as 1.0 × BMR, to heavy exertion, the factor by which the BMR is multiplied must be the average of the composite. Table 15 estimates that average as 1.78 for "moderate work" and 2.1 for "heavy work." The BRM of a 20-year-old weighing 65 kilograms is 70 (see table 5), thus 70 × 1.78 × 24 hours = 2,990 calories per day; 70 × 2.1 × 24 = 3,528. Compare especially table 10, "energy requirement of a subsistence farmer."

10. BRM = 68.

11. BMR = 55. Compare table 14, "energy requirement of a rural woman in a developing country."

12. The minima are from table 48, assuming a weight of 55.5 kilograms and a BMR factor of 1.64 for the boy and a weight of 36 kilograms and the same BMR factor for the girl. These figures are, however, probably somewhat low: compare table 26, "example of the calculation of the daily energy expenditure of a ten and a half year old boy in a developing country," where the BMR factor figured is 1.71, reflecting three hours of "moderate" and one hour of "heavy" activity a day. The higher figures are composites drawn from older estimates of the energy requirements of children and adolescents given in Gallant, *Risk and Survival*, 62–63 and 73. Compare also the remarks of Foxhall and Forbes, "Σιτομετρία," 49 n. 26.

13. First son: (2,990–3,530 calories/day × 365 days = 1,091,350–1,288,450 calories/year) + father: (2,900–3,380 × 365 = 1,058,500–1,233,700) + mother: (2,170–2,400 × 365 = 792,050–876,000) + second son: (2,650–2,900 × 365 = 967,250–1,058,500) + daughter: (1,950–2,350 × 365 = 711,750–857,750) = 4,620,900–5,314,400 calories/year.

14. Note also Clark and Haswell, *The Economics of Subsistence Agriculture*, table 5, offering much lower caloric requirements per day.

15. Figure taken from Foxhall and Forbes, "Σιτομετρία," 45–46, who cite the FAO, *Food Composition Tables for the Near East*, table 2, item 1. Clark and Has-

well, *The Economics of Subsistence Agriculture*, 57–58, assume that even subsistence cultivators would have milled off the bran from the wheat, resulting in a 10 percent reduction in caloric content. The assumption seems unjustifiable (it was obviously written before the American health craze for eating bran): compare Watt and Merrill, *Composition of Foods*, table 1, items 2,430–2,434; FAO, *Food Composition Tables for the Near East*, table 1, item 115. See also White, "Food Requirements and Food Supplies," on the nutritive value of foods prepared from grains under conditions prevalent in antiquity.

16. This total is somewhat high compared with the estimates of other scholars: compare Foxhall and Forbes, "Σιτομετρία," 51, 55–56, 62–63, and table 1; Evans, "Wheat Production," 432–33; this may be a function of assuming body weights for the hypothetical family members that are too high for antiquity. As noted, however, such a worst-case scenario should not affect the validity of the model for the purposes of this exercise.

17. Ratios: Varro 1.44.1; Col. 3.3.4, and compare Cic. *Verr.* 3.112. Problems: see below, n. 20.

18. Col. 3.3.4, Spurr's translation. This is a highly problematic passage; see, recently, De Martino, "Produzione di cereali," 245–46; Ampolo, "Le condizioni materiale," 20–24; Evans, "Wheat Production," 429–31; De Martino, "Ancora sulla produzione," 241–62; Garnsey and Saller, *The Roman Empire*, 79–82; Sallares, *The Ecology of the Ancient Greek World*, 374, 497 n. 243.

19. That is, one liter of wheat weighs about .772 kilogram on average: Foxhall and Forbes, "Σιτομετρία," 43–44; 8.62 liters equal one *modus*: Duncan-Jones, "The Choenix," 51–52; thus .772 × 8.62 = 6.654 kilograms.

20. Sowing rates: Varro 1.44.1; Col. 2.9.1; Pliny *NH* 18.198. The first two authors, however, note that the rate of sowing may in fact vary according to the local conditions, an observation confirmed in modern literature; see Peterson, *Wheat*, 238–39, for factors that would affect the rate of sowing. Compare also Halstead, "Traditional and Ancient Rural Economy," 85; Gallant, *Risk and Survival*, 46–49, for sensible cautions about assuming constant sowing rates. However, the sowing rate used here, about 133 kilograms per hectare, is a fairly high one for a dry climate: see Peterson, *Wheat*, 238–39, who gives sowing rates for arid and semiarid regions of between one-quarter bushel (15 pounds) to two bushels of seed and notes that one bushel (60 pounds = 27 kilograms) per acre (= .4 hectare) is most common. That rate works out to 67.5 kilograms per hectare or about 17 kilograms per *iugerum*, slightly more than half the rate given in the agricultural writers. Compare the highest rates in the examples cited by Gallant, *Risk and Survival*, 48 and table 3.1. Thus the model proposed here is not likely to be underestimating the sowing rate

for wheat; if farmers reduced it, they might have obtained greater seed-to-crop ratios in their harvests, but overall they would have produced a smaller amount of grain.
21. This figure refers to "naked" wheat, which was measured after threshing had removed its hulls. In Latium and elsewhere, *far* or emmer (triticum dicoccum), a hulled wheat (that is, one for which threshing would not remove the hulls from the kernels, requiring that it be stored in its hulls and then parched to remove them just before cooking) served as the principal crop of small farmers: Jasny, *The Wheats of Classical Antiquity*, 117, 120. The hulls made emmer bulkier than naked wheats and, because emmer was sown in its hulls, sowing ratios for it were approximately double those of naked wheats: Varro 1.44.1; Col. 2.9.1; Pliny 18.61, 198. Hence one would expect the volume of grain produced to have been double that for naked wheat as well. Unfortunately, the weight of a modus of emmer once freed of its hulls, and therefore its caloric content, is quite difficult to determine: Jasny, *The Wheats of Classical Antiquity*, 155–59; De Martino, "Ancora sulla produzione," 256–60. For the purposes of this exercise, however, what matters is whether the weight of emmer once the kernels were removed from their hulls would fall below the 66.5 kilograms per *iugerum* estimate used here. Ampolo, "Le condizioni materiale," 23–24, offers the most pessimistic analysis of the yields of emmer. Using 4:1 and 3:1 crop-to-seed ratios, he estimates that emmer would yield 525 kilograms per hectare or 131 kilograms per *iugerum* after seed for the following year's crop had been set aside. He then assumes that 70 percent of that weight would remain after milling, leaving a usable net of 367 kilograms per hectare or 92 kilograms per *iugerum*. De Martino, "Ancora sulla produzione," 256–62, arrives at a minimum figure of 125 to 145 kilograms per *iugerum*, but it is unclear whether he excludes future seed from this total or what he assumes the extraction rate after milling to be. Jasny, *The Wheats of Classical Antiquity*, 144, and compare 155–57, argues that for a given volume of emmer in the hull, the kernels will represent 75 percent of its weight. There is no reason to believe, therefore, that recasting the food component of the model in terms of emmer would require a greater amount of food to be produced by the family and so necessitate increasing the amounts of land and labor calculated here.
22. Col. 2.12.1; White, "The Productivity of Labour," and compare Evans, "*Plebs Rustica*," 137. Note that nine or ten *modi* of emmer require, according to Col. 2.12.1, the same number of days as five *modi* of ordinary wheat; compare Col. 2.6.1 for *semen* [sic. *adoreum*].
23. Col. 11.2.46; other writers indicate that some areas could require as many as

Notes to Page 68

nine plowings: Pliny *NH* 18.181; Pliny *Ep.* 5.6.10. See also Duncan-Jones, *The Economy of the Roman Empire*, 327–35, for analysis of Columella's figures.

24. Columella allows one and a half days for reaping; Varro, one: 1.50.3. Spurr, *Arable Cultivation*, 138 n. 19, doubts that such estimates were realistic and proposes two days per *iugerum*, which is followed here. Compare, too, Peterson, *Wheat*, table XXI, who gives forty-six man-hours as the time required to harvest an acre of wheat by hand in 1829 (about .62 acres = one *iugerum*), which works out to about 28.5 man-hours per *iugerum*, or about three man-days if nine man-hours = one man-day.

25. Spurr, *Arable Cultivation*, 138, but compare Duncan-Jones's estimate: *The Economy of the Roman Empire*, 329. However, Duncan-Jones based his estimate on the differences in the proportions of a wheat crop awarded to a tenant according to whether he helped with the threshing or not in Cato *Agr.* 136, and since it is by no means certain that Cato is referring to tenancy in this passage, the estimate is without foundation: see De Neeve, *Colonus*, 202 n. 5, for a survey of opinion on the meaning of *politio* in the passage.

26. These include gleaning, one-half day; transport of the grain to the threshing floor, one-half day; collecting straw from the field, one day; storage of the grain and seed collection, one day; ditch digging and maintenance, one day: see Spurr, *Arable Cultivation*, 138. I omit Spurr's day for manuring, since there is considerable doubt whether subsistence cultivators had access to adequate supplies of dung: White, *Roman Farming*, 125–44, especially 144; Brunt, Review of White, 156–58. Compare the 63.8 man-days per hectare of wheat among modern peasants in the Peruvian Andes: Brush, "The Myth of the Idle Peasant," 71, table 4-2.

27. Gallant, *Risk and Survival*, 76, drawing on a number of earlier studies to arrive at a composite figure. Compare also the figures given in Shanin, *The Awkward Class*, 81, table 5.III note a; 104, table 6.IV note a. In Roman colonial charters the labor of boys fifteen and older is treated as equivalent to that of an adult male for purposes of the corvée: see the *lex Coloniae Genetivae* (Bruns, *FIRA* no. 21, chap. 98) and the *lex Irnitana* (González, "The Lex Irnitana," chap. 83). The lower labor-potential factor assigned to adult and adolescent women derives from the assumption that women will also be required to do housework, prepare food, offer child care, and perform other tasks that will come at the expense of the amount of time they can work in the fields. In few if any cases do differences in strength between men and women affect their ability to perform agricultural work. Compare Gray, *A History of Agriculture*, 548–49, who notes testimony from the antebellum South that under the gang system female slaves might be rated at three-quarters hands, but were

sometimes "expected to bear the same part in the labors of the field as the men." See also Conrad and Meyer, *The Economics of Slavery*, 62. On women's ability to plow, see below at n. 187.
28. Col. 2.12.8–9.
29. The total potential labor in the family is $1 + 1 + .9 + .7 + .5 = 4.1$.
30. The surplus of labor in this model should not be taken to mean that the "underemployment" it represents was the same thing as idleness. As Brush, "The Myth of the Idle Peasant," 61–64, points out, "underemployment" really reflects a particular definition of "work." Many traditional activities among subsistence cultivators, such as religion or handicrafts, fall outside this definition and yet absorb considerable amounts of the participants' time. Where a shortage of labor for the fields arises, such as through the removal of a family member, these sorts of activities may be curtailed in order to meet the need for a specific amount of agricultural labor for a family to survive.
31. See Chapter 2 at n. 140.
32. Grain could be stored in the ear and then threshed in the winter when the demand for labor had slackened: Col. 2.20.4; Spurr, *Arable Cultivation*, 75–76.
33. See Chapter 2 at n. 127.
34. Beans, in addition to being a rich source of protein, supply lysine, a complementary amino acid for wheat.
35. See Ampolo, "Le condizioni materiale," 30–31 and the table at 16; André, *L'Alimentation*, 35–42; Spurr, *Arable Cultivation*, 103–16; Garnsey, "The Bean: Substance and Symbol." Evans, "*Plebs Rustica*," 147–59, strangely neglects to consider them in his otherwise exhaustive analysis of the nutritional content of an Italian peasant diet.
36. In fact, beans may have required less labor to grow: seven or eight days per *iugerum* according to Col. 2.12.2 (Spurr, *Arable Cultivation*, 107–11), and this figure includes three hoeings which might be omitted: Col. 2.11.6–7.
37. Col. 2.12.1–6. Only a crop of sesame, at fifteen days per *iugerum*, requires substantially more labor than the ten and a half Columella estimates for wheat. On the other hand, note that Columella estimates the labor needed for a *iugerum* of barley at six and a half days, a *iugerum* of chickpeas at six days, and a *iugerum* of lentils at eight.
38. On vines and olives, see Ampolo, "Le condizioni materiale," 31–33. On the vegetable garden, see especially Ps.-Virgil, *Moretum*; Col. 10; White, *Roman Farming*, 246–47. On the diet of smallholders, see generally Evans, "*Plebs Rustica*," 134–73, and Gallant, *Risk and Survival*, 72–79. The latter attempts to model the role each of these would have played in a peasant family's diet and

the labor required to grow the amounts required. Gallant's calculations are vitiated, for the present purposes, by his serious overestimation of the labor required to cultivate legumes.

39. Two *iugera* seems to have been viewed as the traditional size of the Roman *heredium*: Varro *RR* 1.10.2; Pliny *NH* 18.7, and compare Mommsen, *Römische Staatsrecht*, 3.1:22–27; Salmon, *Roman Colonization*, 21; White, *Roman Farming*, 48. Note also Pliny's remark, *NH* 19.50, that in early Rome *heredium* was used in the sense of *hortus*, "garden": so Mommsen, *History of Rome*, 1:239. Note, too, Crawford's suggestion that a farm of two *iugera* was at some point arbitrarily set as the floor for the fifth class: "The Early Roman Economy," 206. It is worth pointing out that this area, about half a hectare or a little less than two acres, seems more than ample: it is about equal to an American football field, including the end zones. Note that Columella estimates the yield of the very worst vineyards as 1 *culleus* or 517 liters per *iugerum* and a more normal minimum crop as three *cullei*: 3.3.10–11.

40. Gallant, *Risk and Survival*, 75–76.

41. Both Frayn, *Subsistence Farming*, 57–72, and Evans, "*Plebs Rustica*," 138–39, stress the importance of wild plants in peasants' diet. Even under the worst-case scenario described here, ample time would exist for this hypothetical family to spend gathering the wild plants and other items it needed as well as the variety of other activities that will have filled the family members' days, on which see Brush, "The Myth of the Idle Peasant."

42. Tibiletti, "Il Possesso dell'*Ager Publicus*," 9–11. Compare White, *Roman Farming*, 273, who assumes a minimum of five *iugera* (that is, about one and a quarter hectares) for a team of oxen.

43. Spurr, *Arable Cultivation*, 138–39. Spurr (n. 16) considers Tibiletti's figures to refer to "the best meadow land," but what Tibiletti seems to be referring to is the amount of labor expended rather than the quality of the land itself: see the preceding note. Hence the assumption used here, that one ox would need four *iugera* of pasture, presumes only that land that could be intensively cultivated was available for this purpose. Where this was not possible, cultivation would be extensive rather than intensive; more land would be required, in other words, to support the animals, but substantially less labor would be expended.

44. Although one might suppose that men would have been necessary for the essential work of plowing, this assumption is unwarranted: see below at n. 187.

45. Compare Roth, *The Logistics of the Roman Army*, 9–10, arguing for an average weight among Roman legionaries in the imperial period of 65.7 kilograms

(145 pounds) and noting evidence suggesting that soldiers at this time were usually taller, and therefore heavier, than civilian men.
46. Spurr, *Arable Cultivation*, 47–48.
47. Scholars have generally assumed that biannual fallowing would have been universally the norm among smallholders in view of the absence of large amounts of fertilizer. The matter is much more complex, however: see Halstead, "Traditional and Ancient Rural Economy," 81–83, for cautions against assuming it as the norm, and Spurr, *Arable Cultivation*, 103–16, on cereal-pulse crop rotation. More to the point, however, land under continuous cropping will never fail to produce a crop simply because of that fact. In the Woburn experiment, yields from plots continuously planted in wheat initially declined over a period of about twenty years and then stabilized: Russell and Voelcker, *Fifty Years of Field Experiments*, 28–29, 236–49; compare the results of the Rothamsted experiment, where an unmanured plot was fallowed every fifth year: Commonwealth Bureau of Soil Science, *Details of the Classical and Long-Term Experiments*, 16, table 2. The real problem with continuous cropping is the gradual infestation of the fields with weeds; the point of fallowing is to eradicate them through plowing when no crop is present: White, *Roman Farming*, 113. On the other hand, a wheat-fallow rotation scheme does not avoid declining yields, and although yields are higher initially, the decline is more precipitous than under continuous wheat; see, for comparison of fallowing with continuous wheat, Clark and Russell, "Crop Sequential Practices," 290–92.
48. Col. 11.2.1–3, 32, 46–47; White, *Roman Farming*, 113, 118; Spurr, *Arable Cultivation*, 25–26.
49. On fallowing as characteristic of large estates rather than small, subsistence farms, see Spurr, "Agriculture and the *Georgics*," 76.
50. Varro 1.20.4, 2.6.5; Pliny *NH* 8.167; Col. 7.1.2; White, *Roman Farming*, 273; Spurr, *Arable Cultivation*, 32.
51. Lirb, "Partners in Agriculture," refuting the claims of Jongman, *Economy and Society of Pompeii*, 83–84, 153, 201, that small farms would not have been able to support oxen; compare White, *Roman Farming*, 345; Gallant, *Risk and Survival*, 51; and Sallares, *The Ecology of the Ancient Greek World*, 312.
52. Pliny 18.178, and compare Horace *Odes* 3.6.38–39. How extensive the practice was is uncertain: Spurr, *Arable Cultivation*, 40, suggests it was rare; compare White, *Roman Farming*, 345; Halstead, "Traditional and Ancient Rural Economy," 84; Gallant, *Risk and Survival*, 51–52; Sallares, *The Ecology of the Ancient Greek World*, 312. Compare also Barker, *A Mediterranean Valley*, 37 and

fig. 18, who notes that farmers in the Biferno valley commonly cultivated their fields by hand as late as the 1970s.

53. Keith Hopkins, in an unpublished paper, argued that seven or eight *iugera* worked by hand would suffice to support a family of 3.25 persons whereas twenty *iugera* would be needed if work animals were kept: cited in White, *Roman Farming*, 336. However, it is unclear how many days of manual cultivation Hopkins assumes would have been required to produce the food the family needed. Gallant, *Risk and Survival*, 51, notes that increases in yields could have offset the greater labor hand cultivation required, based on Lewis, "Plow Culture and Hoe Culture," a study of Mexican farmers. However, although Lewis found much higher yields where hoe cultivation rather than plow cultivation was practiced, the difference, he noted, was "primarily due to greater soil fertility and different type of corn used": "Plow Culture and Hoe Culture," 124.

54. Finley, *The Ancient Economy*, 106: "Modern studies show that the smaller the holding, the greater the number of man-hours expended per acre. What else can a peasant householder do? Since he cannot fire members of his family, ... he must keep them busy somehow; in jargon, his aim is to 'maximize the input of labour.'"

55. Peterson, *Wheat*, 257–58; Jordan and Shaner, "Weed Control," 266–70, 274–75.

56. Gallant, *Risk and Survival*, 49–50; compare Spurr, *Arable Cultivation*, 58–64. Roman agricultural writers were well aware of the advantages intensive cultivation brought to smaller farms: Col. 1.3.9; Pliny *NH* 18.35.

57. Garnsey, "Where Did Italian Peasants Live?," and above, Chapter 2, n. 123.

58. Thus they would need to farm 41.6–47.9 *iugera*. At 19.5 man-days per *iugerum*, 811–933 man-days would be necessary. Adding 100 man-days for garden, orchard, and vineyard and 80 man-days for pasture would bring the total to 991–1,113 out of an available total of 1,189–1,312. Loss of one son would result in 899–992 man-days available and 895.5–1006 required.

59. Frank, *Economic Survey of Ancient Rome*, 5:141 and n. 4; De Martino, "Produzione di cereali," 245–46, and "Ancora sulla produzione," 241–44; Garnsey and Saller, *The Roman Empire*, 79–82; Wallace, "Diskussion," 557. *Contra*, however, Sallares, *The Ecology of the Ancient Greek World*, 374; Duncan-Jones, *The Economy of the Roman Empire*, 49, 377–78; Evans, "Wheat Production," 429–34.

60. See Delano-Smith, *Western Mediterranean Europe*, 195–97; De Martino, "Ancora sulla produzione," 242–54; and Spurr, *Arable Cultivation*, 84–88, for comparative data.

Notes to Page 74

61. Gallant, *Risk and Survival*, 77, where the mean is approximately 674 kilograms per hectare for wheat, which works out to 168.5 kilograms per *iugerum*. Recall that the model outlined here assumes 99.75 kilograms total yield per *iugerum*. Note too Gallant's statement that data collected from modern small farmers around the Mediterranean suggest a consensus that "a farm of 3–6 hectares seems to be about the size peasants consider to [be] sufficient for supporting a subsistence farm." *Risk and Survival*, 84, and compare 82–83. This would imply that conventional wisdom among his informants considered yields as low as those postulated in the model presented here unlikely, for the model assumes cultivation of 7.75–8.5 hectares (= 31–34 *iugera*, including arable, gardens, vineyards, orchard, and pasture) would be required to yield the food the family needed. Note, too, the yield data in Russell and Voelcker, *Fifty Years of Field Experiments*, 28–29, and compare 27 for sowing amounts.
62. Above, n. 47. Pessimism about manure: Brunt, Review of White, 156–58.
63. Evans, "Wheat Production," 431–35, argues that Caesar's *lex Campania*, which allotted ten *iugera* to fathers of three, demonstrates that yields in the *ager Campanus* were on the order of 1:4, based on the assumption that ten *iugera* represented the amount of land required to raise the minimum amount of food necessary to feed a family of five and that half the land would be fallowed annually. Since the *ager Campanus* was among the most fertile in Italy (Cic. *Leg. Agr.* 1.21, 2.80; Col. 3.8.4; Pliny *NH* 18.109–11; Strab. 5.4.3), one might then claim the yields on average land were considerably lower. Yet neither of Evans's assumptions is certain: on fallowing in general, see above, n. 47; on continuous cropping in the *ager Campanus*, note Jongman, *Economy and Society of Pompeii*, 104: "Fallowing was probably unnecessary," and compare 101–3. On the other hand, Caesar may have intended to be generous to the beneficiaries of his law; hence the amount of land awarded could have represented substantially more than what was required for bare survival. One must bear in mind that the "marginality" of Mediterranean small farming is less the result of limitations intrinsic to this form of production than of unequal social relations to which producers were subjected: note Horden and Purcell, *The Corrupting Sea*, 272, "Thus 'subsistence' is not a way of life: it is a state passed through in particular cases of crisis; and it is the widespread result of many systems of political control," and compare the analysis of Gallant, *Risk and Survival*, 185–96, for Greece in the Hellenistic era.
64. This assumes that the amount of land needed to qualify as an *assiduus* would have been quite small, as in fact appears likely: see above, Chapter 2 at n. 178, and below, n. 68.

Notes to Page 75

65. See Foxhall, "The Dependent Tenant," for a discussion of the complexities.
66. If one son, representing 290–320 man-days of labor, left, the family will have needed to cultivate 9.8–11 fewer rented *iugera* and so had to expend 191–214.5 fewer man-days of labor.
67. Possibly the total available labor would have been greater: the daughter would have been in her early teens by the time the second son was old enough to be conscripted and so her contribution to the family's pool of labor would have been greater, seven-tenths rather than one-half a man-day. Father, mother, and daughter together would have disposed of 696–768 total man-days per year. Food consumption would have gone up slightly as the daughter aged, so that between 639.5 and 702 total man-days of labor would have been required to feed the family. But even so, the surplus of available over required labor would have been uncomfortably small.
68. Note, for example, the seven *iugera* distributed to citizens by M.' Curius Dentatus following his conquests of the Sabines and the Praetutii in the mid-third century: Val. Max. 4.3.5; Pliny *NH* 18.18; Col. praef. 14, and compare 1.3.10 and Plut. *Mor.* 194 E; Front. *Strat.* 4.3.12; compare *de Vir. Ill.* 33.6; Forni, "Manio Curio Dentato," 197; Taylor, *Voting Districts*, 60; or the farm of the consul of 256, M. Atilius Regulus, also allegedly a mere seven *iugera*: Val. Max. 4.4.6. Earlier, plebeians had each received seven (or perhaps only four) *iugera* of the *ager Veientanus*: Livy 5.30.8, and compare Diod. 14.102.4 with Ogilvie, *Commentary on Livy*, 693; and the same amount under a *lex Licinia*: Col. 1.3.10, and compare Pliny *NH* 18.18. Note, too, the six *iugera* thought to have been allotted to the Latin colonists at Cosa: Potter, *Roman Italy*, 106, as well as other colonial allotments made to Roman citizens in the second century, ranging from five to ten *iugera*: Livy 39.44.10, 55.7; 40.29.1–2. For even smaller distributions, note the two *iugera* given to each member of the maritime colony established at Anxur (Tarracina) in 329 (Livy 8.21.11), as well as to those participating in the Latin colony at Labici in 418 (Livy 4.47.7). Settlers at Satricum, also a Latin foundation, in 385 got two and a half *iugera*: Livy 6.16.6–7, and compare 36.11. And those who settled on confiscated Latin territory in 340 each got between two and three-quarters and three *iugera*: Livy 8.11.14. Note also the tradition of Romulus's original distribution of two *iugera* to each citizen as *heredium*: Varro *RR* 1.10.2, Pliny *NH* 18.7, and see also Salmon, *Roman Colonization*, 21. Compare the complaints of the tribunes Sextius and Licinius in 369 (Livy 6.36.11); and the supposed four *iugera* farm of L. Quinctius Cincinnatus, dict. 458 (Livy 3.26.8; Pliny *NH* 18.20).
69. For example, Brunt, *Italian Manpower*, 194–95; Hopkins, *Conquerors and Slaves*, 21; Drummond, *CAH*², 7.2:121. Compare Evans, "Wheat Production," 431–

35, who is more optimistic, however, in *"Plebs Rustica,"* 159–63; Frayn, *Subsistence Farming*, 90–91; Frayn, "Subsistence Farming in Italy," 15. *Contra*, however, White, *Roman Farming*, 345–47, based on the unpublished paper by Hopkins cited above, n. 53, and with references to earlier discussions. On comparative farm sizes, see Gallant, *Risk and Survival*, 82–86.

70. Brunt, *Fall of the Roman Republic*, 248–49, and compare *Italian Manpower*, 194–95; Stockton, *The Gracchi*, 15–16; Hopkins, *Conquerors and Slaves*, 21, 24.
71. De Neeve, *Colonus*, and especially 119–21 for a summary; criticism, however, in Brunt, *Fall of the Roman Republic*, 248 n. 31. De Neeve himself suggests that tenancy arose in the course of the second century: *Peasants in Peril*, 30–31. Sharecropping: De Neeve, *Peasants in Peril*, 16. Foxhall, "The Dependent Tenant," 98, claims that "it is now well established that references to agricultural tenancy go back well into the Republican period," but the works to which she refers to substantiate this statement themselves mostly cite De Neeve's study. Johne, Köhn, and Weber, *Die Kolonen in Italien*, 48–51, claim that extensive small-scale tenancy existed in the second century B.C. on the basis of a fragment from Saserna (= Col. 1.7.4), but see the sensible cautions of Scheidel, *"Coloni* und Pächter," 339–41, and Scheidel, *Grundpacht und Lohnarbeit*, 35–38.
72. De Ligt, "Tenancy under the Republic," has recently challenged De Neeve on this point, but his arguments are not persuasive: see Appendix 3.
73. See Chapter 2 at n. 178.
74. Brunt, *Italian Manpower*, 394–95.
75. App. *BC* 1.7. And in view of what is now known of the relationship between the lowering of the property qualification for service in the legions and the republic's monetary history, it is no longer possible to claim that the property needed to qualify as an *assiduus* in the second century was significantly lower than in the third; see Chapter 1 at n. 58. There is no reason to suppose, therefore, that before the Hannibalic War men poor enough to become tenants or sharecroppers would have been exempted from the draft.
76. Jones, *Later Roman Empire*, 149; Southern and Dixon, *The Late Roman Army*, 69.
77. Upton, *Charles XI and Swedish Absolutism*, 73–74; Lindegren, "The Swedish 'Military State,'" 321–22, 329.
78. Even if, as is sometimes supposed, colonists would have had to become tenants or sharecroppers in these new settlements in order to survive, their former landlords would still have suffered inconvenience and hardship when they departed.
79. On enslavements, see Harris, *War and Imperialism*, 59–60, for sources and

references to earlier discussions; on the substitution of slaves for *nexi*, see Finley, *Ancient Slavery*, 83–86; Cornell, *Beginnings of Rome*, 333, 393–94; Cornell, *CAH*², 7.2:413–14. Claims that slaves would not have been cost-effective on grain-growing estates constitute no impediment to this theory in view of Spurr's refutation of this contention on the one hand and the unprovability of the notion that the wealthy would not have begun producing wine and oil at this point on the other. On the date of the *lex Poetilia*, see Cornell, *Beginnings of Rome*, 462 n. 13. On *nexum* as a source of agricultural labor in the early republic, see ibid., 283.

80. Tibiletti, "Ricerche di storia agraria Romana," 227–31; Gabba, "Sulle strutture agrarie," 22–29. See also De Neeve, *Peasants in Peril*, 30. Note also the theory of communal landholding in early Rome advanced by Mommsen, *History of Rome*, 1:238–40, and Westrup, *Introduction to Early Roman Law*, 47, 56. Compare Brunt, *Fall of the Roman Republic*, 246; Frayn, *Subsistence Farming*, 102; Crawford, *The Roman Republic*, 102–3; Potter, *Roman Italy*, 118 and n. 50.

81. Livy 37.57.7–8; 40.34.2; 39.55.7; 40.29.1–2.

82. Compare Salmon, *Roman Colonization*, 72, on the use of communal land to supplement allotments for citizen colonies; on the size of allotments in Latin colonies as an inducement for Roman citizens, see ibid., 24, and Brunt, *Italian Manpower*, 193, who suggests the dangers of living on the frontier, although it is difficult to see why Bononia, for example, would have been considered a more dangerous place requiring greater inducements for settlers than Mutina: Livy 39.2.5; 41.16.7–8. Salmon also notes that the availability of *ager publicus* to supplement the small allotments in Roman colonies would have tended to keep the colonists from moving into higher census categories and voting centuries in the *comitia centuriata*: *Roman Colonization*, 72. Of course, the allied citizens who participated in Latin colonies were not giving up Roman citizenship, so no compensation for this would have entered into their calculations, and the allotments they received might give a better idea of what average-sized holdings in Italy were where no public land supplemented private arable.

83. Varro *RR* 1.2.9; Livy 6.35.5. The questions of this law's authenticity and, if it is genuine, its exact provisions are controversial: see Oakley, *Commentary on Livy*, 1:654–59, for discussion and references to earlier scholarship. I find persuasive Oakley's arguments for some type of limit passed in this period, even if this was not the 500 *iugera* maximum that later Roman tradition claimed.

84. Garnsey, "Non-Slave Labour," 37.

85. On the need of large farms for the labor of their neighbors, see Rathbone,

"The Development of Agriculture," 12–17. On sharing of resources, see Lirb, "Partners in Agriculture," 293, and compare Foxhall, "The Dependent Tenant," 107.

86. Tibiletti, "Il Possesso dell'*Ager Publicus*," 182–90, argues for the collection of a tax on any *ager publicus* that was farmed; *contra*, Burdese, *Studi sull'Ager Publicus*, 63–70. See, further, Gabba, *Appiani, Bellorum Civilium Liber Primus*, 14, for scholarship on these questions prior to 1958; more recently, Nicolet, *Tributum*, 81–82.
87. Livy 31.13.7.
88. App. *BC* 1.7; Nicolet, *Tributum*, 81–82. Other scholars hold, however, that these sums are anachronistic for the middle republic and that Appian has retrojected them from the empire.
89. The payment of *tributum* makes little difference to the model, since, as noted in Chapter 2, this was not particularly onerous for *assidui* in the lowest census classes: see above at n. 156.
90. Cornell, *Beginnings of Rome*, 268–71, 378, 380–81; Harris, *War and Imperialism*, 60–61, 64–65.
91. Horden and Purcell, *The Corrupting Sea*, 178.
92. On storage, see Chapter 1, n. 72.
93. Garnsey, *Famine and Food Supply*, 53–56, and compare De Ligt, "Demand, Supply, Distribution I," 43–45; and for Greece, Gallant, *Risk and Survival*, 94–101, 113–96.
94. Horden and Purcell, *The Corrupting Sea*, 272–73: "In Mediterranean conditions . . . overproduction is the only safe plan. So there is no incompatibility between protestations of autarky and producing surpluses for storage, exchange or redistribution: the latter is the way the former are made plausible. . . . The surplus was the unpredictable residue of sensibly ambitious production when times were not as bad as they could have been."
95. On labor for wealthier neighbors in Italy, Rathbone, "The Development of Agriculture," 12–17; on kinship and social networks as risk buffers in Greece, Gallant, *Risk and Survival*, 153–69.
96. See generally Horden and Purcell, *The Corrupting Sea*, 175–224.
97. On wild plants, see above, n. 41. Compare also Gallant, *Risk and Survival*, 134–36, on mercenary service in Greece.
98. Compare Hopkins, *Conquerors and Slaves*, 4.
99. Compare Evans, *War, Women and Children*, 106–7. In the antebellum southern United States, six seems to have been the youngest age at which slave children were considered at all useful in fieldwork: Gray, *A History of Agriculture*, 549; Conrad and Meyer, *The Economics of Slavery*, 62.

100. See Chapter 2 at n. 186.
101. For example, Hopkins, *Conquerors and Slaves*, 24–25; Crawford, *The Roman Republic*², 99–100; Nicolet, *CAH*², 9:619. Brunt, *Italian Manpower*, seems to suggest that marriage among soldiers was common in the second century but not in the first: 640–41, and compare 140, 247; see also Harmand, *L'Armée et le soldat*, 427–28. Both Delbrück, *History of the Art of War*, 1:277, and Adcock, *The Roman Art of War*, 12, suppose that the *triarii* were *patres familiae*, as does Evans, *War, Women and Children*, 113, who points to the existence of a separate census category for widows and orphans (for example, Livy 3.3.9, *Per.* 59) as confirmation. He believes that it was the deaths of so many husbands and fathers at war that made widows and orphans numerous enough to warrant their being counted separately. But Evans has apparently failed to understand that the difference in men's and women's ages at first marriage and high natural mortality will by themselves have produced a large number of widows and orphans: see below, n. 159.
102. Most of the passages are highly rhetorical: Polyb. 11.28.7, and compare Livy 28.28.7; Sall. *Iug.* 41.8; Livy 21.41.16, 42.34.3–5; Plut. *TG* 9.5, on which compare Brunt, *Italian Manpower*, 76; and from a century later note Cic. *Phil.* 14.38; Dio 38.9.3.
103. Peter, *HRR* 1.308, Tubero frag. 4 (= Gell. *NA* 10.28.1); Livy 22.57.9, 25.5.8, 27.11.5, 27.11.15, and see Appendix 4.
104. Saller's study is based on a statistical tabulation of the ages at which a deceased's commemorators on his tombstone cease to be parents or other kin and begin to be wives: *Patriarchy*, 25–41, especially 37–38. Compare also Saller, "Men's Age at Marriage," 21–34.
105. Saller, *Patriarchy*, 27 n. 52; see further Rosenstein, "Marriage and Manpower," 180–82. Saller's inscriptions also derive largely from urban contexts: *Patriarchy*, 27. However, it should not on that account be supposed that conclusions based upon these data are invalid for Italy's rural population. Urban populations before the modern era required a constant influx of new residents to increase or even simply to remain stable, and these came primarily from the countryside: see, on Rome, Morley, *Metropolis and Hinterland*, 39–46, 50–54, and for Italian towns during the republic, Hopkins, *Conquerors and Slaves*, 66, and compare 68–69; for preindustrial cities generally, Bairoch, "Urbanization and the Economy," 261–62. Any argument for a strong cultural divergence between Italy's urban and rural populations in a matter so basic as marriage customs is therefore prima facie weak. Compare North, "Religion and Rusticity," refuting similar claims for a strong dichotomy between urban and rural religious practices.

106. On Africanus: Sumner, *Orators in Cicero's Brutus*, 34–36.
107. So Münzer, *Römische Adelsparteien*, 104–7, 171, 267–68, 278; Syme, "Marriage Ages for Roman Senators," 319–26 (= *Roman Papers*, 6:233–40).
108. Hajnal, "European Marriage Patterns in Perspective"; Hajnal, "Two Kinds of Pre-industrial Household Formation"; Laslett, "Family and Household," 526–27; Treggiari, *Roman Marriage*, 400. Variations in the pattern occur, but these mainly affect the age at which women marry.
109. So Brunt, *Italian Manpower*, 138.
110. On the declining number of Italians in the imperial army, see Forni, *Il Reclutamento*; Forni, "Estrazione etnica," 382–83.
111. Saller, *Patriarchy*, 28–29, tables 2.2.b,c.
112. Brunt, *Italian Manpower*, 138–40.
113. Saller, *Patriarchy*, 52, table 3.1.e, and compare table 3.2.e. See further below for the implications of this fact for the question at hand.
114. Compare Hopkins, "On the Probable Age Structure," 247.
115. Although Hajnal, "European Marriage Patterns in Perspective," 116–22, has suggested that marriage ages for both men and women in the Middle Ages and for women in antiquity were significantly lower than those in the early modern period, Saller's data show that this is untrue for the early and high empire, at least for men. Women's median age at first marriage is put by Saller at about twenty (*Patriarchy*, 37), which is lower than that for many early modern European populations but similar to examples from the later Middle Ages: see Livi Bacci, *The Population of Europe*, 102–7. However, as Livi Bacci suggests (ibid., 107), these early marriages in the later Middle Ages may have come in response to the large number of deaths from plague at this time as well as the removal of many of the obstacles to marriage that these deaths brought about.
116. Scholars universally ascribe the measure to Augustus: so, recently, Wells, "Celibate Soldiers," 185; Phang, *Marriage of Roman Soldiers*, 345, with full bibliography.
117. See now the perceptive analysis of Phang, *Marriage of Roman Soldiers*, 344–81.
118. Dio 54.25.5–6; Wells, "Celibate Soldiers," 181–86.
119. The fact that soldiers got round this regulation by common-law marriages is no more an argument against this interpretation than is the failure of Augustus's legislation on marriage and adultery to increase the population or prevent infidelity an argument against the latter as the emperor's aims in passing those laws. The ineffectiveness of his solution in each case does not mean that the emperor was not attempting to address what he perceived as a genuine problem.

Notes to Page 85

120. Livy 22.11.9. The levy of 217 was also exceptional, Livy reports, because *even* freedmen with children had been included (22.11.8: *libertini etiam quibus liberi essent et aetas militaris in verba iuraverant*).
121. Gran. Licin. 14F (= 33.26–27Cr.), and compare Brunt, *Italian Manpower*, 399 n. 3. Rutilius's decision to fix the limit at thirty-five cannot be due to the fact that no one below this age could have completed sixteen years of service and for that reason been exempt from the levy. As is well known, in emergencies the maximum number of years of military service the republic could require of citizens went up to twenty: Polyb. 6.19.4. Consequently, the crisis Rome faced in 105 would have obligated anyone who by thirty-five had already served sixteen years to serve for another four. Similarly, on any plausible assumption about the size of the citizen population in 105 and the average number of legions the republic had fielded annually over the previous sixteen years, only a minority of men will have had to serve as many as sixteen years in order to fill the legions. Therefore, plenty of *assidui* ought to have been available whose *stipendia* fell below that number in 105: on the number of legions in the late second century, see Brunt, *Italian Manpower*, 429–30 and table XIII; on the proportion of *assidui* among the citizen body, see Appendix 5. Moreover, Rutilius's edict is paralleled by the decision in 217 to require only freedmen below the age of thirty-five to present themselves for duty as marines. Here there can be no question of time-expired veterans since Rome rarely manned fleets between the end of the First Punic War and the outbreak of the Second, while *liberti* did not, of course, serve in the legions. So few if any of them over thirty-five can have completed sixteen years of service.
122. While the senate's decree in 171 permitting the consuls at their discretion to enlist soldiers under fifty-one may seem an obvious counterexample (Livy 42.31.4), this step was taken to secure as many former centurions as possible for the war against Persius and implies little about the maximum age of ordinary recruits: Livy 42.33.4. In many cases, those who had reached the rank of centurion by the age of thirty may have elected to continue their military service, as Sp. Ligustinus allegedly did: Livy 42.34.5–11. The decree also enabled the consuls to accommodate older volunteers who were lured by the prospect of an easy victory and bountiful loot, a way of allowing them to dispense patronage, in other words: compare Livy 42.32.6.
123. Polyb. 6.21.7–10, 22.1, 23.1, 24.1; Livy 8.8.5–8. Note that age was a prime criterion in assigning recruits to their legions: Polyb. 6.20.3. Marchetti's curious notion—see *Histoire économique*, 217—that the *principes* and *triarii* in a

legion came exclusively from the first census class derives from a misreading of Polyb. 6.23.15–16. The following discussion also accepts the existence of *velites* (or a similar force of light-armed troops under a different name) as an integral part of the legions prior to the Hannibalic War: see Walbank, *Commentary on Polybius*, 1:701–2, for discussion.

124. Meyer, *Kleine Schriften*, 2:198. See also Fraccaro, *Opuscula*, 4:42–43.
125. Indeed, the phrase might suggest that these men were barely twenty: Craig Williams, *Roman Homosexuality*, 19, and compare Oakley, *Commentary on Livy*, 2:469.
126. Polyb. 6.21.10. On the increase in the size of a legion after 200, see Afzelius, *Römische Kriegsmacht*, 48–61; Brunt, *Italian Manpower*, 671–76. Roth's claim, "The Size and Organization of the Imperial Legion," 346–48, that republican legions in this period had no fixed size is unconvincing: see Chapter 4, n. 5.
127. Note that the "female" Coale-Demeny tables do not refer to a single, specific sex: Parkin, *Demography and Roman Society*, 83, 102. For a discussion of the use of the Coale-Demeny[2] model life tables for studies of ancient populations, see Parkin, *Demography and Roman Society*, 67–90, although see now Scheidel, "Roman Age Structure," and Sallares, *Malaria and Rome*, 118, 272–78, for important cautions about assuming that any model life table can ever represent the real age structure of an ancient population.
128. At $e_0 = 25$ and $r = 0.0$, those seventeen and older compose about 63 percent of the total male population. Each 1 percent of the total male population thus is the equivalent of 100/63 or 1.587 percent of a number that represents 63 percent. One hundred percent of the total male population is thus 1.587 × 300,000 = 476,100. If those between thirty and forty-six are 21 percent of this total, then .21 × 476,100 = 99,981. 38,400/100,000 = 38 percent. On the Roman population before 218, see Brunt, *Italian Manpower*, 44–60. His conclusions are strongly challenged, however, by Lo Cascio ("The Population of Roman Italy," 166–71, and "Recruitment and the Size of the Roman Population," 119–33), who argues that the Roman adult male population would have been considerably higher, around 500,000, in 225 B.C. If he is correct, then naturally the percentage of men over thirty called to serve would have been dramatically smaller.
129. The figure 280,000 is deliberately low as an estimate of the citizen population in the early decades of the second century: see below, n. 152. On the average number of legions in these years, see Brunt, *Italian Manpower*, 423.
130. If 280,00 men were seventeen and older, the total male population would

be about 444,360; if those from thirty to forty-six = 21 percent, then there were 93,316 of them. 106,560 man-years/3 years = 35,520. 35,520/93,316 = 38 percent.
131. The actual total was twenty-five, but these included two legions of slave volunteers. The number of legions in this year is somewhat problematic: see Toynbee, *Hannibal's Legacy*, 2:524–27, and compare 650–51, for discussion; Marchetti calculates twenty-seven legions total, believing that two additional legions were with the fleet in Sicily: *Histoire économique*, 59, 61–63. But Marchetti's reconstruction depends on accepting the tradition that the *legiones Cannenses* were debarred from active service, which is debatable; see Brunt, *Italian Manpower*, 652–55.
132. Brunt, *Italian Manpower*, 418, table X.
133. Again, assuming $e_0 = 25$ and $r = 0.0$. 230,000 citizens represent a total male population of about 365,010; if men from thirty to forty-six were 21 percent, then they numbered approximately 76,652. Because these calculations are solely for illustrative purposes, the likely distortions in the age structure of the male citizen population that resulted from the defeats in the war's early years are ignored here. For the estimate of 230,000 citizens at this date, see below, n. 147.
134. 31,240/76,652 = 41 percent.
135. Caution on this point is especially important in light of Lo Cascio's strong challenge to conventional estimates of the citizen population in the middle republic; see above, n. 128. His argument that the adult male citizen population in 225 B.C. numbered around 500,000 implies a far greater number of citizens in this period than the figures used here, and if he is correct, the percentage of the population conscripted must be correspondingly reduced.
136. Nuptuality for the Roman population is impossible to calculate for any period. Suggestive comparisons are to be found in marriage rates of between about 70 and 80 percent for England, France, and Germany in the seventeenth and eighteenth centuries: Livi Bacci, *The Population of Europe*, 94–95, table 5.1, and compare the same author's remark that in Europe, generally speaking, the higher the average age at first marriage within a population, the greater the percentage of never-marrieds: ibid., 101.
137. Note the hired help that was to maintain the consul Regulus's farm in his absence (see Chapter 1 at n. 6); only the accidents that caused the arrangement to go wrong led to a crisis for the family, not the arrangement itself. And the senate's expedient for manning the fleet in 210 assumes that the ownership of at least one slave was common among citizens in the third census class: Livy 24.11.7, and compare Rathbone, "The Census Qualifications," 138. On

the frequency of slave ownership among members of the third census class, see Chapter 5 at n. 99.
138. On the number of a woman's kin, however, see below at n. 157.
139. Compare the exemptions enjoyed by married men during the revolutionary and Napoleonic eras: Forrest, *Conscripts and Deserters*, 48–49, 53–54.
140. Saller, *Patriarchy*, table 3.1.e.
141. Ibid., table 3.1.e, although it is important to note that Saller's model assumes a stationary population, that is, one that is neither growing nor shrinking (ibid., 42), and this assumption may be unwarranted for the middle republic: see Chapter 5 at n. 25.
142. As for example in the draft for the Union army during the United States Civil War (Geary, *We Need Men*, 66, and compare Murdock, *One Million Men*, 78–80), or in Sweden during the seventeenth century (Parker, *The Military Revolution*, 53).
143. See Appendix 5.
144. If we assume an adult male citizen population in which again, $e_0 = 25$ and $r = 0.0$, the total male population is 476,100 (see above, n. 128); of these, men between seventeen and thirty make up 22.58 percent, or 107,503.
145. To yield a total of 187,200 man-years, 23,400 men will each have had to serve eight years on average.
146. Brunt, *Italian Manpower*, 418, table X. This assumes that because, in a legion of 4,500 men, including cavalry, 3,600 men or 80 percent were *velites*, *hastati*, and *principes*, in a legion with a total complement of 3,200 men the same ratio will have obtained: thus $.80 \times 3,200 = 2,560$.
147. Brunt estimates the number of *iuniores* in 212 at 195,000 men (*Italian Manpower*, 66), which implies an adult male population of about 279,000. Since in the age structure represented by Coale-Demeny[2] Model West 3 Female men from seventeen to forty-six years of age represent 44 percent of the total male population, $100/44 = 2.273$. Thus $195,000 \times 2.273$ yields a total male population of 443,235, of whom 63 percent, or about 279,000, were over seventeen. This seems far too large, considering the losses Rome had suffered in the war to this point. The number of citizens assumed here derives from subtracting Brunt's estimate of 50,000 Romans killed between 218 and the end of 215 (*Italian Manpower*, 419–20) from the 285,000 citizens on the rolls in 218, leaving aside the Campanians, and also assuming that another 5,000 might have been lost between 214 and 212. However, the number of citizens may have been substantially greater than this estimate if Lo Cascio's arguments are sound: see above, n. 135.
148. $230,000 \times 1.587 = 365,010$; $.23 \times 365,010 = 83,952$. Of course, the dramatic

losses among men in the seventeen-to-thirty-year-old age-cohort during the war's early years render the assumption of a "normal" age structure very questionable. But because the example here is for illustrative purposes only, this problem can be left aside here. Also, *assidui* wealthy enough to serve in the cavalry have been left out of the reckoning here, although they naturally would have been included among the total of *iuniores* liable for service. But cavalry made up so small a percentage of a legion's forces that its inclusion in these and the calculations that follow would not significantly alter the conclusions to which they point.

149. See Appendix 5.
150. Livy 27.9.7–14. Note, however, that on Lo Cascio's higher estimates of the citizen population in 225, on which see above, n. 128, the proportion of men between seventeen and thirty that Rome would have had to draft will have been about half as large. Even so, this still represents an exceptionally high rate of mobilization; see his tables, "Recruitment and the Size of the Roman Population," 135–37. However, the even higher ratio of soldiers to civilians that Brunt's figures entail is not, *pace* Lo Cascio, an argument against them or in favor of his own estimate of the citizen population since none of the nations to which he compares Rome's ratio made extensive use of slave labor, which unquestionably affected the number of free men who could be removed from the civilian economy for military service. It is significant, in this regard, that one other "slave society," the Confederate States of America, mobilized about 75 percent of all free, adult males between seventeen and fifty years of age in its war for independence; see below at n. 194.
151. Note, for example, that during the Napoleonic era, the minimum height requirements for French draftees were progressively relaxed from about five feet in the 1790s until men only four feet eight inches were being conscripted during the final, desperate levies of the empire. Throughout, the government pressured examiners to grant medical exemptions sparingly: Forrest, *Conscripts and Deserters*, 44–46. Paraguay, during its catastrophic war against Argentina and Brazil (1864–70), was forced to resort to even more extreme measures: John Williams, *The Rise and Fall of the Paraguayan Republic*, 214–17.
152. In general, see Brunt, *Italian Manpower*, 71–75, for discussion of the census returns in this period and probable population size. Several of the returns that Livy transmits are significantly greater than the 280,000 figure used here, but Brunt argues that they must be roughly correct because the one reliable census figure we possess, that of 313,000 or, as corrected by Brunt, 346,000, cannot be understood except against a series of prior census returns well above 280,000. The figure of 280,000 is used here because an unknown num-

ber of men in the first decades of the second century may have been falling below the minimum requirement for service in the legions. Thus a deliberately low base figure for *assidui* avoids counting those *proletarii* as eligible for the legions. Of course, if as Lo Cascio's argument ("Recruitment and the Size of the Roman Population," 123–33) implies, the citizen population in these decades was much greater than Livy's census figures would suggest, then the proportion of men who would have had to be drafted would have been much lower than what is suggested here if most citizens qualified as *assidui*.

153. On length of service in this period, see Appendix 6, and compare Hopkins, *Conquerors and Slaves*, 32–35, on the effect of length of service on the proportion of men who will have had to serve.

154. Note, interestingly, the complaint voiced much later by the author of the Ps. Sall. *ad Caes*. 1.8.6: *Item ne, uti adhuc, militia iniusta aut inaequalis sit, cum alii triginta, pars nullum stipendium facient.*

155. As the consul Regulus's family did (see Chapter 1 at n. 6). Note that during the Hannibalic War, the *patres* assumed that citizens in the third census class would possess a slave: Livy 24.11.5–9, 26.35.3. See also Livy 1.43.1–5 for his understanding of the property qualifications for the various census classes and the discussions of Rathbone, "The Census Qualifications," and Lo Cascio, "Ancora sui Censi Minimi," of these figures. On the distribution of slaveholding among the Romans and Italians, see Chapter 5 at n. 99.

156. Lirb, "Partners in Agriculture," 284–86, and above, n. 93.

157. As Evans, *War, Women and Children*, 113–14, seems to do. His curious notion (ibid., 103) that endogamy between cousins was common in this period and so would have remedied labor shortages caused by conscription is highly implausible. See Saller and Shaw, "Close-Kin Marriage," and Treggiari, *Roman Marriage*, 109–19. Note, too, Livy, frag. 12 (Mueller) from book 19, reporting that an attempt to arrange a match in violation of an old custom forbidding marriages between kin closer than the seventh degree shortly before the Second Punic War resulted in a riot, an event hardly consonant with the notion that endogamy among cousins was commonly practiced at Rome. Although much of the evidence on this topic concerns members of the upper class, we have no reason to assume that marriage practices among smallholders differed to any great extent.

158. For one model of the structure of a woman's kin, see Saller, *Patriarchy*, tables 3.1.a and b, although again it is important to note Saller's assumption of a stationary population. If Italy's population was growing significantly during the middle republic, on the possibility of which see Chapter 5 at n. 25, then presumably the average number of kin would have been larger as well.

159. Evans, *War, Women and Children*, 113, suggests that many widows will have remarried in order to keep their farms going, but the same difference in age at marriage that caused wives generally to outlive their husbands also worked to reduce significantly the pool of potential husbands, including widowers, available for all women seeking husbands, including widows, to marry. Data from Roman Egypt indicate that there "about half the women in their later thirties were no longer married"; see Saller, *Patriarchy*, 68. The high rate of mortality among soldiers had a similar effect, as is discussed in the following chapters. The existence of many fewer potential husbands than eligible wives must also have forced a considerable percentage of women to remain single, and they presumably could have constituted a source of additional labor on many small farms.

160. Martin, "The Construction of the Ancient Family," argues on the basis of evidence drawn from Asia Minor during the imperial period that a "family" might often have comprised more members than simply the parents and children that Saller's models assume through the inclusion of more distant relatives or even people with no kinship ties at all.

161. Compare for Macedonian soldiers, Hamilton, "The Hellenistic World," 171.

162. Chayanov, *The Theory of Peasant Economy*, xvi, 81.

163. These needs of course would include money to purchase items a family could not make or grow itself, such as salt or certain tools, and possibly access to facilities that were too expensive for them to own, such as olive presses or draft animals, for which labor might be exchanged. Similarly families might work in exchange for foods that they could not grow on their own.

164. Some areas also may have seen an increasing number of large estates that may in some cases have monopolized large tracts of public land, although the rate at which "plantation agriculture" was expanding during the second century was probably much more modest than usually supposed: see Chapter 1 at n. 13.

165. I am assuming for the mother a daily requirement of between 2,082 and 2,434 calories per day or 759,930 and 888,410 calories per year and for the daughter 2,160 and 2,500 calories per day or 788,400 and 912,500 calories per year.

166. So, for example, a yield ratio of 1:6 and a sowing rate of 33.25 kilograms per *iugerum* nets 166.25 kilograms per *iugerum* after seed is deducted and means that the produce of between about 2.75 and 3.25 *iugera* would feed the mother and daughter or between about five and a half and six and a half *iugera* if half the crop went to a sharecropper.

167. On the age of women at first marriage during the imperial era, see Treggiari,

Roman Marriage, 409–10; Saller, *Patriarchy,* 25–41, especially 36–37; Shaw, "The Age of Roman Girls." However, it remains quite possible that women's age at first marriage declined somewhat during the middle republic owing to the relative scarcity of potential husbands caused by the high mortality among men serving in the military; see Chapter 5 at n. 13.

168. Daughters married *sine manu* remained in the *potestas* of their fathers and consequently were *sui heredes* entitled to an equal share of the estate upon his death. In developed law, beginning early in the third century B.C., even daughters who had married *cum manu* were apparently recognized as heirs in the first instance along with their brothers. See Johnston, *Roman Law in Context,* 51–52, for a brief overview of testamentary disposition and intestate succession with further bibliography.

169. So Brunt, *Italian Manpower,* 148–54; other examples in Harris, "Child Exposure," 11 n. 94, who offers good reason to disagree. Saller, "Slavery and the Roman Family," 70–71, argues only that under the empire female infanticide brought the numbers of potential brides and grooms back into balance. Death due to childbirth will not have greatly reduced the number of women seeking husbands since little supports the notion that such deaths would have been demographically significant (Parkin, *Demography and Roman Society,* 103–5) and, obviously, prospective brides in most cases had yet to undergo the rigors of giving birth.

170. Harris, "Child Exposure," 5–6, and compare 1–22, for a general discussion of the practice under the empire. In fact, very little can be said with certainty about the practice one way or another in either period: Scheidel, "Progress and Problems in Roman Demography," 44–45.

171. Dio 54.16.2: $\epsilon\pi\epsilon\iota\delta\acute{\eta}\ \tau\epsilon\ \pi o\lambda\grave{\upsilon}\ \pi\lambda\epsilon\hat{\iota} o\nu\ \tau\grave{o}\ \check{a}\rho\rho\epsilon\nu\ \tau o\hat{\upsilon}\ \theta\acute{\eta}\lambda\epsilon o\varsigma\ \tau o\hat{\upsilon}\ \epsilon\mathring{\upsilon}\gamma\epsilon\nu o\hat{\upsilon}\varsigma\ \mathring{\eta}\nu$. See Harris, "Child Exposure," 11 n. 94.

172. Brunt, *Italian Manpower,* 151–52.

173. Compare Harris, "Child Exposure," 13–14.

174. Above, n. 167.

175. Treggiari, *Roman Marriage,* 32–34, noting that there is no good reason to believe that marriage *cum manu* is in any sense the "original" form of marriage at Rome and marriage *sine manu* a later development.

176. Phillips, "Roman Mothers"; Hallett, *Fathers and Daughters,* 259–63; Dixon, *The Roman Mother,* 210–32.

177. On the first point: Saller, "Slavery and the Roman Family," 70; on the second, see the following chapter at n. 149.

178. See above, n. 27.

179. Thus $.5 \times 290 = 145$ man-days; $.5 \times 320 = 160$ man-days.

180. The assumption that women could not successfully run a farm in their husbands' absence is common among scholars: for example, Hopkins, *Conquerors and Slaves*, 24–25, 30; compare Krause, *Witwen und Waisen*, 2:154–56. Note, especially, Evans, *War, Women and Children*, 146 n. 15: "Most women would have found it difficult if not impossible efficiently to use the typical Roman plough, or sole-ard. This was designed in such a way that it had a pronounced tendency to catch in the soil, and it required a constant and very heavy downward pressure to drive a straight furrow with an even depth."
181. On the agrarian role of women in antiquity generally, see now the fine studies of Scheidel, "The Most Silent Women," and "Feldarbeit von Frauen."
182. As White, *Roman Farming*, 176, notes, "In order to enable the plough to bite into the soil, the pointed share has a slight downward curvature at the point. The plough thus has a natural tendency, when in motion, to dig itself into the ground as it cuts its way through. To maintain an even keel, different types of plough require different techniques; in the sole-ard an even depth of cut is achieved by means of downward pressure on the stilt by the ploughman, who can also increase the pressure when necessary by using his foot." Compare White, *Agricultural Implements*, 138–39. However, note the photograph in Barker, *A Mediterranean Valley*, fig. 19 on p. 38, of a farmer in the 1970s using oxen and a sole-ard-type plow, who does not seem to be straining unduly. Note, too, Spurr's discussion, *Arable Cultivation*, 28–34, of the range of different types of plows that fall under this rubric, each of which might have its own operational characteristics.
183. Col. 1.9.2–3; compare Delano-Smith, *Western Mediterranean Europe*, 199.
184. Stanley, *Mothers and Daughters*, 14–17; Scheidel, "The Most Silent Women," 208–10.
185. *Reminiscences of Georgia*, 227, quoted in Gray, *A History of Agriculture*, 549. Compare Eckberg, *French Roots*, 157, who asserts that many female slaves in colonial Illinois would have done heavy fieldwork alongside their male counterparts. Note, too, the recollections of Mrs. Jeffers of life on her farm after her husband departed for service in the Confederate army: "There was no income from the farm and no way to get anything from it except I worked it myself. I plowed in the field day after day." United Confederate Veterans, *Confederate Women of Arkansas in the Civil War*, 155.
186. Segalen, *Love and Power in the Peasant Family*, 105–8, especially 107.
187. Dohr, *Die italienischen Gutshöfe*, 138–39.
188. Scheidel, "The Most Silent Women," 205–7.
189. "Notes of a Tour into the Southern Parts of France, &c.," in *The Papers of Thomas Jefferson*, 11:415, quoted in Eckberg, *French Roots*, 139.

190. Davis, *Land and Family in Pisticci*, 94–95, on the dishonor to males in this southern Italian town of allowing their wives to work in the fields; compare Erdkamp, "Agriculture, Unemployment," 571. Interestingly, little evidence supports the assumption that in antiquity a woman's chastity was strongly linked to the honor of her male relatives, although this connection is common in Mediterranean societies in more recent eras: Treggiari, *Roman Marriage*, 311–12 and n. 268, where she suspects that the practice may have been introduced through the influence of Islam.
191. See the classic study of Boserup, *Woman's Role in Economic Development*, 15–80; Sachs, *Gendered Fields*, 67–101.
192. Compare the judgments of scholars who have examined the question of why men seem to have acquired primary responsibility for plowing from the earliest phases of the Neolithic era's agricultural revolution: see the survey of opinions in Stanley, *Mothers and Daughters*, 38–42, 775–95, and compare Benería and Sen's critique in "Accumulation, Reproduction" of Boserup's analysis of the reasons for the sexual division of agricultural labor.
193. Scheidel, "The Most Silent Women," 211. Compare the attitude of modern women on small farms in Germany (Inhetveen and Blasche, "Women in the Smallholder Economy," 30–31), or those of American women in the colonial and revolutionary eras (Ulrich, *Good Wives*, 35–50; Norton, *Liberty's Daughters*, 15, 225). Note too the widows who kept their farms going with the help of children and hired labor in early modern Spain: Vassberg, *The Village and the Outside World*, 29. The case of Regulus's wife and children, noted at the beginning of this study, might seem to contradict this assertion, but one must remember that this tale represents the reactions of the wife of a consul—and so presumably a woman of very high status herself—to the crisis set in motion by the death of the farm's steward: Chapter 1 at n. 6. Considerations of honor presumably weighed far less heavily upon the wives of smallholders who also, unlike the wives of consuls, will probably not have been unused to fieldwork.
194. The figure of 75 percent is generally accepted among historians of the American Civil War: for example, Faust, *Mothers of Invention*, 30, although no records survive. For discussion, see Moore, *Conscription and Conflict*, especially at 357–58, for total numbers mobilized.
195. Otto, *Southern Agriculture*, 32–33.
196. On the Roman Republic: Brunt, *Social Conflicts*, 18–19; Brunt, *Italian Manpower*, 67; Finley, *Ancient Slavery*, 84; compare Harris, "Roman Warfare," 499. On the Confederacy, see Otto, *Southern Agriculture*, 33–34.
197. Rable, *Civil Wars*, 321 n. 27; Faust, *Mothers of Invention*, 32, 99; Otto, *Southern*

Agriculture, 35. On yeomen's wives performing fieldwork in the antebellum period when necessity compelled this, see Cecil-Fronsman, *Common Whites,* 144; Bynum, *Unruly Women,* 45; Hahn, *Roots of Southern Populism,* 30. One may compare, too, the Swedish parish of Bygdea in the early seventeenth century. As a result of a massive number of deaths among the men the parish sent to fight in Gustavus Adolphus's wars, "no adult male labour was available on many farms and . . . women were in charge of a growing number of farms . . . the responsibility for maintaining production fell mainly on the women of the parish." See Lindegren, "The Swedish 'Military State,'" 317.

198. Bynum, *Unruly Women,* 47–49; Hahn, *Roots of Southern Populism,* 29–31.
199. Livi Bacci, *Population and Nutrition,* 23–62.
200. Ramsdell, *Behind the Lines;* Rable, *Civil Wars,* 78–81, 111; Cecil-Fronsman, *Common Whites,* 212; Bynum, *Unruly Women,* 121–29. However, one should note that much of the evidence for hardship comes from the letters women wrote to governors and other state officials seeking the discharges of their husbands or sons from the army because of difficulties and privations they were suffering at home. Those women who were coping successfully with the demands the war had placed upon them naturally are underrepresented in such collections.
201. The cause of the inflation was a scarcity of goods which was due less to a failure of production than distribution. The South produced the food it needed to feed its armies and civilian population but was unable to transport enough of it to those in need to prevent scarcity, hunger, and, among the civilian population, rising prices. Otto, *Southern Agriculture,* 34–40; Ramsdell, *Behind the Lines,* 19–20, 44–45, 51, 75, 85, 88–89, 113, 115–16; Rable, *Civil Wars,* 96; Beringer, *Why the South Lost the Civil War,* 12–13.
202. On the financial aspects of the crisis of the Hannibalic War, see Nicolet, "A Rome pendant la seconde guerre punique"; on the coinage, Crawford, *Coinage and Money,* 52–62.
203. Koistinen, *Beating Plowshares into Swords,* 76–77.
204. Ford, *Origins of Southern Radicalism,* 52–57, 72–89; Hahn, *Roots of Southern Populism,* 29–34; Ash, *When the Yankees Came,* 4; Ash, *Middle Tennessee Society Transformed,* 19–22.
205. On the integration of peasants in the Roman Empire into urban economies, see De Ligt, "Demand, Supply, Distribution I," and "Demand, Supply, Distribution II."
206. On the question of southern women's responses to the war, see generally Rable, *Civil Wars;* Faust, *Mothers of Invention;* Berlin, "Did Confederate Women

Lose the War?" On the failure of morale, see Wiley, *The Road to Appomattox*, 43–75; Beringer, *Why the South Lost the Civil War*, 336–67; Rable, "Despair, Hope, and Delusion."
207. On Union strategy, see Grimsley, *The Hard Hand of War*.
208. On military pay and donatives in the second century, see Brunt, *Italian Manpower*, 394, table IX, and 411–13, and Harris, *War and Imperialism*, 102–3.
209. Col. 2.1.3, and compare Spurr, *Arable Cultivation*, 121. Compare also the remarkable yields obtained in the Butser experiment from fields sown after many years of fallow (Reynolds "Deadstock and Livestock," 106–10), although the periods of fallow in this case were far in excess of anything that would have obtained in the circumstances here envisioned. For a description of the Butser experiment, see Reynolds, *Iron Age Farm*.
210. On mortality among soldiers, see the following chapter.
211. On the composition of Roman dowries, see Treggiari, *Roman Marriage*, 340–50, especially 248–49 on the inclusion of land.
212. Note the local tribunals implied in Plut. *Cato Mai.* 1.4, 3.1. Although the stories of Cato's pleadings in them may be invention, the settings themselves are likely to have been real enough if a later writer could imagine Cato developing his skill as an advocate in them. See Astin, *Cato the Censor*, 8.
213. See above, Chapter 1 at nn. 13 and 63.
214. For example, once a recruit's enrollment was complete, he was required to assemble at a particular place at a specific time in the future; see examples above in Chapter 2, n. 118. The interval allowed for travel to and from home: Polyb. 6.26.1, and compare 21.5–6.
215. Livy 33.29.2–4; compare Briscoe, *Commentary on Livy XXXI–XXXIII*, 303.
216. Livy 43.11.10, 14.5–10.
217. Val. Max. 9.3.7.
218. Livy 43.14.7, and compare 11.10.

Chapter 4

1. On the Confederate States, see Chapter 3 at n. 194. Oddly, no comprehensive study of mobilization rates exists. Andreski, *Military Organization and Society*, the nearest thing available, is highly schematic and insufficiently grounded in historical evidence. Figures for early modern European states are collected in Corvisier, *Armies and Societies in Europe*, 113, although these must be treated with caution because, where they can be checked, they are often inaccurate or misleading: see, for example, Tulard, *Dict. Napoléon*, 475, for France 1800–

1815; Roberts, *The Swedish Imperial Experience*, 45, for Sweden 1700–1708; Ingrao, *The Hessian Mercenary State*, 132, for eighteenth-century Hesse-Cassel; Ziechmann, *Panorama der fridericianischen Zeit*, 393, for eighteenth-century states generally. Mobilization rates in less complex societies: Keeley, *War before Civilization*, 33–36. For comparable Roman figures for this period: Lo Cascio, "Recruitment and the Size of the Roman Population," 136.

2. On the number of deaths during the Hannibalic War, see Brunt, *Italian Manpower*, 422.

3. So Rathbone, in his important study of the Latin colony of Cosa, "The Development of Agriculture," 18–19, only offers his estimate of a 7 percent mortality rate as a heuristic device to explain the steady decline in the population of Cosa in the century since its founding: "Each year . . . Cosa faced a population decline of only 0.5 percent, equivalent to a simple average of thirteen families. . . . If we conjecture that the death of a smallholder on campaign had a one-in-four chance of causing his family to abandon his allotment, we need only posit an annual casualty rate on campaign of just over 7 percent to explain this demographic decline completely." However, his speculation is accepted by Morley, *Metropolis and Hinterland*, 102–3. A highly uncritical treatment of the sources vitiates the figures offered by Gabriel and Metz, *From Sumer to Rome*, 83–91. See recently Sabin, "The Face of Roman Battle," 5–6, for a discussion of selected figures.

4. An earlier listing in Frank, *Economic Survey of Ancient Rome*, 1:110, also reproduced in Toynbee, *Hannibal's Legacy*, 2:71–72.

5. The sizes of the armies are taken from Afzelius, *Römische Kriegsmacht*, 34–79; Brunt, *Italian Manpower*, 242, table XI. Roth, "The Size and Organization of the Imperial Legion," 346–48 (but compare Roth, *The Logistics of the Roman Army*, 19–20), argues that there was no standard size for Roman legions during the republican period. The evidence he cites, however, can be explained in other ways (for example, the legions that Caesar, at *BC* 3.106.1, describes as containing only 3,200 men are explicitly said to have been depleted by wounds and the difficulty of their march; Caesar's remark implies nothing about these legions' normal complement). The evidence from Livy, discussed by Afzelius, *Römische Kriegsmacht*, 49–50, demonstrates that, at least for the period 200–168, the legions regularly contained 5,200 infantry and 300 cavalry except in rare instances of special need. In reckoning the percentages in Table 2, therefore, 5,500 has been taken as the standard number of Roman citizens in a legion save for the legions in Greece from 171 to 168, which contained 6,000 infantry each: Livy 42.31.2; 43.12.4; 44.21.8. The number of *socii* are as given in Afzelius, *Römische Kriegsmacht*, 78–79.

6. (55,281/24)/(627,800/24) = .088. The assumption that in each battle Roman forces were at their full strength can of course be challenged. For one instance, see Walbank's discussion, *Commentary on Polybius*, 2:584–85 (and compare De Sanctis, *Storia*², 4.1:76 n. 159), on the number and composition of Flamininus's forces at Cynoscephalae. Men on garrison duty of course ran no risk of death in combat, so subtracting them from the number of combatants might raise the death rate for a battle, but since what we are after is an overall death rate for all conscripts, not simply those who participated in battles, it is proper to include them in the count.
7. Brunt, *Italian Manpower*, 694–97.
8. Thus Antony mustered his soldiers during his retreat from Parthia and discovered that he had lost 20,000 men: Plut. *Ant.* 50.1. And Caesar seems to have had a clear idea of how many of his soldiers were killed and wounded following his engagements, for example: *BC* 1.46.5; 2.35.5; 3.53.3; *BH* 31.10.
9. Note, famously, the account books that the enemies of L. Cornelius Scipio Asiagenus demanded to inspect in the senate in 187. The sum in question had come from Antiochus for the troops' pay; see Polyb. 23.14.7–8; Livy 38.55.10–11; and Walbank, *Commentary on Polybius*, 3:245, for further sources and discussion of the incident. And Tiberius Gracchus as quaestor thought it worth risking his life in 136 to return to Numantia following Mancinus's surrender and withdrawal from that city to recover his ledgers, which had fallen into the hands of the Numantines, in order to prevent his enemies in Rome from criticizing his administration of that office: Plut. *TG* 7.1–2. On magistrates' financial records, see also Livy 30.38.1; Cic. *Verr.* 2.1.36–37, 57; *Fam.* 128 (5.20).2, and compare *Att.* 120 (6.7).2. On the administration of logistics: Roth, *The Logistics of the Roman Army*, 256–59.
10. On the food supply of Roman armies, see now Erdkamp, *Hunger and the Sword*.
11. A soldier's monthly ration was 26.9 kilograms of wheat: Foxhall and Forbes, "Σιτομετρία," 86; 2,690 kilograms = 5,945 pounds. On the soldiers' diet, see Roth, *The Logistics of the Roman Army*, 18–44.
12. See Culham, "Archives and Alternatives," 104–5, although it must be added that our only extant *commentarii*, those of Caesar, rarely provide figures for the numbers of his soldiers who died in battles.
13. Evidence for collection of Roman dead following victories: Livy 23.46.5; 27.2.9; 27.42.8; 39.21.7. Note, too, the unpopularity of Lucullus among his soldiers owing to a failure to bury Roman dead following a defeat: Plut. *Pomp.* 39.1; compare also Caes. *BG* 1.26.5; Phlegon of Tralles, *FGrH* 257 F36 III.2.
14. See below, n. 128. One can compare the extraordinary organizational system

Notes to Pages 112–13

of the Aztecs, the only other preindustrial state to have engaged in as long and as successful a period of conquests: Hassig, *Aztec Warfare*, 27–109.

15. Livy 22.56.1–2. Whatever uncertainty there may have been in his count will have been due precisely to the fact that the army's organizational structure had broken down temporarily: *ad decem milia militum ferme esse incompositorum inordinatorumque*.
16. Val. Max. 2.8.1.
17. Val. Max. 2.8.1; Oros. 5.4.7. On the dating of the law, see, for example, Rotondi, *Leges Publicae*, 279, who puts the law circa 179 in connection with the awarding of triumphs for minor (or, indeed, no) victories in Liguria in this period, on which see Livy 40.38.8–9, 59.1, and compare Cic. *Brut.* 255.
18. Livy 33.22.7.
19. Livy 38.46.7.
20. Livy 35.6.9–10.
21. Livy 33.25.9.
22. Of the battles listed in Table 2, note those in 196, 193, 191, 190, and 185 as well as the victory of Minucius Rufus in 197: Livy 33.22.7–8. Compare also Livy 22.49.16–17; 38.46.7.
23. For a list of passages, see Afzelius, *Römische Kriegsmacht*, 48–61, and compare 62–77.
24. Note, interestingly, Lucian, *Hist. Conscr.* 20, who writes as if official reports of casualties from an army's officers would have been available to historians in the second century A.D. Whether he knew this to be true or merely presumed it was so is unfortunately unclear. On casualty reports during this period, see Mattern, *Rome and the Enemy*, 105–6, whose assessment of their reliability, however, seems overly pessimistic. The figures for Romans and Italians slain in Boudicca's revolt—70,000 in Tacitus *Ann.* 14.33 but 80,000 in Dio 62.1.1—represent those who died in a general massacre, where an accurate count would have been highly unlikely, rather than a pitched battle. When the latter occurred in the course of the revolt, the 400 Roman troops reported slain (Tacitus *Ann.* 14.37) are quite in line with the extent of casualties suffered in several Roman victories reported in Table 2.
25. Livy 42.60.1, and compare Plut. *Mor.* 197F.
26. For example, Polyb. 15.18.3, and compare Livy 30.16.10, 37.3; App. *Lib.* 54; Dio 17.57.82; Polyb. 18.1.13, and compare Livy 32.33.3; Polyb. 18.44.6, and compare Livy 33.30.5; Livy 38.9.9; 38.11.4, and compare Polyb. 21.32.5; Polyb. 21.43.10, and compare Livy 38.38.7. Other examples in Walbank, *Commentary on Polybius*, 3:133.
27. On the fate of deserters: Livy 30.34.13.

28. For example, Livy 34.16.7; 37.60.3–6.
29. Brunt, *Italian Manpower*, 694–97, and compare 419.
30. For the exaggeration of numbers of enemies killed, note Livy's well-known strictures on the claims of Valerius Antias, for example, 30.19.11; 33.10.8; 36.38.6–7; 38.23.8, and compare 23.26.25. And note, too, Livy's remarks on the different figures his sources preserved for the number of men Scipio transported to Africa in 204: Livy 29.25.1–4.
31. Polyb. 3.117.4; Livy 22.49.15, but compare the figures in App. *Han.* 25: 50,000; Plut. *Fab.* 16.8: 50,000; Quint. *Inst.* 8.6.26: 60,000. On Polybius's source, see Walbank, *Commentary on Polybius*, 1:440.
32. Livy 30.35.3, and compare Polyb. 15.14.9; App. *Lib.* 48.
33. Livy 22.7.2–4, although according to De Sanctis, *Storia*^{2}, 3.2:113, and Brunt, *Italian Manpower*, 419 n. 1, this figure includes prisoners of war.
34. Polyb. 3.84.7, 85.1–2. On the Punic sources for these figures, see De Sanctis, *Storia*^{2}, 3.2:113; Walbank, *Commentary on Polybius*, 1:420.
35. Livy 27.1.13.
36. Plut. *Aem.* 21.3, and, on Posidonius (Jacoby, *FGrH* 2B 169), compare 19.4. On the Roman casualties, see also Livy 44.42.8.
37. Livy 42.60.1; Plut. *Aem.* 9.2, but compare *Mor.* 197F where Plutarch gives a combined total of 2,800 for both Roman dead and prisoners of war. Because at *Aem.* 9.2 he puts the number of the latter at 600, the figure of 2,800 at *Mor.* 197F may mean that Plutarch here followed a source that, like Livy's, put the Roman dead at 2,200.
38. Plut. *Aem.* 19.4–5.
39. On *auxilia* see Afzelius, *Römische Kriegsmacht*, 90–98. Compare also the Gallic forces with P. Cornelius Scipio in the Po Valley in 218: Polyb. 3.65.5, 67.1–3; Livy 21.46.5, 48.1–4. Much greater losses noted among Spanish *auxilia* than for Romans and Italian allies in 181: 200 Romans, 830 Latins, and 2,400 *auxilia*: Livy 40.32.7; and in 180: 472 Romans, 1,019 Latins, and again 3,000 auxiliaries: Livy 40.40.13. Compare, however, the limited Spanish losses in 185 with the Roman and Italian losses: Livy 39.31.15. On the Roman forces at Callinicus, see Livy 42.58.11–59.11, and compare 60.8–9 and App. *Mac.* 12. Many of those fighting on the Roman side at Cynoscephalae were Greeks; presumably then they composed at least a portion of the 700 killed in that battle: see Walbank, *Commentary on Polybius*, 2:584–85, for discussion of the composition of Flamininus's forces. Note, too, the twenty-five soldiers from Eumenes' forces who died at Thermopylae in 190 (Livy 37.44.2) and the preponderance of local *auxilia* who died in Claudius Centho's defeat in 170 (Livy 43.11.11).

40. Compare Brunt, *Italian Manpower*, 663, on the casualty figures for Roman losses in Spain between 154 and 133.
41. Below, n. 57.
42. On Greek perceptions of the Celts, see Momigliano, *Alien Wisdom*, 59–63; on their frequent employment as mercenaries by Hellenistic powers, G. T. Griffith, *Mercenaries of the Hellenistic World*, 252–53.
43. For example, Sall. *Cat.* 53.3: *sciebam . . . facundia Graecos, gloria belli Gallos ante Romanos fuisse*. On the image of the Gauls generally, see Kremer, *Das Bild der Kelten*; on warfare against the Gauls, J. H. C. Williams, *Beyond the Rubicon*, 175–77; and on the senate's preparations in 225, Polyb. 2.22.7–24.17.
44. Note the emphasis on Gallic courage and determination in Polybius's account of the battle at the Telemon River in 225: 2.30.7. Note, too, the 1,500 killed Polybius reports the Romans suffering at Zama (Polyb. 15.14.9, and compare Livy 30.35.5; Appian *Lib.* 48 puts the Roman losses at 2,500 killed), about 8.5 percent if Brunt's estimate of the size of Scipio's army at 17,600 men is correct: Brunt, *Italian Manpower*, 672–73, but see Lazenby, *Hannibal's War*, 203, for a significantly higher figure.
45. Cannae: Polyb. 3.117.6, and on the fighting, see especially 3.116.10–11; Trasimene: Polyb. 3.85.5, and compare 3.84.2–8 on the battle itself. Calculations for Trasimene assume that Hannibal's army numbered about 50,000 as it did at Cannae (Polyb. 3.114.5), although as Walbank, *Commentary on Polybius*, 1:439, notes, Hannibal probably left some of these troops to guard his camp, which would raise the percentage of those killed in combat.
46. On Polybius's pro-Carthaginian Greek sources for the Hannibalic War, see Walbank, *Commentary on Polybius*, 1:28–29. Note that the Roman tradition, as preserved in Livy 22.52.6, put Hannibal's losses at Cannae at 8,000 and 2,500 dead at Trasimene: Livy 22.7.5.
47. Above at n. 20.
48. Polyb. 35.4.1–2, compare 35.1.1–6, 4.3–14; App. *Iber.* 49; Livy *Per.* 48; Oros. 4.21.1; Val. Max. 3.2.6b; and on the reputation of Spanish soldiers among later authors, compare Florus 1.22.38; 2.10.3.
49. On the characterization of easterners as lacking martial prowess, see Petrochilos, *Roman Attitudes to the Greeks*, 46–48, 93–104; Wheeler, "The Laxity of Syrian Legions," 237–46.
50. Livy 39.20.7–9.
51. Scheidel, "Finances, Figures and Fiction," quotations from 223, 237. See also Duncan-Jones, *Money and Government*, 16–19.
52. See the figures in Table 2 under "number killed" marked "<", ">" and, "*f(erme)*."

53. The figures include four multiples of 30, one of 400, as well as possibly one multiple of 10 (see under 168). In addition, there are two multiples of 200, four of 500, two of 600, one of 700, and twelve that may be classified as "other." Few of these appear in Dreizehnter's discussion of conventional figures for groups of men in Livy: *Die rhetorische Zahl*, 96–102.
54. See also under 193, 191, 181, 176, and note, too, Livy 31.33.10, from Polybius.
55. Above, n. 22.
56. Krentz, "Casualties in Hoplite Battles," 13–14 and the table on 19. Compare Vaughn, "The Identification and Retrieval of the Hoplite Battle-Dead." See also Sabin, "The Face of Roman Battle," 5–6, who takes Krentz's findings as likely approximations of Roman deaths in battles in this period.
57. There are 16,443 killed out of 391,500 in victories and 33,837 out of 209,800 in defeats. These calculations count the two battles fought in Spain in 185 separately but do not include the draw in 194.
58. The missing defeats are the consul Hostilius's repulse in Macedonia in 170, lost owing to the lacuna in Livy's book 43 (Livy 43.11.9; Plut. *Aem.* 9) and the defeat suffered by Sempronius Tuditanus in Spain in 196 (Livy 33.25.8–9). Of those defeats listed in appendix 1 of Rosenstein, *Imperatores Victi*, all can be found in Table 2: no. 8 = 199; no. 19 = 196; no. 29 = 194; no. 61 = 186; nos. 16 and 22 = 170; no. 60 = 169. In addition, Table 2 also includes figures for three initial defeats which were subsequently redeemed by victories, and which, therefore, are not included in Appendix 1: 190, 185, and 171. On the victories, see Table 3.
59. As Table 3 indicates, several legions were involved in more than one pitched battle in a particular year, for example, the legions stationed in Gaul in 196. Because of this, a "victorious legion" is defined as an actual legion and its allied contingent that won a single, discrete victory in a battle (whether by itself or as part of a two-legion consular army) in a year; likewise, in the case of "defeated legions." Under this system of reckoning, the two actual legions in Gaul in 196 commanded by Marcellus, having fought three separate battles in which they were defeated once and won the other two engagements, represent four "victorious legions" and two "defeated legions" in Table 3. The other two "victorious legions" in Gaul listed under that year were those commanded by Marcellus's colleague. They fought alongside Marcellus's forces in one of the latter's three battles and shared in the victory, yielding a total of six "victorious legions" in Gaul for that year. Under this system of reckoning, the total number of "victorious legions" in Table 3 is not the same as the total number of actual legions levied in the period 200–168 that won *one or more* victories. These legions numbered 80.

60. Over 10 percent: 193 and 173.
61. On the number of citizens in a legion, see above, n. 5; the figure for the average number of *socii* accompanying them is more problematic since the sizes of allied contingents attested in Livy in this period range between 5,250 and 10,500: Afzelius, *Römische Kriegsmacht*, 62–63. However, the average of the seventy-seven allied contingents whose size Livy gives is 6,688: Baronowski, "Roman Military Forces," 194, which may be rounded up to 6,700 for the present purposes.
62. 63 legions × 12,200 = 768,600 men × .042 = 32,281 deaths. This calculation assumes that each legion was at full strength when it entered combat, which cannot have been true where legions fought more than one major battle in a year, and is unlikely to have regularly been true in most other cases. But for the purposes of this illustration, this presumption may be accepted.
63. 63 legions × 12,200 = 768,600 × .0265 = 20,368.
64. (55,281, from Table 2, + 20,368 =) 75,649/(12,200 men × 110 [94 "victorious legions" plus 16 legions that suffered a defeat] = 1,342,000) = .056.
65. As indicated earlier, details of the consul Mancinus's defeat in 170 are lacking owing to the loss of the relevant part of Livy's text: compare Plut. *Aem.* 9.3. Note, however, that the *legati* sent by the senate to investigate conditions in Macedonia did not mention the casualties his army had suffered in this battle as a reason for its shortage of men: Livy 43.11.9–10. Sempronius's defeat in 196 apparently involved only Italian troops because the legions in both Spains had been demobilized in 197. The *patres* sent 8,400 *socii* to replace them, apparently 4,200 for each province: Livy 32.28.11. New legions went out only with the praetors of 196, one of whom replaced Sempronius: 33.26.3–4; compare Afzelius, *Römische Kriegsmacht*, 35–36. Each praetor received, in addition to one legion, a contingent of 4,300 allies, and this apportionment of forces suggested to Afzelius, reasonably enough, that each praetor was to command 5,500 Romans and 7,900 allies (infantry and cavalry, together), the usual ratio of legionaries to allies in this period: *Römische Kriegsmacht*, 66–67, and compare 79. If this is correct, then Sempronius commanded at most 4,200 Italian allies, and if he lost 16 percent of them, his casualties are likely to have amounted to no more than about 700 men.
66. Note, for example, the clash between Carthaginian and Roman cavalry near the mouth of the Rhone in 218 (Polyb. 3.45.2, and compare Livy 21.29.3), that between Roman and Macedonian cavalry in 200 (Livy 31.33.10), and the minor battles Flamininus fought against the Spartans in 195 (Livy 34.24.3, 28.3–11, 37.6–8, 39.1–13).

67. That is, eighty actual legions won one or more victories (above, n. 59) and sixteen actual legions suffered defeats.
68. Again for the purposes of this exercise, one may assume that these 139 legions were at full strength throughout the year, which obviously cannot have been true of legions that also fought one or more pitched battles in any single year. Thus 139 × 12,200 = 1,695,800 × .01325 = 22,469.
69. Deaths: 75,649 in defeats and victories (from n. 64 above) + 22,469 in minor battles + 1,647 from the draw in 186 = 99,765.
70. Afzelius, *Römische Kriegsmacht*, 34–79; Brunt, *Italian Manpower*, 422–24.
71. 99,765/3,847,140 = .0259.
72. Afzelius, *Römische Kriegsmacht*, 75–77.
73. Brooks, *Civil War Medicine*, 126–27; Brooks draws here on Dyer, *A Compendium of the War of the Rebellion*, 18, a more accurate compilation than United States, Surgeon-General's Office, *The Medical and Surgical History of the War of Rebellion*, on overall losses due to wounds and disease. Crimean War: Aldea and Shaw, "The Evolution of the Surgical Management," 561.
74. Hodge, "On Mortality," 225–26.
75. Corvisier, *L'Armée française*, 674–81. Note Duffy's comment, *Military Experience in the Age of Reason*, 245–46, that these gunshot and shrapnel wounds represent only those that their victims survived; consequently, sword wounds are overrepresented as a proportion of the total since, in Duffy's view, they were generally more survivable than cannon and gunshot wounds. See also Hodge, "On Mortality," 255–56, on the minuscule number of bayonet compared with gunshot wounds suffered by British forces at the battle of Inkermann in 1854. This is not, of course, to claim that the bayonet was not a weapon of considerable tactical importance even though it accounted for so few wounds; see Paddy Griffith, *Forward into Battle*, 20–30, especially 27.
76. Brooks, *Civil War Medicine*, 127: "Of the 246,712 cases of wounds reported in the Medical Records by weapons of war, 245,790 were shot wounds and 922 were saber and bayonet."
77. Duffy, *Military Experience in the Age of Reason*, 247, and compare Brooks's description of the damage caused by minié balls during the Civil War: *Civil War Medicine*, 75.
78. Quoted in Parker, *The Army of Flanders*, 168.
79. Brooks, *Civil War Medicine*, 75.
80. See ibid., 88, on wounds to the extremities, and 75–88, on conditions in operating rooms.
81. Keegan, *Face of Battle*, 113.

Notes to Pages 126–27

82. Salazar, *Treatment of War Wounds*, 17.
83. Polyb. 2.30.8, 33.3–5; 3.114.2, although the story is doubted by De Sanctis, *Storia*², 3.1:306–7 n. 119, and Walbank, *Commentary on Polybius*, 1:209. On Gallic swords generally, see Pleiner, *The Celtic Sword*, who discusses Polybius's claims about its supposed weakness at 157–64.
84. On the penetrating power of a composite bow, see Patterson, "The Archers of Islam," 83–86, and note the descriptions in Yadin, *The Art of Warfare in Biblical Lands*, 82–83. On English longbows, see Hardy, *Longbow*, especially 53–54.
85. Livy 31.39.12.
86. On the kinds of wounds suffered in hand-to-hand infantry combat in antiquity, see also Hanson, *The Western Way of War*, 210–18.
87. Livy 31.34.4–5.
88. Polyb. 6.22.1–23.16.
89. Polyb. 2.30.8; 6.23.1–5; compare the discussion in Goldsworthy, *The Roman Army at War*, 209–12, 218–19. Note, too, the interesting analysis of Gabriel and Metz, *From Sumer to Rome*, 58–59, 65, 70–72, on the effectiveness of the protection that shields afforded.
90. Again, descriptions of this confrontation are well known: Polyb. 18.29.1–30.11 on the array of spearpoints a phalanx presented; Plut. *Aem.* 19.1–2 on the appearance of the Macedonian phalanx, and compare Polyb. 29.17.1. Note also the effectiveness of the *scutum* against javelins (Livy 34.39.1–6), although admittedly in this case these were not being hurled with full force. On the other hand, one may suspect that the very high numbers of wounded that Alexander the Great's Macedonian infantry suffered in some battles, ten and twelve times the number of those killed in action, were in part due to the very small shields that his phalangites carried: figures and sources in Engels, *Alexander the Great*, 151.
91. Tetanus: Gabriel and Metz, *From Sumer to Rome*, 97. These statistics need to be treated with considerable caution since I cannot find their source. Gas gangrene: Aldea and Shaw, "The Evolution of the Surgical Management," 561, citing R. H. Major, *Fatal Partners*, but he, at 281, states that in the First World War, "In some engagements between 7 and 12 percent of the wounded developed gas gangrene."
92. Major, *Fatal Partners*, 281–84; Aldea and Shaw, "The Evolution of the Surgical Management," 561; Gabriel and Metz, *From Sumer to Rome*, 97. Compare Hare, "The Antiquity of Diseases," 115–16.
93. On the limited manuring possible in a Mediterranean climate, see White, *Ro-*

man Farming, 136–37; Brunt, Review of White, 156–58; Spurr, Arable Cultivation, 126–31.

94. On gangrene, Aldea and Shaw, "The Evolution of the Surgical Management," 561; on septicemia, Gabriel and Metz, From Sumer to Rome, 98 (however, I cannot find the source for this last statistic).

95. On the lack of doctors and hospitals in Roman armies in this period, Jacob, "Le Service de santé"; Harmand, L'Armée et le soldat, 201–5. But see, now, Salazar, Treatment of War Wounds, 76–79, on doctors and the treatment of war wounds in the middle and late republic, and in particular 94: "In the case of practical skills in treating wounds one can assume that a large number of people possessed them, mostly among those who had served in armies for a long time and among the rural population."

96. We are ignorant of just how soldiers would have treated wounds in this period, although see, generally, Salazar, Treatment of War Wounds, 9–54. Certainly, many of the techniques later used by the medical corps of the imperial legions were quite effective: Davies, Service in the Roman Army, 214–20, and compare Celsus De Med. 5.26.21–24, 7.5.1–5, 7.33.1–2, but how much of this can be assumed to have been common practice in the armies of the middle republic cannot be known. Still, it is worth pointing out that some of the simpler treatments recommended by Celsus, such as bathing a wound in wine or vinegar, would not have been beyond the ability of ordinary soldiers to perform and would have been quite effective in preventing infection. The polyphenol that occurs naturally in wine is a very powerful bactericide, while washing wounds in wine, according to Majno, The Healing Hand, 186–88, has been the most common type of treatment for wounds since the Greeks.

97. Servilius: Livy 45.39.14–19; Plut. Aem. 31.2; Marius: Sall. Iug. 85.29; Cato: Plut. Cato Mai. 1.5.

98. Livy 44.42.8; 28.34.2.

99. Sources for Alexander in Engels, Alexander the Great, 151. Caesar's ratios: 70 killed, 600 wounded: BC 1.46.5; 1,000 killed, 500 wounded: BH 31.10; compare BC 2.35.5: 600 killed, 1,000 wounded; and BC 3.53.3: every man in a cohort wounded. Compare, too, the 5 Romans killed and 100 wounded in a victory over Tigranes: Plut. Luc. 28.6.

100. Or at least their historians never report them doing so, perhaps for ideological reasons: compare Rosenstein, Imperatores Victi, 116–24.

101. As occurred in the aftermath of Cannae: Livy 22.51.5–9.

102. For British forces, see Hodge, "On Mortality," 233: 16,312 officers and men killed and 70,078 wounded in battles, a ratio of 1:4.3; for Union forces, see

Brooks, *Civil War Medicine*, 126–27: 67,058 officers and men killed in battle, 246,712 wounded, a ratio of 1:3.7.
103. On comparative rates of death among wounded, see above at n. 73.
104. For outright deaths in combat, see above, n. 69.
105. Livy 25.26.7–15; 28.46.15; 29.10.1–3; 41.5.11, 6.6.
106. App. *Pun* 99; *Iber.* 78. Compare also the illness among Lucullus's soldiers in Spain in 151 due, according to Appian, to an unusual diet that produced dysentery: *Iber.* 54. Note, too, that on completing his retreat from Parthia Antony mustered his army and found that he had lost 24,000 men on the march, of whom more than half had perished from disease: Plut. *Ant.* 50.1; compare Florus 2.20.8–10 and Reinhold, *From Republic to Principate*, 62, for further discussion.
107. Livy 23.34.11, 40.2; 26.26.4; 28.24.1; 34.10.5; 39.56.2; note, too, the serious illness of the son of the censor Q. Fulvius Flaccus while serving in Illyria in 172 (Livy 42.28.10–11) and the death of the proconsul Sex. Iulius Caesar from disease in 90 at the siege of Asculum (App. *BC* 1.48).
108. Brooks, *Civil War Medicine*, 128, and compare 106. For the Boer War, see Prinzing, *Epidemics Resulting from Wars*, 290.
109. Major, *Fatal Partners*, 239; compare Colin Clark, *Population Growth and Land Use*, 119–20 and table III.24, and Prinzing, *Epidemics Resulting from Wars*, 286–301, for specific numbers of deaths due to disease among soldiers in this period.
110. Major, *Fatal Partners*, 240; compare Prinzing, *Epidemics Resulting from Wars*, 296–99.
111. Grmek, *Diseases in the Ancient Greek World*, 89, 336.
112. Sallares, *The Ecology of the Ancient Greek World*, 241, 266–68.
113. Zinsser, *Rats, Lice and History*, 119–24; Littman and Littman, "The Athenian Plague," 266–75; Littman, "The Plague at Syracuse"; Poole and Holladay, "Thucydides and the Plague," 291–92; Sallares, *The Ecology of the Ancient Greek World*, 264–65.
114. Whether this epidemic was due to murine typhus, caused by *Rickettsiae typhia* (= *R. mooseri*), or epidemic typhus, caused by *R. prowazeki*, is unclear. Zinsser, *Rats, Lice and History*, 241, identifies it as epidemic typhus, whereas Hare, "The Antiquity of Diseases," 118, claims it as murine typhus and states that the first clear outbreaks of epidemic typhus were in Italy in 1505 and 1528. This is not to say, however, that typhus did not exist prior to the sixteenth century. It may well have existed in its nonepidemic, endemic form for centuries prior to this date before conditions arose that allowed epidemic typhus to develop and infect large human populations: Zinsser, *Rats, Lice and History*,

236–39. Whether either form of typhus was known in antiquity is unclear: Grmek, *Diseases in the Ancient Greek World*, 304, and Zinsser, *Rats, Lice and History*, 117–19; both refute claims of typhus in reference to specific cases described in Greek medical writers, although Grmek, *Diseases in the Ancient Greek World*, 348, remarks in passing that the disease was known to them, without, however, citing any specific case.

115. On the absence of smallpox in the Hippocratic corpus, see Grmek, *Diseases in the Ancient Greek World*, 89, and compare 86; Jackson, *Doctors and Diseases*, 23. Note also that the rash of elevated pustules on the mummy of the Pharaoh Rameses V may or may not have been caused by smallpox: Sandison and Tapp, "Disease in Ancient Egypt," 44. The epidemic at Syracuse: Diod. 2.14.70; Littman, "The Plague at Syracuse." Littman and Littman, "The Athenian Plague," 266–75, and Sallares, *The Ecology of the Ancient Greek World*, 244–54, have recently argued at length for smallpox as the cause of the plague at Athens. Scholars however are by no means anywhere near a consensus; see recent discussions in Morens and Littman, "Epidemiology of the Plague at Athens," and Bellemore and Plant, "Thucydides, Rhetoric and the Plague at Athens," both with references to earlier scholarship and diagnoses. See also Bellemore, Plant, and Cunningham, "Plague of Athens—Fungal Poisoning?," who suggest "alimentary toxic aleukia" as the cause, and compare Boucher's letter to the *Journal of the Neurological Sciences*, reviving the claims of Kobert that the cause was contamination by ergot of rye, another fungal infection.

116. Littman and Littman, "Galen and the Antonine Plague"; Sallares, *The Ecology of the Ancient Greek World*, 254–55.

117. On the impact of typhus as well as other diseases on armies, see Prinzing, *Epidemics Resulting from Wars*, especially 106–64; Zinsser, *Rats, Lice and History*, 159–65.

118. Malaria, although epidemic in the ancient world, is not a disease that significantly affects military mortality except in those cases in which armies composed of soldiers who had no experience of the disease fought in areas where it was prevalent, such as in Cuba during the U.S. war with Spain at the end of the nineteenth century or northern Greece during the conflict there in the First World War: Sallares, *The Ecology of the Ancient Greek World*, 279, and compare 271–81. Discussion of malaria in Italy: Brunt, *Italian Manpower*, 611–24, and see now Sallares, *Malaria and Rome*. Malaria was quite prevalent in Sardinia: ibid., 90–93, but after the disease had devastated a Roman army in 234 (Zon. 8.18), the *patres* apparently stationed troops there as infrequently as possible in this period: see Table 3 and compare Strabo 5.2.7. In general, the prevalence of malaria in Italy and elsewhere varies significantly according

to topography and ecology; see Sallares, *Malaria and Rome*, 43–114. Because Rome drew its soldiers from throughout the peninsula, soldiers' resistance to the disease will have varied considerably according to its prevalence or absence in their native regions. Consequently, malaria must have had a significant but unquantifiable impact on military mortality when soldiers from areas where the disease was largely absent and who were therefore vulnerable to infection (ibid., 42, 86) were stationed in areas where the disease was endemic. See, for example, Caesar's legions in Apulia in 48 B.C.: Caesar *BC* 3.2, and compare Sallares, *Malaria and Rome*, 263–68.

119. Hare "The Antiquity of Diseases," 118; Grmek, *Diseases in the Ancient Greek World*, 348; Jackson, *Doctors and Diseases*, 53, 103.

120. Brooks, *Civil War Medicine*, 8, 108–9. Compare the lack of sanitation at Verdun during the siege in 1792 that gave rise to a severe outbreak of dysentery, and similar conditions and results in British concentration camps during the Boer War: Prinzing, *Epidemics Resulting from Wars*, 92–93, 294–95.

121. According to Brooks (*Civil War Medicine*, 127, and compare 118), 27,050 Union soldiers died of typhoid and another 37,794 of diarrhea or dysentery (the two terms were used more or less interchangeably: 115, and compare 113–17) out of a total of 71,173 deaths from disease. Although Brooks elsewhere (126) puts Union deaths due to disease at 224,586 men, the discrepancy appears to result from the fact that in many cases muster reports simply indicate that a soldier had died of disease without further specifying its nature; compare fig. 20 on p. 107.

122. Note the outbreak of disease at the siege of Syracuse in 212 or 211: Livy 25.26.7–15. In 206 Hannibal remained inactive throughout the summer (Livy 28.12.1), suggesting that the epidemic began in his encampments and spread from there to the Romans: Livy 28.46.15, 29.10.1–3. For the outbreaks of disease at the sieges of Carthage and Numantia, see above, n. 106. Note, too, the outbreaks of disease that occurred in the camp of the Roman army in the course of its five-month siege of Agrigentum in 262 (Polyb. 1.19.1, and compare 1.17.6–18.11); in Massilia during Caesar's siege in 49 (Caesar *BC* 2.22.1), although Caesar attributes the Massilians' illness in part to eating old stockpiles of wheat (on which it is interesting to compare the fungal poisoning claimed by Bellemore et al., above, n. 115, as the cause of the plague at Athens); and in Pompey's camp during the siege of Dyrrachium in 47 (Caesar *BC* 3.49.2). The single exception to this pattern, which occurred in 178, represents a case of an epidemic among the civilian population breaking out in the army rather than as a result of factors intrinsic to military service. Italy in these years lay in the grip of a major epidemic, and the numbers of sick

and dead among the civilian population was so large that they slowed significantly the levying of the legions in 181 and 180 and again in 174 (Livy 40.19.3, 26.5–6, 36.14, 42.6; 41.21.5–6, 10–11; compare also the pestilence expiated in 187: Livy 38.44.7). It is also possible that sieges of large cities like Syracuse or Carthage brought soldiers drawn from rural backgrounds in Italy into contact with urban population concentrations large enough to have allowed some diseases to become endemic in them, meaning that while the urban population had developed an immunity through repeated exposure, the disease might become epidemic among those drawn from populations that had not. Yet the fact that during the siege of Syracuse, at least, sickness afflicted both sides (albeit the Romans more severely than the Syracusans) seems to tell against such a hypothesis: Livy 25.26.7.

123. Kipple et al., *Cambridge World History of Human Disease*, 1008–9.
124. Roth, *The Logistics of the Roman Army*, 7–44.
125. Schulten, *Numantia*, 4:63, and compare 4:53. As Schulten concedes, however, Marx and others translated *latrina* as "bath" rather than "latrine."
126. Davies, *Service in the Roman Army*, 211.
127. Schulten, *Numantia*, 4:63.
128. Polybius's long description of their camps is the clearest indication of how striking and unusual Roman practices were; see now Eckstein, "Physis and Nomos," 175–85, on Polybius's appreciation for the high degree of discipline and organization the Romans brought to their war making. Compare also Philip V's remarks (Livy 31.34.7–8) and Pyrrhus's (Plut. *Pyrr.* 16.5), although note that according to Front. *Str.* 4.1.14 (compare Livy 35.14.8), the Romans learned the art of castrametation from Pyrrhus. On water, see Roth, *The Logistics of the Roman Army*, 36. Compare the contaminated water that caused outbreaks of typhoid fever in the British army during the Boer War: Prinzing, *Epidemics Resulting from Wars*, 290–91.
129. App. *Pun.* 99.
130. Note, for example, Sall. *Iug.* 44.4, discussing operations in Africa in 110: *Nam Albinus . . . plerumque milites stativis castris habebat nisi cum odor aut pabuli egestas locum mutare subegerat*; and Caes. *BG* 8.52.2: *Ipse [Caesar] tantum itinerum faciebat, quantum satis esse ad mutationem locorum propter salubritatem existimabat.*
131. Compare Proctor, *Hannibal's March*, 17.
132. Parker, *The Army of Flanders*, 168–69; Corvisier, *L'Armée française*, 677–80, although at 671–73, he notes the evidence for disease and guesses that more French soldiers died of this in the eighteenth century than from enemy action.
133. Capua: Livy 23.18.9–16, 45.4–6, and compare 7.38.5 and Strabo 5.4.13; Asia

Minor: Livy 39.6.7–9; Sall. *Cat.* 11.5–6, and compare Plut. *Sull* 25.2; App. *Mith* 61. Note too the charges of corrupting the discipline of his troops at Syracuse in 205/4 leveled against Scipio Africanus: Livy 29.19.11–13, and compare 21.13 and, in general, Wheeler, "The Laxity of Syrian Legions," 239.

134. Note, for example, that even in Greece, Flamininus's troops remained in encampments during the winter of 197/6: Livy 33.29.2–6. The punishment meted out to the legions defeated by Pyrrhus in 280, to spend the winter under canvas (Front. *Strat.* 4.1.24), simply means that they were forbidden to build themselves more substantial barracks, as Marcellus's soldiers did in 215/14 (Livy 23.48.2), not that Roman legions normally passed the winter in towns. On sanitation in Roman towns during the imperial period, see Scobie, "Slums, Sanitation, and Mortality"; Laurence's critique of Scobie, "Writing the Roman Metropolis," is not convincing: see Scheidel, "Germs for Rome." Conditions in ancient cities during the second century B.C. would scarcely have been any better and were often worse.

135. Note that as an added disgrace the senate forbade the soldiers from the *legiones Cannenses* serving in Sicily during the Hannibalic War to winter in towns or construct camps within ten miles of one: Livy 26.1.10. This could suggest an expectation among the *patres* that soldiers would ordinarily do so, but it more likely means only that the senate recognized that these troops constituted a de facto garrison for the island and wanted to ensure that they did not enjoy the comforts that usually accompanied such duties.

136. Roman practice in the fourth century may have been different, however; see Oakley, *Commentary on Livy*, 2:365–66. Note also that Afzelius assumed that Rome regularly placed garrisons in southern Italy as well as Sicily and Sardinia in the second century; see below, n. 143. And Orosius at 5.11.4 mentions a garrison of 30,000 Roman soldiers at Utica who died of plague in 125.

137. See Kallet-Marx, *Hegemony to Empire*, 22–29, on the nature of Rome's *imperium* in this period.

138. Richardson, *Hispaniae*, 16–17; Richardson, *The Romans in Spain*, 11–16; Knapp, *Aspects*, 15–28, and compare 159.

139. Lindegren, "The Swedish 'Military State,'" 317; Parker, *The Military Revolution*, 53, with further bibliography at n. 30; Major, *Fatal Partners*, 98–100. On diseases in Central Europe during the Thirty Years' War, see Prinzing, *Epidemics Resulting from Wars*, 25–77. See, in general, Flinn, *The European Demographic System*, 52–53, on early modern armies as transmitters of disease.

140. Scheidel, *Measuring Sex, Age and Death*, 117–24, quotation from 124. Compare the conclusions of Balfour, "Statistical Data for Forming Troops," 194–97,

that in the first half of the nineteenth century British troops serving in Britain in peacetime enjoyed rates of mortality not much different from their civilian counterparts, although the rates of those serving abroad were significantly higher.

141. Note Caesar's remark at *BC* 3.2.3 that *gravis autumnus in Apulia circumque Brundisium ex saluberrimis Galliae et Hispaniae regionibus omnem exercitum valetudine temptaverat*, and compare 3.87.2 and Brunt, *Italian Manpower*, 620, for discussion, implying that despite ten years of war in Gaul that included lengthy winter encampments every year and several protracted sieges, disease had not presented a serious problem to his forces during that time. Compare the good health of Caesar's forces besieging Dyrrachium: Caesar *BC* 3.49.5. These cases strongly suggest the existence of well-developed practices that usually were able to keep illnesses arising from poor sanitation at bay. Compare also Harmand, *L'Armée et le soldat*, 205–9, for a generally positive assessment of sanitation in Roman armies of the first century B.C., with references to earlier scholarship.

142. Note, for example, the epidemic that broke out in the army during the Gallic campaign of 224 as a result, apparently, of heavy rains: Polyb. 2.31.10. The inclement weather may have seriously impeded the legions' mobility and so compelled them to remain in camp for extended periods where an inability to control runoff from the latrines caused by the deluge may have contaminated water supplies. Caesar likewise connects the illness that affected his troops in 48 with a period of bad weather in Italy (*BC* 3.2.3), and a retreat under harsh conditions through Media along with serious shortages of food and water formed the background to the heavy losses to disease in Antony's legions (Plut. *Ant.* 45.4–50.1; Dio 49.28.3–29.1; and above, n. 106).

143. Thus $116{,}580 \times .019 = 2{,}215 \times 33$ years $= 73{,}096$. $116{,}580 \times .026 = 3{,}031 \times 33 = 100{,}026$. This calculation does not take into account the possibility that significant numbers of Italian *socii* served as garrison troops in Sardinia, Sicily, and southern Italy, as Afzelius, *Römische Kriegsmacht*, 75–77, suggests.

144. Deaths from combat: $109{,}742$ (above, n. 69) $+ 73{,}096 = 182{,}838$ total deaths. $109{,}742 + 100{,}026 = 209{,}768$. $182{,}838/3{,}847{,}140$ total man-years of service $= .0475$. $209{,}768/3{,}847{,}140 = .0545$.

145. $116{,}580$ conscripts annually $\times .015 = 1{,}749 \times 33 = 57{,}707$ total "natural" deaths. $182{,}838$ total deaths among conscripts from all causes $- 57{,}707 = 125{,}131$ deaths beyond natural mortality; $209{,}768 - 57{,}707 = 152{,}061$. $125{,}131/3{,}847{,}140 = .0325$; $152{,}061/3{,}847{,}140 = .0395$.

146. See Brunt, *Italian Manpower*, 426–29, on the number of legions levied in this period. The one conflict in northern Italy, Ap. Claudius Pulcher's supposed

Notes to Pages 136–38

victory over the Salassi in 143, cost Rome 5,000 dead according to Oros. 5.4.7, and compare Dio frag. 74.2.

147. App. *Iber.* 45: 6,000 dead; *Iber.* 46: 4,000; *Iber.* 56: 6,000; *Iber.* 56: 9,000; *Iber.* 58: 7,000. On the inflation of Appian's figures, see Brunt, *Italian Manpower*, 663. Schulten's estimate of 150,000 to 200,000 Romans and Italians lost in the Spanish wars (*Geschichte von Numantia*, 5) is clearly much too high.

148. If we assume, once again, that all legions contained 5,500 citizens and 6,700 allies for a total of 12,200 conscripts per legion, 12,200 × 6.5 legions = 79,300 men. 79,300 × .0475 mortality rate × 34 years = 128,070; 79,300 × .0545 × 34 = 146,943. 79,300 × .0325 excess mortality rate × 34 = 87,627; 79,300 × .0395 × 34 = 106,500.

149. Rome's armies, on the assumptions used here, required 3,847,140 total man-years of service from 200 through 168 and 2,696,200 man-years between 167 and 133 for a total of 6,543,340. If each legionary or allied conscript served twelve years on average, then 545,278 men would have had to serve between 200 and 133 if none of them died during his enlistment. But if between 278,891 and 358,059 soldiers died during this period, then Rome's armies would have required a total of between 824,169 and 903,337 men over the course of these sixty-seven years. 278,891/824,169 = .338; 358,059/903,337 = .396.

150. So, for example, 12,200 men × .90 = 10,980. 10,980 × .0265 mortality rate in battle = 291 deaths per "victorious legion." 291 × 63 "victorious legions" = 18,331 deaths + 55,281 deaths from Table 2 = 73,612 total deaths, as compared with 75,644 deaths in the preceding calculations.

151. The other very high casualty rates occurred in Gaul and Liguria where in contrast to its practice prior to the Second Punic War Rome seems not to have employed local auxiliary forces in this period, probably on account of Gallic and Ligurian support for Hannibal.

152. Even if one were to put the Roman and Italian deaths in these battles at, say, 25 percent of soldiers involved, rather than the 44 to 50 percent given in Table 2, total number of combat deaths becomes: 55,281 deaths from Table 2 − 12,700 deaths + (26,800 men × .25 = 6,700) = 49,281.

153. On the "proverbial" character of 300, see Duncan-Jones, *Money and Government*, 16–19.

154. Multiples of 300 occur in 196, 191, 190, 185, and 173 and represent 13,200 deaths in eleven legions. Omitting them from the reckoning reduces the total deaths in Table 2 to 42,081 and adds eleven legions to those represented on Table 3, eight "victorious" and three defeated. If the average loss in "victo-

rious legions" was 2.65 percent, then $(76 + 8 =) 84 \times (12{,}200 \times .0265 = 323)$ = 27,132 deaths. Assuming that the number of deaths in the three defeated legions reflected the average in other defeated legions for which figures are preserved, then $(12{,}200 \times .16 = 1{,}952) \times 3 = 5{,}856$. Total unreported deaths would thus be $27{,}132 + 5{,}856 = 32{,}988$. The total from Tables 2 and 3 would then be $42{,}081 + 32{,}988 = 75{,}069$ as against 75,649 in the calculations above.

155. For a balanced appraisal of the annalists' general reliability, see Briscoe, *Commentary on Livy XXXI–XXXIII*, 11–12.
156. Livy 32.6.5–8.
157. App. *Iber.* 41–42.
158. Livy 40.50.6–7.
159. Livy 40.39.3; Richardson, *Hispaniae*, 102.
160. Triumphs for C. Calpurnius Piso and L. Quinctius Cripinus in 184; ovations for M. Fulvius Nobilior in 191; L. Manlius Acidinus Fulvianus in 185; and A. Terentius Varro in 182. See also Richardson, *Hispaniae*, 95–101.
161. In addition, note Livy's skepticism about exaggerated reports of fighting in 171 in Greece and the casualties resulting that he found in some of his sources: Livy 42.66.9–10, and compare 64.1–66.8. See also Luce's vigorous defense of Livy's abilities in his discussion of the historian's account of the trial of the Scipios: *Livy*, 92–104, especially 104.
162. Livy 32.6.8: *Ceteri Graeci Latinique auctores quorum quidem ego legi annales . . .*
163. See Luce, *Livy*, 64–65, on this fabrication; on Antias's alleged mendacity generally, 148–49, but compare the much harsher judgment in Oakley, *Commentary on Livy*, 1:89–92.
164. Even when Livy claims that consuls accomplished nothing worthy of note, this may only indicate that they fought no important battles, not that their provinces were peaceful: Harris, *War and Imperialism*, 258–59.
165. Certainly, the situation may be different for battles during the Second Punic War. De Sanctis, for example, sees Livy's accounts of fighting around Nola in 216–214 and in Spain from 217 to 211 as exaggerated accounts of minor fighting if not complete inventions: *Storia*^{2}, 3.2:225 n. 47; 236 n. 73; 244 n. 104; 249 n. 119. But in such cases it is not difficult to see a desire to glorify Rome's resistance to Hannibal as the annalists' motivation at a stage of the struggle when the prestige of Roman arms badly needed glorifying. Eagerness to burnish the image of a commander like Marcellus, to set him up as Rome's counterpart to Hannibal, may also have led to fabrication. Note De Sanctis's verdict on Livy's account at 27.12.7–14.15 of the clash between these two generals at Canusium in 209: *Storia*^{2}, 3.2:455–56, and compare Münzer,

RE, 3:2752–53; T. Broughton, *MRR*, 1:289 n. 5. But in the second century, when the republic's legions and its generals were going from victory to victory, similar embellishment would have been superfluous.

166. The Gallic war ended in 191 B.C.; subsequent campaigns in northern Italy were directed against the Ligurians, who generally proved to be much less formidable opponents, although this was not always the case, as the 3,000 or more Romans and Italians who died fighting against them in 173 discovered.

167. Losses for these six legions (Table 2: 200, 193, 191) total 8,484 men out of 79,600.

Chapter 5

1. See Chapter 3 at n. 123.
2. Livy 8.8.11–13.
3. The position of the *triarii* in the line of battle did not, of course, reduce their risk of death from disease or complications arising from those wounds they did sustain.
4. It is at least suggestive on this point that the allied contingents accompanying the legions were also organized in maniples: Polyb. 18.28.10 with Walbank's note, *Commentary on Polybius*, 2:586; compare Livy 8.8.14 with Oakley's note, *Commentary on Livy*, 2:475.
5. Saller, *Patriarchy*, 52, table 2.e, although if the population was growing, as is argued below, the proportion of only children will have been smaller.
6. Although Saller argues that most families would have been small and nuclear, it is well to recall that, at least in Asia Minor during the empire, a "family" might include a variety of members other than parents and children, including both other sorts of kin and nonrelated adults: Chapter 3, n. 160. One should not therefore presume that in every case a nuclear family would have represented the total universe of possible workers on a family farm.
7. Johnston, *Roman Law in Context*, 51–52, for a brief overview of testamentary disposition and the rules governing intestate succession. On the frequency with which ordinary Romans made wills, see Champlin, *Final Judgments*, 55 with n. 49, and Cherry, "Intestacy and the Roman Poor," both arguing that intestate succession was very common.
8. Again, we know nothing about laws governing succession to estates among Rome's Italian allies or in Latin colonies; possibly they operated under principles similar to those of an earlier stage of Roman law, embodied in the

Twelve Tables, under which the *gens* rather than cognate relatives inherited in default of agnate heirs.
9. Chapter 3 at n. 80.
10. No evidence survives, of course, to indicate how public land was apportioned within a community. For the procedure in pre-Soviet Russia, see Shanin, *The Awkward Class*, 36–37.
11. Brunt, *Italian Manpower*, 422, for the period 218–203; compare Toynbee, *Hannibal's Legacy*, 1:473–77. Brunt's figure depends on a low estimate for the number of deaths in 218–216, below any derived from the figures given in the sources. But if the arguments outlined here on the reliability of second-century casualty figures are correct, then the case for accepting at least some of the figures preserved in our sources for deaths in these battles is stronger than Brunt would allow.
12. Livy 31.13.6. Clearly this land did not lack heirs, for no one would buy it without clear title. Possibly the current owners did not want to work the land themselves, perhaps due to inconvenient locations that hindered efficient farming.
13. Compare Flinn, *The European Demographic System*, 21–22, 53–54.
14. As happened, for example, after the Black Death: see below, n. 23. Compare also De Neeve, *Peasants in Peril*, 35 n. 90.
15. Compare Livi Bacci, *Population and Nutrition*, 101–2, for a brief survey of the rise in wages following crises of mortality in medieval and early modern Europe.
16. See further below at n. 54. On child exposure during the empire, see Harris, "Child Exposure," with additional discussion of the problem in Scheidel, "Quantifying the Sources of Slaves," 164–67, and Harris's response in "Demography, Geography," 73–74. How common child exposure was during the middle republic is unclear.
17. Easterlin, "Population Change and Farm Settlement," especially 63–68.
18. That such considerations would not have been alien to the mind-set of a farmer in antiquity is suggested by Hesiod's well-known injunction, *Op.* 376–78 (trans. Lattimore): "One single-born son would be right to support his father's / house, for that is the way substance piles up in the household; / if you have more than one, you had better live to an old age."
19. Morley, *Metropolis and Hinterland*, 39, 46, and in general 33–50; on mortality in Rome, see now Scheidel, "Germs for Rome." Morley's calculations and, in particular, their implications for the overall population of Italy are strongly criticized by Lo Cascio, "Recruitment and the Size of the Roman Population,"

Notes to Pages 145–46

113–19, and "Populazione e risorse agricole," 223–29. However, Lo Cascio does not question Rome's dramatic growth in this period.

20. Sources Chapter 4, n. 122. Note, too, the epidemics of 165 and 142: Obseq. 13, 22.
21. On colonization, see Salmon, *Roman Colonization*, 95–109; Toynbee, *Hannibal's Legacy*, 2:192–210. Individual migration to the Po Valley may also have brought significant numbers from central and southern Italy to the region: Broadhead, "Migration and Transformation."
22. Note the colonies of Sipontium and Buxentum, founded in 194, which the consul of 186 found abandoned: Livy 34.45.2–3, 39.23.3–4. The cause may not have been that the settlers could not make a go of it but simply that they had more attractive options elsewhere, for example in the Po Valley. On appeal of this region to settlers, see Broadhead, "Migration and Transformation," 147–50.
23. Herlihy, *The Black Death*, 46–49; Gottfried, *The Black Death*, 135–40; Ziegler, *The Black Death*, 230–58.
24. See Scheidel, "Progress and Problems in Roman Demography," 52–57, for a fine summary and analysis of the debate over this problem.
25. Brunt, *Italian Manpower*, 69–81, especially tables VI and VII.
26. This accepts the figure for the number of citizens in this period transmitted in Livy *Per.* 60; see Brunt, *Italian Manpower*, 77–78.
27. See Appendix 7.
28. For the formula used in calculating the geometric rate of a population's annual growth, see, for example, Sanders, *Basic Demographic Measures*, 65–66.
29. 5,500 citizens multiplied by an average of roughly 6 legions per year from 167 through 124 = 33,000 citizens under arms per year. If the total mortality rate among legionaries was the same as that for the army as a whole, that is, 4.9 to 5.7 percent per year, then 1,617 to 1,881 citizens died each year, yielding a total of 71,418 to 82,764 deaths over 44 years. If, however, deaths among citizens fell below the proportion of a Roman army that legionaries typically composed, that is, 40 percent, due to fewer citizens being killed in combat (see, for example, Table 2 nn. 4 and 5), then the overall mortality rate among citizens would have been lower. Assuming that only 25 rather than 40 percent of those who died were citizens means the overall mortality rate for citizens might have been only 3.6 to 4.3 percent and consequently total deaths would have been 52,272 to 62,436. Again, if the overall mortality rate was only 75 percent of the earlier period (see Chapter 4 at n. 148), then the total number of deaths might have been as low as $(52{,}272 \times .75) = 39{,}204$. Inevitable mortality of 1.5 percent per year over 44 years reduces

the range of these figures by 21,780 to 60,984 at the high end and 17,424 at the bottom.

30. Livi Bacci, *The Population of Europe*, table 1.1 for examples. See also Gregg, *Population Growth*, 54–63.
31. Brunt, *Italian Manpower*, 77: "The census returns obviously do not seem to justify the view that Roman territory was suffering from depopulation. . . . Probably, however, [the census return for 135] comprised a much larger proportion not only of *proletarii* but of freedmen. . . . Probably then the totals returned by the censors conceal a marked decline in *assidui*"; compare 74: "The rate of natural increase since 203 had been just over 1 per cent. . . . But the natural increase must have been much less, since an unknown but probably significant proportion of citizens were freedmen, newly manumitted."
32. See Chapter 1 at n. 39.
33. For example, Brunt, *Italian Manpower*, 74: "The return of 168 is proved by its consistency with those that follow to have been more or less correctly transcribed."
34. This interpretation of the census figures for this period is implicit in Frank, "Roman Census Statistics," and is now argued forcefully by Lo Cascio in a variety of publications; see especially "The Size of the Roman Population," "Recruitment and the Size of the Roman Population," and "Populazione e risorse agricole." For other suggestions along these lines, see n. 64 below.
35. A good introduction to and discussion of nonparity-specific fertility limitation (also called "natural fertility") and the demographic transition is Frier, "Natural Fertility." See also Frier, "More Is Worse," 142–49.
36. Coale, "The Decline in Fertility in Europe," 1–10; Wrigley, "Fertility Strategy," 197–209, especially 209. On the mechanisms that control pretransitional populations, Wrigley and Schofield, *Population History of England*, 458–66.
37. As suggested to me by Walter Scheidel in a private communication.
38. See below for discussion.
39. Brunt, *Italian Manpower*, 72–74 and table VII. Brunt regards the census of 168 as unusually accurate compared with the two preceding it for which figures are reported, and so the rate of increase since 188 is calculated on the basis of this return. However, as a result of the possibility of corruption in the figures Livy preserves for the returns for 178 and 173 and his failure to record a figure for 183, it is not possible to determine whether a significant falloff in the rate of increase occurred between 183 and 178 as the hypothesis offered here would predict.
40. Brunt, *Italian Manpower*, 67–68 (emphasis added).
41. Ibid., 269–77.

42. Ibid., 276 n. 2.
43. Chapter 4, n. 122.
44. Livy 28.46.15, and compare 29.10.1.
45. Livy 27.23.6–7.
46. Brunt, *Italian Manpower*, 273–75.
47. However, one may readily concede that in particular regions like Bruttium or Campania the war may have prevented the sowing of crops or caused shortages of food for reasons specific to the conduct of operations there. See also the recent debate over the impact of the war on agriculture and the long-term prospects of the civilian population of southern Italy between Cornell, "Hannibal's Legacy," and Erdkamp, *Hunger and the Sword*, 270–96.
48. On a shortage of recruits, see Rosenstein, "Marriage and Manpower," 178–79, 186–87.
49. Livy 26.28.13.
50. See Appendix 5. On the proportion of Romans who never married, see also Chapter 3, n. 136.
51. See Treggiari, *Roman Marriage*, 44–45.
52. Livy 39.3.4–6, 41.8.6–12, and compare 38.36.5–6, 42.8.3.
53. On the *ius migrationis*, see Sherwin-White, *Roman Citizenship*, 34, 110; and Brunt, *Fall of the Roman Republic*, 95–96, although the latter does not believe the Roman population would have been substantially increased by immigration from Latin colonies: *Italian Manpower*, 72.
54. On this phenomenon in early modern Europe, see Flinn, *The European Demographic System*, 21.
55. Coale, "The Decline in Fertility in Europe," 8–10; on the complex contraceptive effects of length of breast-feeding and variations in suckling intensity during that time, see James Wood, *Dynamics of Human Reproduction*, 338–70. For examples of other means by which preindustrial couples might limit their fertility, see Wrigley, "Family Limitation," 264–66.
56. On breast-feeding: Soranus *Gyn.* 2.17–18, 36–40, 46–48, although the extent to which the practices he advocates prevailed among ordinary women in Italy during the middle republic is quite uncertain. One may compare, however, the requirement imposed upon Egyptian wet nurses by their employers to abstain from sexual intercourse during their employment (Bradley, "Sexual Regulations," 322), owing to beliefs about the detrimental effects of sexual relations on lactation. See Soranus *Gyn.* 2.19; Galen *de Sanitate Tuenda* 1.9.4–6; and Burguière, Gourevitch, and Malinas, *Soranos d'Éphèse, maladies des femmes*, 2:96 n. 152, for similar injunctions in other ancient medical writers; Bradley, "Sexual Regulations," 322; and Bagnall and Frier, *The Demography of*

Roman Egypt, 149, for additional discussions of the practice of postpartum abstinence by nurses and nursing mothers in Egypt and elsewhere.

57. Legal sources put the minimum age of marriage at twelve; Treggiari suggests that girls might marry soon after menarche, which she puts around thirteen: *Roman Marriage*, 39–42. However, Shaw's finding that during the empire women in the western provinces tended to marry in their late teens strongly suggests that physical maturity was not the sole factor governing when young women were felt to be ready to wed: "The Age of Roman Girls"; see also Saller, *Patriarchy*, 36–37.
58. Infanticide under such circumstances might also be expected to decline, but as there is little evidence for its extensive practice, it seems unwise to ascribe more than a marginal role to its decrease. See above, n. 16.
59. Malthus, *An Essay on the Principle of Population*, 243: "The havoc made by war in the smaller states of Italy, particularly during the first struggles of the Romans for power, seems to have been still greater than in Greece. Wallace . . . observes, 'On an accurate review of the history of the Italians during this period, we should wonder how such vast multitudes could be raised as were engaged in those continual wars till Italy was entirely subdued.' . . . But these wonders will perhaps be sufficiently accounted for, if we suppose, what seems to be highly probable, that the constant drains from wars had introduced the habit of giving nearly full scope to the power of population; and that a much larger proportion of births and of healthy children were rising into manhood . . . than is usual in other states not similarly circumstanced." On the stability of population in early modern Europe, see Flinn, *The European Demographic System*, 18–22.
60. On the colonies and distributions, see Chapter 2 at n. 186.
61. Certainly, Rome's *socii* contributed a substantial proportion of the colonists, but even so the republic's population will have declined under a "no-growth" demographic regime.
62. Desire for land in this period as a motive for Roman warfare: Harris, *War and Imperialism*, 60.
63. Sources for the legislation in T. Broughton, *MRR*, 1:225.
64. Evans, "Resistance at Home," 129–31; and, independently, Rathbone, "The Italian Countryside," 19; Lo Cascio, "Populazione e risorse agricole," 230–31; Paterson, "Hellenistic Economies," 376. See also now Morley, "The Transformation of Italy," 59–62. Compare Erdkamp, "Agriculture, Unemployment," 560.
65. However, merely the ordinary trickle of smallholders with too many sons or who had lost their farms owing to misfortunes of one sort or another cannot

have enabled Gracchus to take the unprecedented step of deposing a fellow tribune whose veto obstructed his law's passage and then secure its enactment without the senate's express approval in defiance of convention, as Horvath, "The Origins of the Gracchan Revolution," 107, supposes. Such a view fails to account for the enthusiasm, indeed the passions, that Gracchus's land law aroused and that evince the genuine social misery behind it.

66. As Last, *CAH*, 9:89, believed, followed by Shochat, *Recruitment and the Programme of Tiberius Gracchus*, 7–8. See also Boren, "The Urban Side of the Gracchan Economic Crisis," 890–902, although his contention of a severe economic slump, a decline in public building, and consequently widespread unemployment among the urban poor just prior to Gracchus's tribunate has been refuted by Coarelli, "Public Building in Rome," 1–9.
67. Diod. 34/35.6.1; App. *BC* 1.13-14, although these passages do not rule out significant support among the urban plebs: Astin, *Scipio Aemilianus*, 345–46; Badian, "Tiberius Gracchus," 717–18.
68. Plut. *TG* 8.7. See Nagle, "The Etruscan Journey of Tiberius Gracchus."
69. Chapter 1 at n. 13.
70. Echoes of contemporary polemic in, for example, Sall. *Iug.* 41; App. *BC* 1.7; Plut. *TG* 8.2–3; on the extent of dispossession by large estates, see Chapter 1, n. 11.
71. So, for example, Astin, *Scipio Aemilianus*, 196; Badian, "Tiberius Gracchus," 684–85.
72. Earl, *Tiberius Gracchus*, 31–40; Badian, "Tiberius Gracchus," 682–90. Compare the criticisms of Brunt, Review of Earl, 192. The argument here assumes that the beneficiaries of the *lex agraria* were exclusively Roman citizens rather than Romans and Italians, as is sometimes maintained, most recently by Richardson, "The Ownership of Roman Land."
73. Plut. *TG* 9.5.
74. As even Brunt, *Italian Manpower*, 76, concedes. See also Rich, "The Supposed Roman Manpower Shortage," 301 n. 65.
75. Evans, "Resistance at Home," 131.
76. The reduction in the minimum census from 4,000 to 1,100 (or, as a marginal correction in the manuscript has it, 1,500) *asses* is inferred from Cic. *Rep.* 2.40, and compare Gell. *NA* 16.10.10; Nonius 228L. Discussion of the date and motives in Rich, "The Supposed Roman Manpower Shortage," 309–16, with references to earlier scholarship; Rathbone, "The Census Qualifications," 139–46. However, the existence of any reduction at this date is strongly disputed by Lo Cascio, "Ancora sui Censi Minimi," 282, 286–88. The existence of a fragment of a speech by the Elder Cato, *de Tribunis Militum*,

which seems to advocate recruiting the plebs into the legions (Malcovetti, *ORF*² 58, M. Porcius Cato, frag. 152), is sometimes taken as indicative of a shortage of recruits: so Evans, "Resistance at Home," 131, but the speech is usually placed around 171; the meaning is ambiguous; and the context very uncertain; see Astin, *Cato the Censor*, 118 and n. 46. On the reluctance of recruits to serve in Spain, see Shochat, *Recruitment and the Programme of Tiberius Gracchus*, 56–60; Rich, "The Supposed Roman Manpower Shortage," 317–18. On increasing numbers of *assidui* claiming to be *proletarii* as the reason for the senate's reduction in the minimum census in 212/11, see Rosenstein, "Marriage and Manpower," 187–88.

77. App. *BC* 1.7, 9–11.
78. Lo Cascio, "Populazione e risorse agricole," 230–31, argues that the terms Appian uses at *BC* 1.7 and 1.11 to describe the situation that Tiberius sought to alleviate, δυσανδρία, and the result that he sought to achieve, εὐανδρία, cannot mean, as they are usually translated, "lack of men" and "abundance of men," since ὀλιγανδρία and πολυανδρία are the proper terms for these conditions. However, this line of reasoning is not persuasive in view of Appian's language at *BC* 1.7: τοὺς δ' Ἰταλιώτας ὀλιγότης καὶ δυσανδρία κατελάμβανε; and 1.9: Τιβέριος Σεμπρώνιος Γράκχος ... δημαρχῶν ἐσεμνολόγησε περὶ τοῦ Ἰταλικοῦ γένους ... φθειρομένου δὲ κατ' ὀλίγον εἰς ἀπορίαν καὶ ὀλιγανδρίαν. These phrases make it difficult to take Appian's use of εὐανδρια at *BC* 1.11 to describe Tiberius's aims as meaning anything other than "abundance of men." Note also Appian's use of πολυπλησία and his judgment of the situation the Romans found themselves in as a result of the repeal of the Gracchan laws: ὅθεν ἐσπάνιζον ἔτι μᾶλλον ὁμοῦ πολιτῶν τε καὶ στρατιωτῶν, both at *BC* 1.27.
79. The figures are listed conveniently in Brunt, *Italian Manpower*, 13; compare 70, table VI. These are of course raw figures and require at least some correction, on which see ibid., 70–71. Compare Earl, *Tiberius Gracchus*, 32; Rich, "The Supposed Roman Manpower Shortage," 303–4.
80. See Chapter 4 at n. 147.
81. Livy *Per.* 59; Suet. *Aug.* 89.2; Gell. *NA* 1.6.1, 6.7, but see McDonnell, "The Speech of Numidicus," who argues that Gellius's identification of the censor in question as Numidicus, who was elected in 102, is correct and that Macedonicus and his son each gave a speech on this topic during his censorship. See also Badian, "Which Metellus?"
82. Plut. *TG* 8.3–4; on the date: Astin, *Scipio Aemilianus*, 308.
83. App. *BC* 1.7. On the evidence of a declining population in this period, see also Astin, *Scipio Aemilianus*, 171–72.

84. Dupâquier et al., *Histoire de la population française*, 2:56–57, 65–68.
85. See Chapter 2 at n. 186.
86. Compare Brunt, *Italian Manpower*, 71: "The revised estimates [of the number of citizens] reach a peak in 146 . . . a decline sets in afterwards. The decline is perhaps to be accounted for by a progressive failure of the censors to register all the citizens; indeed, if the new peak in 124 is authentic, their failure was considerable."
87. Above, n. 76; on growing poverty among the citizens, see further below.
88. Compare the size of the allotments the colonists received at Bononia in 189, Aquileia in 181, and Luna in 177: Livy 37.57.8; 40.34.2; 41.13.5.
89. On Russia, see Shanin, *The Awkward Class*, 63–66; on the western United States, above at n. 17; and for antiquity, compare Hesiod, *Op.* 379–80.
90. Forrest, *Conscripts and Deserters*, 53–54.
91. Shanin, *The Awkward Class*, 1–2.
92. Ibid., 96–121.
93. Ibid., 85–88.
94. If we assume that the parents kept a quarter of the family's wealth for themselves and gave a quarter to each of three children when they wed. If the parents kept a larger share, the portions the children obtained would have been correspondingly reduced. For the pressures on a farmer with too many sons to provide for, see also the well-known warning of Hesiod, *Op.* 376–78, quoted above, n. 18.
95. Saller, *Patriarchy*, 51, table 3.1.d.
96. Ibid., 46.
97. Compare the hypothesis of Easterlin, discussed above at n. 17.
98. Below, n. 105.
99. Livy 24.11.5–9; 26.35.1–36.12.
100. Livy 1.43.5, expressed, as usual in Livy, in sextantal *asses* rather than the libral *asses* that were in use in 214 B.C.: see Rathbone, "The Census Qualifications," 138; Lo Cascio, "Ancora sui Censi Minimi," 284–85.
101. Compare Dohr, *Die italienischen Gutshöfe*, 35, 139–40.
102. For the sake of comparison, note that allotments to equestrians in Latin colonies in the second century ranged from 30 to 140 *iugera*: Livy 35.40.6; 40.34.2, and compare 35.9.7–9; 37.57.8.
103. Compare App. *BC* 1.7, 10, noting complaints about slaves' exemption from military service.
104. Compare Rathbone, "The Italian Countryside," 19.
105. On the lack of colonies between 181 and 128, see Chapter 2, n. 185. The de-

cision to end colonization in Italy, to the extent it was a conscious one, was probably due to lack of threats from enemies within Italy after that date and perhaps political jealousy within the aristocracy over the office of founding commissioners, as Salmon, *Roman Colonization*, 112–13, suggests. He also supposes that aristocrats preferred to exploit Rome's public land for their private benefit rather than found colonies on it, but this seems less likely in view of the lack of evidence for the growth of great slave-run estates in this period. Moreover, if, as suggested above, the *patres* gauged population trends among the citizens by the census returns, then the fact that the number of citizens these reported had not been growing in the years after 416 will have made the wisdom of new colonies appear less certain.

106. See MacMullen, "How Many Romans Voted?" We have no way of estimating the number of recipients Tiberius envisioned for his land allotments, but it is instructive to compare the somewhat better known scale of his brother's distributions. Gaius Gracchus's supporters passed legislation to establish at least three colonies, and the size of one of them, Junonia, is known: 6,000 settlers drawn from all over Italy, a number larger, we are told, than the enabling legislation had called for: App. *BC* 1.24. These figures may be compared with Livius Drusus's proposal in 122 for twelve colonies, probably to be composed exclusively of citizens, containing 3,000 colonists each: Plut. *CG* 9.2; App. *BC* 1.23; Salmon, *Roman Colonization*, 118–19. The scale of Drusus's proposal, in other words, amounted to considerably less than 10 percent of the citizenry if these numbered, as Brunt argues, 476,000 in 124 B.C. (*Italian Manpower*, 77–79), while the proportion of recipients among the citizens that Gaius foresaw for allotments in his colonies seems far below even that mark.

107. Val. Max. 4.4.8; Plut. *Aem.* 5.4, 28.7. Compare the extended family in which M. Licinius Crassus, cos. 70, was raised in the early first century (Plut. *Crass.* 1.1), where Plutarch's emphasis on the small house in which the future triumvir lived with his wife, his parents, his two brothers and their wives suggests that here, too, the reason was the lack of sufficient family wealth to set up the sons and their spouses in independent households in the appropriate style: Crook, "*Patria Potestas*," 117–18. However, not all extended families among the nobility need have been due to poverty: for example, Cato the Elder and his son: Plut. *Cato Mai* 24.1–2; Cn. Plancius: Cic. *Planc.* 29. I thank Professor Myles McDonnell for allowing me to see a copy of his paper touching on this topic in advance of publication.

108. Plut. *TG* 9.4–5.

109. App. *BC* 1.7.

110. App. *BC* 1.7 ascribes this tendency to the fact that free workers were liable to conscription, which certainly could also have been a factor. Rathbone, "The Slave Mode of Production in Italy," 164, 167–68, notes that the "industrial" employment of slaves long predates the appearance of the villa system and makes the attractive suggestion that the villas of the mid-first century B.C. represent the introduction of a previously developed urban version of the "slave mode of production" into the countryside.
111. Plut. *TG* 14.1; compare Astin, *Scipio Aemilianus*, 350–51, for additional sources and discussion of the ancient traditions on the purposes to which Tiberius intended to put this money.
112. Livy 40.38.6. The denomination of the units of money involved is not specified, but the fact that they were in silver implies that they were *denarii*.
113. Compare, cautiously, Salmon, *Roman Colonization*, 25; note similar steps by Antiochus the Great in planting a colony at Lysimachia: App. *Syr.* 1.1.
114. Livy *Per.* 58.
115. So Earl, *Tiberius Gracchus*, 94–95, although his interpretation is excessively cynical. See also Astin, *Scipio Aemilianus*, 351.
116. As Bernstein, *Tiberius Sempronius Gracchus*, 208–9, believes.
117. Broadhead, "Migration and Transformation," 159–65.
118. Rightly stressed by Harris, *War and Imperialism*, 58–65, 101–4.
119. Compare Wrigley and Schofield's "West African Situation," in *Population History of England*, xxiv.

Appendix 1

1. Dumont, *Servus*, 57–71.
2. Restrictions on freedmen tribal enrollment: Livy *Per.* 20. See Treggiari, *Roman Freedmen*, 42–47, for further discussion of this complex problem and additional bibliography.
3. Dumont claims that "le plus que parfait ne marque ici que l'antériorité logique d'une mesure générale par rapport à des exceptions": *Servus*, 59.
4. Treggiari, *Roman Freedmen*, 46.
5. *de Orat.* 1.38, a passage containing considerable hyperbole. Compare, too, *de Vir. Ill.* 57.3.
6. Scheidel, "The Slave Population of Roman Italy," 132–33.

Appendix 2

1. See Michels, *Calendar of the Roman Republic*, 16–18.
2. For example, De Sanctis, *Storia*², 3.2:84–85, 115–17, 130–31; Walbank, *Commentary on Polybius*, 1:412–13, 438; Sumner, "The Chronology of the Outbreak," 12 n. 38; Derow, "The Roman Calendar, 190–168 B.C.," 348; Brind'Amour, *Le Calendrier romain*, 157.
3. Derow, "The Roman Calendar, 218–191 B.C.," table at 272–73, and compare 274–81.
4. Ovid *Fasti* 6.763–70; Derow, "The Roman Calendar, 218–191 B.C.," 266–67.
5. Polyb. 14.2.1–2; Soltau, *Römische Chronologie*, 194–96, quotation from 195.
6. "In the Sahel the winter . . . lasts from October to April . . . April and May are transition months." Knox, *The Climate of the Continent of Africa*, 44, and compare the tables at 49–51, where the break in mean temperatures and rainfall between March and April is clear. Note also Great Britain, Naval Intelligence Division, *Tunisia*, 77: "Spring is only a short transition period between winter and summer. It is usually fresh, and April and May are often cold"; and 72–73: "The prevailing winds, which depend upon the pressure distribution, also change with the seasons. In winter (October to April) westerly winds . . . prevail in most parts of Tunisia and bring rain. . . . In summer (May to September) the winds blow towards the Saharan low pressures." Compare Great Britain, Naval Intelligence Division, *Algeria*, 1:116: "Spring is generally considerably cooler than autumn on the coast, April and May often being quite cold, and really hot weather is seldom experienced before the beginning of July," and Griffiths, *Climates of Africa*, 39 and table XXXIV.
7. Casson, *Ships and Seamanship*, 270–71.
8. Compare Walbank, *Commentary on Polybius*, 2:428; Pédech, *Méthode*, 464, especially n. 190 in reference to Polyb. 39.5.1.
9. See Marchetti, "La Marche du calendrier romain," 478–80, who argues for about eighty-five days; compare Briscoe, *A Historical Commentary on Livy XXXIV–XXXVII*, 19–20.
10. Derow, "The Roman Calendar, 218–191 B.C.," 274–81.
11. Polyb. 3.100.6–8; Desy, "Il grano dell'Apulia."
12. *Fasti* 6.767–68.
13. Polyb. 5.101.6.
14. Walbank, *Commentary on Polybius*, 1:433; compare Erdkamp, *Hunger and the Sword*, 132 n. 31.
15. Polyb. 3.56.3; Livy 21.38.1. Walbank, *Commentary on Polybius*, 1:392, suggests

that Polybius's general statement of the march's duration comes from the inscription Hannibal set up on the Lacinian promontory.

16. Polyb. 3.54.1, and compare Livy 21.35.6; Walbank, *Commentary on Polybius*, 1:390, for discussion.
17. Proctor, *Hannibal's March*, 40–45, 77.
18. Walbank, *Commentary on Polybius*, 1:365.
19. Polyb. 3.40.6: ἤδη δὲ τούτων συνῳκισμένων . . .
20. Asc. 3C, and compare Eckstein, "Two Notes," 255–72.
21. See, in general, De Sanctis, *Storia*² 3.1:241–60, especially 259.
22. Morgan, "Calendars and Chronology."
23. Ibid., 93.
24. Livy 43.15.1.
25. Livy 43.14.2–4.
26. Livy 43.14.9–10.
27. Sal. *Hist.* 2.98.4M (= 2.82.4 McGushin); compare Proctor, *Hannibal's March*, 56.
28. Polyb. 1.14.2, 6.19.5–7; Afzelius, *Römische Kriegsmacht*, 34–47.
29. Zon. 8.10.
30. Compare the legions of 292, Chapter 2 at n. 24.
31. Polyb. 1.17.9: ἀκμαζούσης . . . τῆς τοῦ σίτου συναγωγῆς; compare Walbank, *Commentary on Polybius*, 1:70.
32. Zon. 8.10; Great Britain, Naval Intelligence Division, *Italy*, 1:529–31, and compare 422.
33. Zon. 8.10.
34. Polyb. 1.24.9–10.
35. Lazenby, *First Punic War*, 67–68, 72; Thiel, *History of Roman Sea-Power*, 187–89; De Sanctis, *Storia*², 3.1:123–26.
36. Chapter 2 at n. 96.
37. Brind'Amour, *Le Calendrier romain*, 176–78; Holzapfel, *Römische Chronologie*, 98–99, and compare 290.
38. On the location: Salmon, *Samnium*, 266.
39. Livy 10.25.10–11.
40. Brind'Amour, *Le Calendrier romain*, 176. For dates down to 329 B.C., see Oakley, *Commentary on Livy*, 2:612–14.
41. Holzapfel, *Römische Chronologie*, 94–98; Soltau, *Römische Chronologie*, 302–3.
42. Holzapfel, *Römische Chronologie*, 99; compare Brind'Amour, *Le Calendrier romain*, 177–78; Fränkel, *Der Amtsantritt der römischen Consuln*, 107; Livy 10.45.11, 46.1–2.
43. Proctor, *Hannibal's March*, 17.

44. Salmon, *Samnium*, 272–73.
45. Great Britain, Naval Intelligence Division, *Italy*, 1:529; compare 426–29.
46. Livy 10.45.11, 46.1.
47. For the earlier date, see Skutsch, *The Annals of Q. Ennius*, 311–13, for discussion and earlier scholarship; on the later, see Frier, *Libri Annales*, 115–17, also with discussion and bibliography.

Appendix 3

1. De Ligt, "Tenancy under the Republic," quotation from 378.
2. Crone, *Pre-industrial Societies*, 42.
3. Fergus Millar, "Political Power in Mid-Republican Rome"; Fergus Millar, "Political Character of the Classical Roman Republic"; compare Crone, *Pre-industrial Societies*, 38–57, 68–69, for typical relations between preindustrial states and the societies they ruled.
4. Compare Harris, *War and Imperialism*, 60–61.
5. Livy 26.35.5.
6. Although poor tenants might not qualify as *assidui* (but see Chapter 2 at n. 178 on the very small amount of land that formed the threshold for assiduate status), they certainly were liable to serve as rowers in the fleet, of which the republic was in desperate need during the war, so much so that the senate was led to press slaves into service: see Appendix 5. On the question of whether tenants might be liable to serve in the legions, compare the interesting point of Badian, "Tiberius Gracchus," 673.
7. App. *BC* 1.7.

Appendix 4

1. Mommsen, *Römische Staatsrecht*, 1:506 n. 2; Brunt, *Italian Manpower*, 399 n. 3. Evidence: Peter, *HRR*, 1:308, Tubero frag. 4 (= Gell. *NA* 10.28.1); Livy 25.5.8, 27.11.5, 27.11.15. Compare Livy 22.57.9, where the implication is that seventeen-year-olds, that is, raw recruits, had not been drafted in 216 for the extraordinary effort at Cannae since the senate imagined that lack of experience was to blame for Rome's earlier defeats: Polyb. 3.106.5 and see Rosenstein, *Imperatores Victi*, 98–99.
2. Compare Bradley, *Slaves and Masters*, 87, on the age at which a master could manumit a slave.

Appendix 5

1. Brunt, *Italian Manpower*, 64–66, 417–20.
2. Rich, "The Supposed Roman Manpower Shortage," 311; see also Chapter 1, n. 60.
3. So, for example, if in the period 200–167 the pool of *assidui* over thirty comprised only about 50,000 men instead of 93,000, then roughly 70 percent of them will have had to serve at some point if each *triarius* served for three years: see Chapter 3, n. 130.
4. See Rosenstein, "Marriage and Manpower," for detailed arguments in support of the following points.
5. Livy 24.11.5–9, 26.35.1–36.12; Brunt, *Italian Manpower*, 65, and compare Libourel, "Galley Slaves," 116–19. Shochat, *Recruitment and the Programme of Tiberius Gracchus*, 66, similarly argues that the senate's decision to compel citizens resident in maritime colonies to serve as rowers in 190 indicates a shortage of *proletarii* at that date: Livy 36.3.4–6.
6. Although Cicero and Dionysius of Halycarnassus claim that the single century of *proletarii* in the *comitia centuriata* already contained a majority of citizens in the reign of Servius Tullius, this statement, if it has any validity at all, clearly reflects conditions in the late republic: Cic. *Rep.* 2.40; Dion. Hal. 4.18.2, 7.59.6.
7. See Chapter 2 at n. 178.
8. Polyb. 2.24.1–17; that the figures Polybius cites in this passage derive from Pictor is clear from Eutrop. 3.5 and Oros. 4.13.6. On the difficult question of the interpretation of the figures, see Walbank, *Commentary on Polybius*, 1:196–203; Brunt, *Italian Manpower*, 44–60; Baronowski, "Roman Military Forces," 181–202, all with references to earlier scholarship. Their conclusions are challenged, however, by Lo Cascio, "Recruitment and the Size of the Roman Population," 127–33.
9. Livy 24.18.7–8: *quibus neque vacatio iusta militiae neque morbus causa fuisset.* See Rosenstein, "Marriage and Manpower," for a detailed analysis of this passage.
10. Brunt, *Italian Manpower*, 64–66. Note, however, Lo Cascio's much higher estimates, "Recruitment and the Size of the Roman Population," 127–33.
11. 285,000 men over seventeen implies a total male population of 452,000 according to the age structure of Coale-Demeny² Model West 3 Female, of whom 44 percent would be between seventeen and forty-six, or 199,000 men.
12. Brunt, *Italian Manpower*, 64–66, claims that so many *iuniores* had failed to serve because they were *proletarii* and so were ineligible to serve in the legions.

But Livy does not say that those who had been excused from serving got their exemptions because they were *proletarii*, and, as discussed above, no more than about 10 percent of all *iuniores* would have been rated as *proletarii* at this point.

13. Note that the crisis following Cannae had already forced the city to levy men of marginal physical strength: Livy 23.25.8, and compare 22.57.9-12, 23.14.2-4.
14. Mommsen, *Römische Staatsrecht*, 3.1:241-44; Brunt, *Italian Manpower*, 391.
15. On the political popularity of consuls who granted exemptions, note Livy 43.14.3.
16. Livy 22.57.9-11; 23.14.2-4.
17. Brunt, *Italian Manpower*, 66; on the date at which the census was lowered, see above, Chapter 1 at n. 58.
18. Livy 25.5.5-9, 27.38.2-5. The decision to fix the minimum census required for military service about 27 percent lower in 212/11 was probably taken because, with *vacationes* no longer easily obtainable and demands for soldiers increasing, some potential recruits began seeking to escape service on the pretext that they were too poor. Compare Rathbone, "The Census Qualifications," 145. Note, too, the lack of men and of money to pay them that caused twelve Latin colonies in 209 to refuse to supply the contingents required of them for the allied components of Rome's armies: Livy 27.9.7, 13, and compare 9.1-10.10.

Appendix 6

1. Steinwender, "Altersklassen und reguläre Dienstzeit," followed by Brunt, *Fall of the Roman Republic*, 267 n. 123, and *Italian Manpower*, 399-401. Compare also Toynbee, *Hannibal's Legacy*, 2:72-80. Little can be built upon the case of Sp. Ligustinus's six years of service from 200 to 195 since his voluntary service with various armies thereafter may have precluded his being drafted in those years: Livy 42.34.5-11.
2. Compare the dismissal of the consul Livius's legions after their victory at the Metaurus in 207 (Livy 28.10.4) and, similarly, the discharge of the legions of most of the consuls who triumphed during the second century down to 167: for example, Marcellus's in 196, Cato's and Flaminius's in 195, Nasica's in 191, Vulso's and Nobilior's in 187, Paullus's in 181, probably Flaccus's in 179, Pulcher's in 177, Gracchus's and Lepidus's in 175, and Paullus's in 167.
3. Livy 26.28.6-13.

4. De Sanctis, *Storia*², 3.2:614–15, table I, and Toynbee, *Hannibal's Legacy*, 2:650–51, table III, reckon that the legions they number as 19, 23, and 24 were each levied in 214. Marchetti, *Histoire économique*, table following p. 94, holds that his legion number 3 (= De Sanctis/Toynbee number 19) was levied in 216, but this depends on his identification (pp. 52–53) of the forces under the consul Varro's command at Tarentum in Apulia at the end of that year with the *legio classica* mentioned at Livy 22.57.8. But this is uncertain: De Sanctis, *Storia*², 3.2:222 n. 38; Brunt, *Italian Manpower*, 648–51, especially 649 n. 5.
5. Livy 39.30.1–31.18; 40.30.1–33.9.
6. Livy 39.38.4–5; 40.35.3–7.
7. Livy 39.38.3–12; 40.36.1–12.
8. Compare App. *Iber.* 78, where in 140 soldiers serving in Spain were replaced after six years there. Taylor, "Forerunners of the Gracchi," 24, suggests that an unknown tribunician law was passed sometime during the middle of the second century limiting overseas service to six years, but this is unlikely.
9. On the demand for discharge: Livy 28.24.7; on their length of service in Spain: Polyb. 11.28.3–6; Livy 28.25.6, 28.13. See Chrissanthos, "Scipio and the Mutiny at Sucro," 174 and 179–80, for discussion.
10. Livy 40.35.6–7. The literary tradition on the mutiny in 206 emphasized the soldiers' indiscipline: for example, Livy 28.24.5–11. Although this is probably false (Chrissanthos, "Scipio and the Mutiny at Sucro," 181), such charges nevertheless may well have circulated and been believed subsequently within the senatorial class.
11. On this and the preceding point about the hardship of serving in Spain, see Smith, *Service in the Post-Marian Roman Army*, 7 n. 6; Astin, *Scipio Aemilianus*, 169–70; Harris, *War and Imperialism*, 44–45.
12. So, acutely, Harris, *War and Imperialism*, 45, who also thinks terms of service of twelve or fourteen years more credible than six or seven, noting the eighteen-year stint in Spain of the soldier mentioned by Lucilius: 490–491M. I accept, along with most scholars, the usual emendation of the figure "six" in Polybius's text at this point to "sixteen": discussion in Walbank, *Commentary on Polybius*, 1:698; see also Harris, *War and Imperialism*, 45; Brunt, *Italian Manpower*, 399–401. Harris also rightly deems unpersuasive Brunt's claim that Polybius's sixteen-year maximum was anachronistic in the second century. This is particularly so since Brunt bases this assertion (at 399) on the fact that Polybius refers at 6.19.2–4 to campaigns, not years of service. Consequently, Brunt claims, this must refer to an earlier period when wars were short, usually lasting no more than six months. But as shown in the first part

of this study, brief campaigns during the summer season had not been a feature of Roman warfare for at least a century and a half when Polybius wrote. Yet Polybius's source for his description of the Roman army in book 6, while outmoded in some respects, surely cannot be almost 150 years out of date. On Polybius's source for this account and the anachronistic elements therein, see Rawson, "Literary Sources," 13–31 (= *Roman Culture and Society*, 34–57).

13. Compare Humphreys, *Anthropology and the Greeks*, 164–68, especially 165: "Alongside the seasonal rhythm of production and destruction in hoplite warfare ran the longer-term cycle of a man's life, with youth as the period of search for wealth abroad, followed by home production in the *oikos* between approximately the ages of thirty and sixty. It was in the young men's world of unofficial enterprises and liminal status, I suggest, that the ethos and structure of the Homeric *hetaireia* continued to flourish." Compare also Ducrey, "Remarques sur les causes du mercenariat," 115–23.

Appendix 7

1. Chapter 4 at n. 144.
2. See Chapter 4, n. 61.
3. Total noncombat mortality: $(73,096-100,026) \times .4 = 29,238-40,010$.
4. For the total number of combat deaths, see Chapter 4 at n. 104.
5. In 181 Livy records 200 Roman and 830 Latin deaths in battle, along with 2,400 *auxilia* killed (40.32.7); the 200 Romans represent about 20 percent of the total Roman and Italian casualties. A battle in the following year claimed 472 Romans, 1,019 Latins, and 3,000 Spanish allies (Livy 40.40.13). Here the Romans constituted about 32 percent of the combined Roman and Italian deaths.
6. Livy 45.42.8, and compare a battle against the Boii in 193, where the consul held the legions in reserve while other forces bore the brunt of the attack (Livy 35.5.1–6). On the other hand, in some cases the legions may have born the brunt of the fighting, as they did at Zama (Polyb. 15.9.6–9, 13.7–14.6). Consequently, it would be unwise to assume that Roman casualties always fell as low as 25 percent of the total.
7. If 8.7 legions on average served each year and each contained 5,500 citizen legionaries, then each year on average 47,850 men served between 200 and the end of 168. 56,674–67,446 total deaths in this period yield an average of 1,717 to 2,043 deaths per year. $1,717/47,850 = .0358$; $2,043/47,850 = .0426$.
8. This figure assumes a total "normal" mortality among all conscripts, both Ro-

Notes to Page 192

man and Italian, of 57,070 from 200 through 168 (see Chapter 4 at n. 144), and that if Romans made up 40 percent of all conscripts their normal mortality would have totaled 23,083. Thus 56,674 less 23,083 = 33,591; 83,907 − 23,083 = 59,824.

9. Livy 30.18.14; Zama: Polyb. 15.14.9, Livy 30.35.5, although compare Appian *Lib.* 48, who puts the Roman losses at 2,500 killed.
10. Livy reports a fourth battle in this period, but it is unlikely to have been more than a skirmish: Livy 30.19.11–12.
11. The size of Scipio's army is disputed. I follow Brunt, *Italian Manpower*, 672–73, who concludes that it numbered somewhat over 16,000 Roman and Italian infantry and 1,600 cavalry. See Lazenby, *Hannibal's War*, 203, however, for a significantly higher estimate.
12. For Scipio's army, see Brunt, *Italian Manpower*, 672–73, 680, and compare Baronowski, "Roman Military Forces," 195–96.
13. 3,200 citizens × 17 = 54,400 × 3 = 163,200 man-years of service. 163,200 × .019 = 3,101 total citizen deaths; 163,200 × .026 = 4,243. On the assumption that Roman combat deaths represented 25 percent of the total and mortality due to disease was 1.9 percent: 1,174 + 3,101 = 4,275; if combat deaths were 50 percent and mortality from disease was 2.6 percent: 2,347 + 4,243 = 6,590.
14. Average number of citizens with the legions per year 203–201: 54,400. 54,400 × .015 = 816 × 3 years = 2,448 ordinary mortality.

Bibliography

Adcock, Frank E. *The Roman Art of War under the Republic.* Martin Classical Lectures, vol. 8. Cambridge: W. Heffer and Sons, 1960.
Afzelius, Adam. *Die römische Eroberung Italiens (340–264 v. Chr.).* Acta Jutlandica 14:3. Copenhagen: Universitetsforlaget i Aarhus, 1942. Reprinted in *Two Studies on Roman Expansion* (New York: Arno Press, 1975).
———. *Die römische Kriegsmacht während der Auseinandersetzung mit den hellenistischen Grosmächten.* Acta Jutlandica 16:2. Copenhagen: Universitetsforlaget i Aarhus, 1944. Reprinted in *Two Studies on Roman Expansion* (New York: Arno Press, 1975).
Aldea, Peter A., and William D. Shaw. "The Evolution of the Surgical Management of Severe Lower Extremity Trauma." In *Lower Extremity Trauma and Reconstruction*, edited by William D. Shaw, 549–69. *Clinics in Plastic Surgery* 13, no. 4 (October 1986).
Ampolo, Carmine. "Le condizioni materiale della produzione. Agricoltura e paesaggio agrario." *Dialoghi di Archeologia*, n.s., 2 (1980): 15–46.
Anderson, John K. *Military Theory and Practice in the Age of Xenophon.* Berkeley: University of California Press, 1970.
André, Jacques. *L'Alimentation et la cuisine à Rome.* Paris: C. Klincksieck, 1961.
Andreski, Stanislav. *Military Organization and Society.* Berkeley: University of California Press, 1968.
Arthur, Paul. *Romans in Northern Campania: Settlement and Land-Use around the Massico and the Garigliano Basin.* Archaeological Monographs of the British School at Rome, no. 1. London: British School at Rome, 1991.
Ash, Stephen V. *When the Yankees Came: Conflict and Chaos in the Occupied South, 1861–1865.* Chapel Hill: University of North Carolina Press, 1995.
———. *Middle Tennessee Society Transformed, 1860–1870: War and Peace in the Upper South.* Baton Rouge: Louisiana State University Press, 1988.
Astin, Alan E. *Cato the Censor.* Oxford: Oxford University Press, 1978.
———. *Scipio Aemilianus.* Oxford: Oxford University Press, 1967.
Azzi, G. "Il clima del grano in Italia." *Nuovi annale del ministro per l'agricoltura,* no. 3 (September 30, 1922): 453–624.

Badian, Ernst. "Which Metellus? A Footnote to Professor Barchiese's Article." *AJAH* 13 (1988 [1997]): 106–12.

———. "Tiberius Gracchus and the Beginning of the Roman Revolution." *ANRW* 1.1 (1972): 668–731.

Bagnall, Roger S., and Bruce W. Frier. *The Demography of Roman Egypt.* Cambridge: Cambridge University Press, 1994.

Bairoch, Paul. "Urbanization and the Economy in Preindustrial Societies: The Findings of Two Decades of Research." *Journal of European Economic History* 18 (1989): 239–90.

Balfour, Edward. "Statistical Data for Forming Troops and Maintaining Them in Health in Different Climates and Localities." *Journal of the Statistical Society of London* 8 (1845): 193–209.

Barker, Graeme. *A Mediterranean Valley: Landscape Archaeology and Annales History in the Biferno Valley.* London: Leicester University Press, 1995.

Barker, Graeme, John Lloyd, and Derrick Webley. "A Classical Landscape in Molise." *PBSR* 46 (n.s. 33) (1978): 35–51.

Baronowski, Donald. "Roman Military Forces in 225 B.C. (Polybius 2.23–4)." *Historia* 42 (1993): 181–202.

Bellemore, J., and I. M. Plant. "Thucydides, Rhetoric and the Plague at Athens." *Ath* 82 (1994): 385–401.

Bellemore, J., I. M. Plant, and L. M. Cunningham. "Plague of Athens—Fungal Poisoning?" *Journal of the History of Medicine and Allied Sciences* 49 (1994): 521–45.

Beloch, Julius. *Römische Geschichte bis zum Beginn der punischen Kriege.* Berlin: De Gruyter, 1926.

———. *Die Bevölkerung der griechisch-römischen Welt.* Historische Beiträge zur Bevölkerungslehre. T. 1. Leipzig: Duncker & Humblot, 1886.

Benería, Lourdes, and Gita Sen. "Accumulation, Reproduction, and Women's Role in Economic Development: Boserup Revisited." *Signs* 7 (1981): 279–98.

Beringer, Richard E. *Why the South Lost the Civil War.* Athens: University of Georgia Press, 1986.

Berlin, Jean V. "Did Confederate Women Lose the War? Deprivation, Destruction, and Despair on the Home Front." In *The Collapse of the Confederacy*, edited by Mark Grimsley and Brooks D. Simpson, 168–93. Lincoln: University of Nebraska Press, 2001.

Bernstein, Alvin H. *Tiberius Sempronius Gracchus: Tradition and Apostasy.* Ithaca, N.Y.: Cornell University Press, 1978.

Boren, Henry C. "The Urban Side of the Gracchan Economic Crisis." *AHR* 63 (1958): 890–902.

Boserup, Ester. *Woman's Role in Economic Development.* London: Allen & Unwin, 1970.

Boucher, M. Letter to the Editor. *Journal of the Neurological Sciences* 150, issue 1001 (September 1997): S172.

Bradley, Keith R. *Slavery and Rebellion in the Roman World, 140 B.C.–70 B.C.* Bloomington: Indiana University Press, 1989.

———. *Slaves and Masters in the Roman Empire.* Collection Latomus, vol. 185. Brussels: Latomus, Revue d'Etudes Latines, 1984. Reprint, New York: Oxford University Press, 1987.

———. "Sexual Regulations in Wet-Nursing Contracts from Roman Egypt." *Klio* 62 (1980): 321–25.

Brind'Amour, Pierre. *Le Calendrier romain: recherches chronologiques.* Ottawa: Editions de l'Universitá d'Ottawa, 1983.

Briscoe, John. *Commentary on Livy, Books XXXIV–XXXVII.* Oxford: Clarendon Press, 1981.

———. *Commentary on Livy, Books XXXI–XXXIII.* Oxford: Clarendon Press, 1973.

———, ed. *Titi Livi Ab urbe condita libri XLI–XLV.* Stuttgart: B. G. Teubner, 1986.

Broadhead, William. "Migration and Transformation in North Italy in the 3rd–1st Centuries BC." *BICS* 44 (2000): 145–66.

Brooks, Stewart. *Civil War Medicine.* Springfield, Ill.: C. C. Thomas, 1966.

Broughton, A. L. "The *Menologia Rustica.*" *CPh* 31 (1936): 353–56.

Broughton, T. Robert S. *Magistrates of the Roman Republic.* 3 vols. Philological Monographs, no. 15. New York: American Philological Association, 1951–86.

Bruns, Carolus, ed. *Fontes iuris romani antiqui.* Tübingen: I. C. B. Mohrius, 1909.

Brunt, Peter A. *The Fall of the Roman Republic and Related Essays.* Oxford: Oxford University Press, 1988.

———. Review of *Roman Farming*, by K. D. White. *JRS* 62 (1972): 153–58.

———. *Italian Manpower.* London: Oxford University Press, 1971.

———. *Social Conflicts in the Roman Republic.* New York: W. W. Norton, 1971.

———. Review of *Tiberius Gracchus: A Study in Politics*, by Donald C. Earl. *Gnomon* 37 (1965): 189–92.

———. "The Army and the Land in the Roman Revolution." *JRS* 52 (1962): 69–86. Reprinted with changes in Brunt, *The Fall of the Roman Republic*, 240–80.

Brush, Stephen B. "The Myth of the Idle Peasant: Employment in a Subsistence Economy." In *Peasant Livelihood: Studies in Economic Anthropology and Cultural Ecology*, edited by Rhoda Halperin and James Dow, 60–78. New York: St. Martin's Press, 1977.

Burdese, Alberto. *Studi sull'Ager Publicus*. Università di Torino. Memorie dell'Istituto Guridico, ser. 2, memoria 75. Turin: G. Giappichelli, 1952.

Burguière, Paul, Danielle Gourevitch, and Yves Malinas, eds. *Soranos d'Ephèse, maladies des femmes*, tome II, livre II. Paris: Les Belles Lettres, 1990.

Burke, Emily. *Reminiscenses of Georgia*. Oberlin, Ohio: J. M. Fitch, 1850.

Bynum, Victoria E. *Unruly Women: The Politics of Social and Sexual Control in the Old South*. Chapel Hill: University of North Carolina Press, 1992.

Calderone, S. "Di un antico problema di esegesi polybiana." *AAntHung* 25 (1977): 383–87.

Calderone, S., I. Bitto, L. De Salvo, and A. Pinzone. "Polibio 1,11,1 sq." *Quaderni Urbinati di Cultural Classica*, n.s., 7 (1981): 7–78.

Camilli, A., L. Carta, T. Conti, A. De Laurenze, and M. De Simone. "Recongizioni nell'Ager Faliscus." In *Settlement and Economy in Italy, 1500 BC–AD 1500: Papers of the Fifth Conference of Italian Archaeology*, edited by Neil Christie, 395–402. Oxbow Monograph 41. Oxford: Oxbow Books, 1995.

Camilli, A., and B. Vitali Rosati. "Nuove Ricerche nell'Agro Capenate." In *Settlement and Economy in Italy, 1500 BC–AD 1500: Papers of the Fifth Conference of Italian Archaeology*, edited by Neil Christie, 403–13. Oxbow Monograph 41. Oxford: Oxbow Books, 1995.

Casson, Lionel. *Ships and Seamanship in the Ancient World*. Princeton: Princeton University Press, 1971.

Cecil-Fronsman, Bill. *Common Whites: Class and Culture in Antebellum North Carolina*. Lexington: University Press of Kentucky, 1992.

Champlin, Edward. *Final Judgments: Duty and Emotion in Roman Wills, 200 B.C.–A.D. 250*. Berkeley: University of California Press, 1991.

Chayanov, Aleksandr Vasilevich. *The Theory of Peasant Economy*. Edited by Daniel Thorner, Basile Kerblay, and R. E. F. Smith. Homewood, Ill.: R. D. Irwin for the American Economic Association, 1966.

Cherry, David. "Intestacy and the Roman Poor." *Legal History Review* 64 (1996): 155–74.

Chrissanthos, Stefan G. "Scipio and the Mutiny at Sucro, 206 B.C." *Historia* 46 (1997): 172–84.

Churchill, J. Bradford. "*Ex qua quod vellent facerent*: Roman Magistrates' Authority over *Praeda* and *Manubiae*." *TAPA* 129 (1999): 85–116.

Clark, A. L., and J. S. Russell. "Crop Sequential Practices." In *Soil Factors in Crop Production in a Semi-Arid Environment*, edited by J. S. Russell and E. L. Greacen, 279–300. St. Lucia, Queensland: University of Queensland Press, 1977.

Clark, Colin. *Population Growth and Land Use*. London: Macmillan; New York: St. Martin's Press, 1967.

Clark, Colin, and Margaret Haswell. *The Economics of Subsistence Agriculture*. 4th ed. London: Macmillan; New York: St. Martin's Press, 1970.

Coale, Ansley. "The Decline in Fertility in Europe since the Eighteenth Century as a Chapter in Human Demographic History." In *The Decline of Fertility in Europe: The Revised Proceedings of a Conference on the Princeton European Fertility Project*, edited by Ansley J. Coale and Susan Cotts Watkins, 1–30. Princeton: Princeton University Press, 1986.

Coarelli, Filippo. "Public Building in Rome between the Second Punic War and Sulla." *PBSR* 45 (n.s. 32) (1977): 1–19.

Commonwealth Bureau of Soil Science. *Details of the Classical and Long-Term Experiments up to 1967: Rothamsted Experimental Station*. Harpenden: Lawes Agricultural Trust, 1970.

Conrad, Alfred H., and John R. Meyer. *The Economics of Slavery, and Other Studies in Econometric History*. Chicago: Aldine, 1964.

Cornell, Tim. "Hannibal's Legacy." In *The Second Punic War: A Reappraisal*, edited by Tim Cornell, Boris Rankov, and Philip Sabin, 97–113. *Bulletin of the Institute of Classical Studies*, suppl. 67. London: Institute of Classical Studies, School of Advanced Study, University of London, 1996.

———. *The Beginnings of Rome: Italy and Rome from the Bronze Age to the Punic Wars, c. 1000–263 BC*. London: Routledge, 1995.

———. "The End of Roman Imperial Expansion." In *War and Society in the Roman World*, edited by John Rich and Graham Shipley, 139–70. London: Routledge, 1993.

———. "Rome and Latium to 390 B.C." In *The Cambridge Ancient History*, 2d ed., vol. 7, pt. 2: *The Rise of Rome to 220 B.C.*, edited by F. W. Walbank, A. E. Astin, M. W. Frederiksen, and R. M. Ogilvie, 243–308. Cambridge: Cambridge University Press, 1989.

———. "The Conquest of Italy." In *The Cambridge Ancient History*, 2d ed., vol. 7, pt. 2: *The Rise of Rome to 220 B.C.*, edited by F. W. Walbank, A. E. Astin, M. W. Frederiksen, and R. M. Ogilvie, 351–419. Cambridge: Cambridge University Press, 1989.

Corvisier, André. *Armies and Societies in Europe, 1494–1789*. Bloomington: Indiana University Press, 1979.

———. *L'Armée française de la fin du XVIIe siècle au ministère de Choiseul. Le soldat*. Publications de la Faculté des lettres et sciences humaines de Paris-Sorbonne. Série "Recherches," tomes 14–15. Paris: Presses universitaires de France, 1964.

Crawford, Michael H. *The Roman Republic*. 2d ed. Cambridge, Mass.: Harvard University Press, 1993.

Bibliography

——. *Coinage and Money under the Roman Republic: Italy and the Mediterranean Economy.* London: Methuen, 1985.

——. "The Early Roman Economy, 753–280 B.C." In *L'Italie préromaine et la Rome républicaine: mélanges offerts à Jacques Heurgon*, 1:197–207. Collection de l'Ecole française de Rome 27. Rome: Ecole française de Rome, 1976.

——. *Roman Republican Coinage.* 2 vols. London: Cambridge University Press, 1974.

——. "War and Finance." *JRS* 54 (1964): 29–32.

Crone, Patricia. *Pre-industrial Societies.* Oxford: Blackwell, 1989.

Crook, J. A. "*Patria Potestas.*" *CQ* 17 (1967): 113–22.

Culham, Phyllis. "Archives and Alternatives in Republican Rome." *CP* 84 (1989): 100–115.

Curti, Emmanuele, Emma Dench, and John R. Patterson. "The Archaeology of Central and Southern Italy: Recent Trends and Approaches." *JRS* 86 (1996): 170–89.

Dalton, G. "How Exactly Are Peasants Exploited?" *American Anthropologist* 76 (1974): 553–61.

D'Arms, John H. *Commerce and Social Standing in Ancient Rome.* Cambridge, Mass.: Harvard University Press, 1981.

David, Jean-Michel. *La République romaine de la deuxième guerre punique à la bataille d'Actium 218–31. Crise d'une aristocratie.* Nouvelle Historie de l'Antiquité 7. Paris: Editions du Seuil, 2000.

——. *The Roman Conquest of Italy.* Translated by Antonia Nevill. Oxford: Blackwell, 1997.

Davies, Roy W. *Service in the Roman Army.* Edited by David Breeze and Valerie Maxfield. New York: Columbia University Press, 1984.

Davis, John. *Land and Family in Pisticci.* Monographs on Social Anthropology, no. 48. London: Athlone Press; New York: Humanities Press, 1973.

Degrassi, Atilius. *Inscriptiones Italiae.* Vol. 13, fasc. 1: *Fasti et Elogia.* Rome: La Libreria dello stato, 1947.

Delano-Smith, Catherine. *Western Mediterranean Europe: A Historical Geography of Italy, Spain and Southern France since the Neolithic.* London: Academic Press, 1979.

Delbrück, Hans. *History of the Art of War within the Framework of Political History.* Translated by Walter J. Renfroe Jr. 4 vols. Westport, Conn.: Greenwood Press, 1975.

De Ligt, Luuk. "Studies in Legal and Agrarian History II: Tenancy under the Republic." *Ath* 88 (2000): 377–91.

——. "Demand, Supply, Distribution: The Roman Peasantry between Town and Countryside II: Supply, Distribution and a Comparative Perspective." *Münstersche Beiträge* 10 (1991): 33–77.

———. "Demand, Supply, Distribution: The Roman Peasantry between Town and Countryside I: Rural Monetization and Peasant Demand." *Münstersche Beiträge* 9 (1990): 24–56.

De Martino, Francesco. "Ancora sulla produzione di cereali in Roma arcaica." *PP*, n.s., 39 (1984): 241–62.

———. "Produzione di cereali in Roma nell'età Arcaica." *PP*, n.s., 34 (1979): 241–53.

Dench, Emma. *From Barbarians to New Men: Greek, Roman, and Modern Perceptions of Peoples of the Central Apennines.* Oxford: Oxford University Press, 1995.

De Neeve, Pieter W. Review of *The Corn Supply of Ancient Rome*, by Geoffrey Rickman. *Mnemosyne* 38 (1985): 443–48.

———. *Colonus: Private Farm-Tenancy in Roman Italy during the Republic and the Early Principate.* Amsterdam: J. C. Gieben, 1984.

———. *Peasants in Peril: Location and Economy in Italy in the Second Century B.C.* Amsterdam: J. C. Gieben, 1984.

Derow, P. S. "The Roman Calendar, 218–191 B.C." *Phoenix* 30 (1976): 265–81.

———. "The Roman Calendar, 190–168 B.C." *Phoenix* 27 (1973): 345–56.

De Sanctis, Gaetano. *Storia dei Romani.* 2d ed. 4 vols. Florence: La Nuova Italia, 1968.

de Ste. Croix, G. E. M. *The Class Struggle in the Ancient Greek World from the Archaic Age to the Arab Conquests.* Ithaca, N.Y.: Cornell University Press, 1981.

Desy, Philippe. "Il grano dell'Apulia e la data della battaglia del Trasimeno." *PP* 44 (1989): 102–15.

Dixon, Suzanne. *The Roman Mother.* Norman: Oklahoma University Press, 1988.

Dohr, Heinz. *Die italischen Gutshöfe nach den Schriften Catos und Varros.* Inaug.-Dissertation, University of Cologne, 1965.

Dreizehnter, Alois. *Die rhetorische Zahl: Quellenkritische Untersuchungen anhand der Zahlen 70 und 700.* Zetemata Heft 73. Munich: Beck, 1978.

Drummond, A. "Sources for Early Roman History." In *The Cambridge Ancient History*, 2d ed., vol. 7, pt. 2: *The Rise of Rome to 220 B.C.*, edited by F. W. Walbank, A. E. Astin, M. W. Frederiksen, and R. M. Ogilvie, 1–29. Cambridge: Cambridge University Press, 1989.

———. "Rome in the Fifth Century I: The Social and Economic Framework." In *The Cambridge Ancient History*, 2d ed., vol. 7, pt. 2: *The Rise of Rome to 220 B.C.*, edited by F. W. Walbank, A. E. Astin, M. W. Frederiksen, and R. M. Ogilvie, 113–71. Cambridge: Cambridge University Press, 1989.

Ducrey, Pierre. "Remarques sur les causes du mercenariat dans la Grèce ancienne et la Suisse moderne." In *Buch der Freunde für J. R. von Salis zum 70. Geburtstag, 12. Dezember 1971*, 115–23. Zurich: Orell Füssli, 1971.

Duffy, Christopher. *Military Experience in the Age of Reason*. London: Routledge & Kegan Paul, 1987.

Dumont, Jean Christian. *Servus. Rome et l'esclavage sous la République*. Collection de l'Ecole française de Rome 103. Rome: Ecole française de Rome, 1987.

Duncan-Jones, Richard. *Money and Government in the Roman Empire*. Cambridge: Cambridge University Press, 1994.

———. *The Economy of the Roman Empire: Quantitative Studies*. Cambridge: Cambridge University Press, 1982.

———. "The Choenix, the Artaba and the Modus." *ZPE* 21 (1976): 43–52.

Dupâquier, Jacques, et al. *Histoire de la population française*. Vol. 1: *Histoire de la population française des origines à la Renaissance*. Paris: Presses universitaires de France, 1988.

Dyer, Frederick H. *A Compendium of the War of the Rebellion*. . . . Des Moines, Iowa: Dyer Publishing Company, 1908.

Dyson, Stephen L. *Community and Society in Roman Italy*. Baltimore: Johns Hopkins University Press, 1992.

Earl, Donald C. *Tiberius Gracchus: A Study in Politics*. Collection Latomus, vol. 66. Brussels-Berchem: Latomus, 1963.

Easterlin, Richard A. "Population Change and Farm Settlement in the Northern United States." *Journal of Economic History* 36 (1976): 45–75.

Eckberg, Carl J. *French Roots in the Illinois Country: The Mississippi Frontier in Colonial Times*. Urbana: University of Illinois Press, 1998.

Eckstein, Arthur M. "Brigands, Emperors, and Anarchy." *International History Review* 22 (2000): 862–79.

———. "Physis and Nomos: Polybius, the Romans, and Cato the Elder." In *Hellenistic Constructs: Essays in Culture, History, and Historiography*, edited by Paul Cartledge, Peter Garnsey, and Erich Gruen, 175–98. Berkeley: University of California Press, 1997.

———. *Senate and General: Individual Decision Making and Roman Foreign Relations, 264–194 B.C.* Berkeley: University of California Press, 1987.

———. "Two Notes on the Chronology of the Outbreak of the Hannibalic War." *RhM* 126 (1983): 255–72.

———. "Polybius on the Rôle of the Senate in the Crisis of 264 B.C." *GRBS* 21 (1980): 175–90.

Engels, Donald W. *Alexander the Great and the Logistics of the Macedonian Army*. Berkeley: University of California Press, 1978.

Erdkamp, Paul. "Feeding Rome or Feeding Mars? A Long-Term Approach to C. Gracchus' *Lex Frumentaria.*" *AncSoc* 30 (2000): 53–70.

———. "Agriculture, Unemployment, and the Cost of Rural Labour in the Roman World." *CQ* 49 (1999): 556–72.

———. *Hunger and the Sword: Warfare and Food Supply in Roman Republican Warfare (264–30 B.C.).* Amsterdam: J. C. Gieben, 1998.

Errington, R. Malcolm. *A History of Macedon.* Translated by Catherine Errington. Berkeley: University of California Press, 1990.

Evans, John K. *War, Women and Children in Ancient Rome.* London: Routledge, 1991.

———. "Resistance at Home: The Evasion of Military Service in Italy during the Second Century B.C." In *Forms of Control and Subordination in Antiquity*, edited by Toru Yuge and Masaoki Doi, 121–40. Leiden: E. J. Brill for the Society for Studies on Resistance Movements in Antiquity, Tokyo, 1988.

———. "Wheat Production and Its Social Consequences in the Roman World." *CQ* 31 (1981): 428–42.

———. "*Plebs Rustica*: The Peasantry of Classical Italy." *AJAH* 5 (1980): 19–47, 134–73.

Faust, Drew Gilpin. *Mothers of Invention: Women of the Slaveholding South in the American Civil War.* Chapel Hill: University of North Carolina Press, 1996.

Finley, Moses I. *Ancient Slavery and Modern Ideology.* New York: Viking Press, 1980.

———. *The Ancient Economy.* Berkeley: University of California Press, 1973.

Flinn, Michael W. *The European Demographic System, 1500–1820.* Baltimore: Johns Hopkins University Press, 1981.

Food and Agriculture Organization of the United Nations. Food Policy and Nutrition Division. *Food Composition Tables for the Near East.* . . . FAO Food and Nutrition Paper 26. Rome: FAO, 1982.

Food and Agriculture Organization, World Health Organization, and United Nations University. *Energy and Protein Requirements: Report of a Joint FAO/WHO/UNU Expert Consultation.* World Health Organization Technical Report Series, no. 724. Geneva: World Health Organization; Albany, N.Y.: WHO Publications Center USA, 1985.

Forbes, Hamish, and Lin Foxhall. "Ethnoarchaeology and Storage in the Ancient Mediterranean: Beyond Risk and Survival." In *Food in Antiquity*, edited by John Wilkins, David Harvey, and Mike Dobson, 69–86. Exeter: University of Exeter Press, 1995.

Ford, Lacy K. *Origins of Southern Radicalism: The South Carolina Upcountry, 1800–1860.* New York: Oxford University Press, 1988.

Forni, Giovanni. "Estrazione etnica e sociale dei soldate delle legioni nei primi tre secoli dell'imperio." *ANRW* 2.1 (1979): 339–91.

———. *Il Reclutamento delle legioni da Augusto a Diocleziano*. Milan: Fratelli Bocca, 1953.

———. "Manio Curio Dentato Uomo Democratico." *Ath*, n.s., 31 (1953): 170–239.

Forrest, Alan I. *Conscripts and Deserters: The Army and French Society during the Revolution and Empire*. New York: Oxford University Press, 1989.

Forsythe, Gary. *The Historian, L. Calpurnius Piso Frugi, and the Roman Annalistic Tradition*. Lanham, Md.: University Press of America, 1994.

Foster, George M. "Introduction: What Is a Peasant?" In *Peasant Society: A Reader*, edited by Jack M. Potter, May N. Diaz, and George M. Foster, 2–14. Boston: Little, Brown, 1967.

Foxhall, Lin. "Farming and Fighting in Ancient Greece." In *War and Society in the Greek World*, edited by John Rich and Graham Shipley, 134–45. Leicester-Nottingham Studies in Ancient Society, vol. 4. London: Routledge, 1993.

———. "The Dependent Tenant: Land Leasing and Labour in Italy and Greece." *JRS* 80 (1990): 97–114.

Foxhall, Lin, and H. A. Forbes. "Σιτομετρία: The Role of Grain as a Staple Food in Classical Antiquity." *Chiron* 12 (1982): 41–90.

Fraccaro, Plinio. *Opuscula*. 4 vols. Pavia: Presso la rivista "Athenaeum," 1956.

Frank, Tenney. "Roman Census Statistics from 225 to 28 B.C." *CPh* 19 (1924): 329–41.

———, ed. *An Economic Survey of Ancient Rome*. 5 vols. Baltimore: Johns Hopkins Press, 1933–40.

Franke, P. R. "Pyrrhus." In *The Cambridge Ancient History*, 2d ed., vol. 7, pt. 2: *The Rise of Rome to 220 B.C.*, edited by F. W. Walbank, A. E. Astin, M. W. Frederiksen, and R. M. Ogilvie, 456–85. Cambridge: Cambridge University Press, 1989.

Fränkel, Arthur. *Der Amtsantritt der römischen Consuln während der Periode 387–532 d. St. Das Verhältness römischen Kalenders zum julianischen während des Zeitraums 440–552 d. St.* Studien zur Römischen Geschichte Heft 1. Breslau: J. U. Kern's Verlag, 1884.

Frayn, Joan M. *Sheep-Rearing and the Wool Trade in Italy during the Roman Period*. ARCA, Classical and Medieval Texts, Papers, and Monographs 15. Liverpool: F. Cairns, 1984.

———. *Subsistence Farming in Roman Italy*. Fontwell, Sussex: Centaur Press, 1979.

———. "Subsistence Farming in Italy during the Roman Period." *G&R* 21 (1974): 11–18.

Frederiksen, Martin W. "Cambiamenti delle strutture agrarie nella tarda Repubblica: La Campania." In *Società romana e produzione schiavistica*, vol. 1: *L'Italia: insediamenti e forme economiche*, edited by Andrea Giardina and Aldo Schiavone, 265–87. Bari: Laterza, 1981.

———. "The Contribution of Archaeology to the Agrarian Problem in the Gracchan Period." *Dialoghi di Archeologia* 4–5 (1970–71): 330–57.

Frier, Bruce W. "More Is Worse: Some Observations on the Population of the Roman Empire." In *Debating Roman Demography*, edited by Walter Scheidel, 139–59. Mnemosyne, Bibliotheca Classica Batava, suppl. 211. Leiden: Brill, 2001.

———. "Natural Fertility and Family Limitation in Roman Marriage." *CPh* 89 (1994): 318–33.

———. *Libri Annales Pontificum Maximorum: The Origins of the Annalistic Tradition.* Papers and Monographs of the American Academy in Rome, vol. 27. Rome: American Academy in Rome, 1979.

Gabba, Emilio. "Sulle strutture agrarie del'Italia romana fra III e I sec. A.C." In *Strutture agrarie e allevamento transumante nell'Italia Romana (III–I sec. A.C)*, by E. Gabba and M. Pasquinucci, 15–73. Biblioteca di studi antichi 18. Pisa: Giardini, 1979.

———. "Considerazioni sulla decadenza della piccola proprietà contadina nell'Italia centro-meridionale del II sec. a.C." *Ktema* 2 (1977): 269–84.

———. "Considerazaioni politiche ed economiche sullo sviluppo urbano in Italia nei secoli II e I a.C." In *Hellenismus in Mittelitalien: Kolloquium in Göttingen vom 5. bis 9. Juni 1974*, edited by Paul Zanker, 315–26. Abhandlungen der Akademie der Wissenschaften in Göttingen, Philologisch-Historische Klasse; 3. Folge, Nr. 97. Göttingen: Vandenhoeck und Ruprecht, 1976.

———. "Origins of the Professional Army at Rome: The 'proletarii' and Marius' Reform." In *Republican Rome, the Army and the Allies*, by E. Gabba, translated by P. J. Cuff, 1–19. Oxford: Blackwell, 1976.

———. *Republican Rome, the Army and the Allies.* Translated by P. J. Cuff. Oxford: Blackwell, 1976.

———. Review of *Hannibal's Legacy*, by A. J. Toynbee. In *Republican Rome, the Army and the Allies*, by E. Gabba, translated by P. J. Cuff, 154–61. Oxford: Blackwell, 1976.

———. "The Roman Professional Army from Marius to Augustus." In *Republican

Rome, the Army and the Allies, by E. Gabba, translated by P. J. Cuff, 20–69. Oxford: Blackwell, 1976.

———. "Motivazioni economiche nell'opposizione alla legge agraria di Tib. Sempronio Gracco." In *Polis and Imperium: Studies in Honour of Edward Togo Salmon,* edited by J. A. S. Evans, 129–38. Toronto: Hakkert, 1974.

———. "Urbanizzazione e rinnovamenti urbanistici nell'Italia centro-meridionale del 1 Sec. a.C." *SCO* 21 (1972): 73–112.

———. *Appiani, Bellorum Civilum Liber Primus. Introduzione, testo critico, e commento con traduzione e indice.* Florence: La "Nuova Italia," 1958.

Gabba, Emilio, and Marinella Pasquinucci. *Strutture agrarie e allevamento transumante nell'Italia romana (III–I Sec. a.C.).* Biblioteca di studi antichi 18. Pisa: Giardini, 1979.

Gabriel, Richard A., and Karen M. Metz. *From Sumer to Rome: The Military Capabilities of Ancient Armies.* Contributions in Military Studies, no. 108. New York: Greenwood Press, 1991.

Gallant, Thomas W. *Risk and Survival in Ancient Greece: Reconstructing the Rural Domestic Economy.* Stanford: Stanford University Press, 1991.

Garlan, Yvon. *Recherches de poliorcétique grecque.* Bibliothèque des écoles françaises d'Athènes et de Rome, fasc. 223. Athens: Ecole française d'Athènes, 1974.

Garnsey, Peter. "The Bean: Substance and Symbol." In *Cities, Peasants and Food in Classical Antiquity: Essays in Social and Economic History,* by Peter Garnsey, edited by Walter Scheidel, 214–25. Cambridge: Cambridge University Press, 1998.

———. *Famine and Food Supply in the Graeco-Roman World: Responses to Risk and Crisis.* Cambridge: Cambridge University Press, 1988.

———. "Non-Slave Labour in the Roman World." In *Non-Slave Labour in the Greco-Roman World,* edited by Peter Garnsey, 34–47. Supplementary volume of the Cambridge Philological Society, no. 6. Cambridge: Cambridge Philological Society, 1980.

———. "Where Did Italian Peasants Live?" *PCPhS* 205 (n.s. 25) (1979): 1–25.

Garnsey, Peter, and Richard Saller. *The Roman Empire: Economy, Society and Culture.* Berkeley: University of California Press, 1987.

Garoufalis, Petros. *Pyrrhus, King of Epirus.* London: Stacey International, 1979.

Geary, James W. *We Need Men: The Union Draft in the Civil War.* Dekalb: Northern Illinois University Press, 1991.

Gelzer, Matthias. *Kleine Schriften.* Edited by Hermann Strasburger and Christian Meier. 3 vols. Wiesbaden: F. Steiner, 1962–64.

Giardina, Andrea, and Aldo Schiavone, eds. *Società romana e produzione schiavistica.* Vol. 1: *L'Italia, insediamenti e forme economiche;* vol. 2: *Merci, mercati e scambi nel*

Mediterraneo; vol. 3: *Modelli etici, diritto e trasformazioni sociali.* Bari: Laterza, 1981.
Goldsworthy, Adrian Keith. *The Roman Army at War, 100 BC–AD 200.* Oxford: Oxford University Press, 1996.
Gomme, A. W., A. Andrews, and J. K. Dover. *A Historical Commentary on Thucydides.* 5 vols. Oxford: Clarendon Press, 1945–81.
González, Julián. "The Lex Irnitana: A New Copy of the Flavian Municipal Law." *JRS* 76 (1986): 145–243.
Gottfried, Robert S. *The Black Death: Natural and Human Disaster in Medieval Europe.* New York: Free Press; London: Collier Macmillan, 1983.
Gray, Lewis C. *A History of Agriculture in the Southern United States to 1860.* Carnegie Institution of Washington. Publication no. 430, Contributions to American Economic History, no. 7. Washington, D.C.: Carnegie Institution of Washington, 1933.
Great Britain. Naval Intelligence Division. *Tunisia.* Geographical Handbook Series, B.R. 523. Oxford: Naval Intelligence Division, 1945.
———. *Italy.* Geographical Handbook Series, B.R. 517, vol. 1. Oxford: Naval Intelligence Division, 1944.
———. *Algeria.* Geographical Handbook Series, B.R. 505, vol. 1. Oxford: Naval Intelligence Division, 1943.
Gregg, David. *Population Growth and Agrarian Change: An Historical Perspective.* Cambridge Geographical Studies 13. Cambridge: Cambridge University Press, 1980.
Griffith, G. T. *Mercenaries of the Hellenistic World.* Cambridge: University Press, 1935.
Griffith, Paddy. *Forward into Battle: Fighting Tactics from Waterloo to Vietnam.* Strettington, Chichester, Sussex: Antony Bird Publications, 1981.
Griffiths, J. F., ed. *Climates of Africa.* World Survey of Climatology, vol. 10. Amsterdam: Elsevier, 1972.
Grimsley, Mark. *The Hard Hand of War: Union Military Policy toward Southern Civilians, 1861–1865.* Cambridge: Cambridge University Press, 1995.
Grmek, Mirko D. *Diseases in the Ancient Greek World.* Translated by Mireille Muellner and Leonard Muellner. Baltimore: Johns Hopkins University Press, 1989.
Gruen, Erich S. "The 'Fall' of the Scipios." In *Leaders and Masses in the Roman World: Studies in Honor of Zvi Yavetz,* edited by I. Malkin and Z. W. Rubinsohn, 59–90. Leiden: E. J. Brill, 1995.
———. *Culture and National Identity in Republican Rome.* Cornell Studies in Classical Philology, vol. 52. Ithaca, N.Y.: Cornell University Press, 1992.

———. *Studies in Greek Culture and Roman Policy.* Cincinnati Classical Studies, n.s., vol. 7. Leiden: E. J. Brill, 1990.

———. *The Hellenistic World and the Coming of Rome.* 2 vols. Berkeley: University of California Press, 1984.

———. *Last Generation of the Roman Republic.* Berkeley: University of California Press, 1974.

Grundy, G. B. *Thucydides and the History of His Age.* 2 vols. Oxford: Blackwell, 1948.

Gualtieri, M., and F. de Polignac. "A Rural Landscape in Western Lucania." In *Roman Landscapes: Archaeological Survey in the Mediterranean Region,* edited by Graeme Barker and John Lloyd, 194–203. Archaeological Monographs of the British School at Rome, no. 2. London: British School at Rome, 1991.

Habinek, Thomas N. *The Politics of Latin Literature: Writing, Identity, and Empire in Ancient Rome.* Princeton: Princeton University Press, 1998.

Hahn, Steven. *The Roots of Southern Populism: Yeomen Farmers and the Transformation of the Georgia Upcountry, 1850–1890.* New York: Oxford University Press, 1983.

Hajnal, J. "Two Kinds of Pre-industrial Household Formation System." In *Family Forms in Historic Europe,* edited by Richard Wall, Jean Robin, and Peter Laslett, 65–104. Cambridge: Cambridge University Press, 1983.

———. "European Marriage Patterns in Perspective." In *Population in History: Essays in Historical Demography,* edited by D. V. Glass and D. E. C. Eversley, 101–43. Chicago: Aldine, 1965.

Hallett, Judith P. *Fathers and Daughters in Roman Society: Women and the Elite Family.* Princeton: Princeton University Press, 1984.

Halstead, Paul. "Traditional and Ancient Rural Economy in Mediterranean Europe: Plus ça Change?" *JHS* 107 (1987): 77–87.

Hamilton, Charles D. "The Hellenistic World." In *War and Society in the Ancient and Medieval Worlds: Asia, the Mediterranean, and Mesoamerica,* edited by Kurt Raaflaub and Nathan Rosenstein, 163–92. Center for Hellenic Studies Colloquia 3. Washington, D.C.: Center for Hellenic Studies; Cambridge, Mass.: Harvard University Press, 1999.

Hanson, Victor Davis. *The Western Way of War: Infantry Battle in Classical Greece.* 2d ed. Berkeley: University of California Press, 2000.

———. *Warfare and Agriculture in Classical Greece.* Rev. ed. Berkeley: University of California Press, 1998.

Hardy, Robert. *Longbow: A Social and Military History.* Cambridge: Stephens, 1976.

Hare, Ronald. "The Antiquity of Diseases Caused by Bacteria and Viruses: A Review of the Problem from a Bacteriologist's Point of View." In *Diseases in Antiquity: A Survey of the Diseases, Injuries, and Surgery of Early Populations,* edited by Don R. Brothwell and A. T. Sandison, 115–31. Springfield, Ill.: C. C. Thomas, 1967.

Harmand, Jacques. *L'Armée et le soldat à Rome, de 107 à 50 avant notre ère.* Paris: A. J. Picard et Cie, 1967.

Harris, William V. "Demography, Geography and the Sources of Roman Slaves." *JRS* 89 (1999): 62–75.

———. "Child Exposure in the Roman Empire." *JRS* 84 (1994): 1–22.

———. "Roman Warfare in the Economic and Social Context of the Fourth Century B.C." In *Staat und Staatlichkeit in der frühen römischen Republik: Akten eines Symposiums, 12.–15. Juli, 1988, Freie Universität Berlin,* edited by Walter Eder, 494–510. Stuttgart: Franz Steiner Verlag, 1990.

———. *War and Imperialism in Republican Rome, 327–70 B.C.* Oxford: Oxford University Press, 1979.

———. *Rome in Etruria and Umbria.* Oxford: Clarendon Press, 1971.

Hassig, Ross. *Aztec Warfare: Imperial Expansion and Political Control.* Civilization of the American Indian Series, vol. 188. Norman: University of Oklahoma Press, 1988.

Herlihy, David. *The Black Death and the Transformation of the West.* Edited and with an introduction by Samuel K. Cohn Jr. Cambridge, Mass.: Harvard University Press, 1997.

Hodge, W. B. "On Mortality Arising from Military Operations." *Journal of the Statistical Society of London* 19 (1856): 219–71.

Hölkeskamp, Karl-J. "Conquest, Competition and Consensus: Roman Expansion in Italy and the Rise of the *Nobilitas.*" *Historia* 42 (1993): 12–39.

Holmes, T. Rice. *The Roman Republic and the Founder of the Empire.* 3 vols. Oxford: Clarendon Press, 1923.

Holzapfel, Ludwig. *Römische Chronologie.* Leipzig: B. G. Teubner, 1885.

Hopkins, Keith. *Conquerors and Slaves.* Sociological Studies in Roman History, vol. 1. Cambridge: Cambridge University Press, 1978.

———. "On the Probable Age Structure of the Roman Population." *Population Studies* 20 (1966): 245–64.

Horden, Peregrine, and Nicholas Purcell. *The Corrupting Sea: A Study of Mediterranean History.* Oxford: Blackwell, 2000.

Horvath, Robert. "The Origins of the Gracchan Revolution." *Studies in Latin Literature and Roman History* 7 (1994): 87–116.

Hoyos, B. D. "Polybius' Roman οἱ πολλοί in 264 B.C." *LCM* 9 (1984): 88–93.

Humphreys, Sarah C. "Diskussion, Sektion VI: Aussenbeziehungen und innere Entwicklung." In *Staat und Staatlichkeit in der frühen römischen Republik: Akten eines Symposiums, 12.–15. Juli, 1988, Freie Universität Berlin*, edited by Walter Eder, 549–51. Stuttgart: Franz Steiner Verlag, 1990.

———. *Anthropology and the Greeks*. London: Routledge & Kegan Paul, 1978.

Ingrao, Charles W. *The Hessian Mercenary State: Ideas, Institutions, and Reform under Frederick II, 1760–1785*. Cambridge: Cambridge University Press, 1987.

Inhetveen, H., and M. Blasche. "Women in the Smallholder Economy." In *Peasants and Peasant Societies, Selected Readings*, edited by Teodor Shanin, 28–34. Oxford: Blackwell, 1987.

Jackson, Ralph. *Doctors and Diseases in the Roman Empire*. London: British Museum Publications, 1988.

Jacob, O. "Le Service de santé dans les armées romaines." *AntClass* 2 (1933): 313–29.

Jameson, Michael. "Agricultural Slavery in Classical Athens." *CJ* 73 (1977–78): 122–45.

Jasny, Naum. *The Wheats of Classical Antiquity*. Johns Hopkins University Studies in Historical and Political Science, ser. 62, no. 3. Baltimore: Johns Hopkins Press, 1944.

Jefferson, Thomas. "Notes of a Tour into the Southern Parts of France, &c." In *The Papers of Thomas Jefferson*, edited by Julian P. Boyd, 11:415–64. Princeton: Princeton University Press, 1955.

Johne, Klaus-Peter, Jens Köhn, and Volker Weber. *Die Kolonen in Italien und den westlichen Provinzen des römischen Reiches: eine Untersuchung der literarischen, juristischen und epigraphischen Quellen vom 2. Jahrhundert v.u.Z. bis zu den Severern*. Schriften zur Geschichte und Kultur der Antike 21. Berlin: Akademie-Verlag, 1983.

Johnston, David. *Roman Law in Context*. Cambridge: Cambridge University Press, 1999.

Jones, A. H. M. *The Later Roman Empire, 284–602: A Social, Economic, and Administrative Survey*. Norman: University of Oklahoma Press, 1964.

———. *Ancient Economic History: An Inaugural Lecture Delivered at University College London*. London: H. K. Lewis, 1948.

Jongman, Willem. *The Economy and Society of Pompeii*. Dutch Monographs on Ancient History and Archaeology, vol. 4. Amsterdam: J. C. Gieben, 1988.

Jordan, L. S., and D. L. Shaner. "Weed Control." In *Agriculture in Semi-Arid Environments*, edited by A. E. Hall, G. H. Cannell, and H. W. Lawton, 266–96. Ecological Studies, vol. 34. Berlin: Springer-Verlag, 1979.

Kallet-Marx, Robert Morstein. *Hegemony to Empire: The Development of the Roman Imperium in the East from 148 to 62 B.C.* Hellenistic Culture and Society 15. Berkeley: University of California Press, 1995.

Keegan, John. *The Face of Battle*. New York: Viking Press, 1976.
Keeley, Lawrence H. *War before Civilization*. New York: Oxford University Press, 1996.
Kent, N. L. *Technology of Cereals: An Introduction for Students of Food Science and Agriculture*. 3d ed. Oxford: Pergamon, 1983.
Keppie, Lawrence. *The Making of the Roman Army: From Republic to Empire*. London: B. T. Batsford, 1984.
Kipple, Kenneth F., et al., eds. *Cambridge World History of Human Disease*. Cambridge: Cambridge University Press, 1993.
Klotz, Alfred. "Eine römische Verlustliste." *RhM* 83 (1934): 251–54.
Knapp, Robert C. *Aspects of the Roman Experience in Iberia, 206–100 B.C.* Anejos de Hispania antiqua 9. Valladolid: Universidad, D.L., 1977.
Knox, Alexander. *The Climate of the Continent of Africa*. Cambridge: University Press, 1911.
Kobert, R. "Ueber die Pest des Thukydides." *Janus* 4 (1899): 240–99.
Koistinen, Paul A. C. *Beating Plowshares into Swords*. Lawrence: University Press of Kansas, 1996.
Krause, Jens-Uwe. *Witwen und Waisen im Romischen Reich*. 4 vols. Heidelberger Althistorische Beiträge und Epigraphische Studien, Bd. 16. Stuttgart: F. Steiner, 1994.
Kremer, Bernhard. *Das Bild der Kelten bis in augusteische Zeit: Studien zur Instrumentalisierung eines antiken Feindbildes bei griechischen und römischen Autoren*. Historia Einzelschriften, Heft 88. Stuttgart: F. Steiner, 1994.
Krentz, Peter. "Casualties in Hoplite Battles." *GRBS* 26 (1985): 13–20.
Kromayer, Johannes. *Antike Schlachtfelder: Bausteine zu einer antiken Kriegsgeschichte*. 4 vols. Berlin: Weidmann, 1903.
Laslett, Peter. "Family and Household as Work Group and Kin Group: Areas of Traditional Europe Compared." In *Family Forms in Historic Europe*, edited by Richard Wall, Jean Robin, and Peter Laslett, 513–64. Cambridge: Cambridge University Press, 1983.
Last, Hugh. "Tiberius Gracchus." In *The Cambridge Ancient History*, vol. 9: *The Roman Republic, 133–43 B.C.*, edited by S. A. Cook, F. E. Adcock, and M. P. Charlesworth, 1–39. New York: Macmillan; Cambridge: University Press, 1932.
Laurence, Ray. *The Roads of Roman Italy: Mobility and Cultural Change*. London: Routledge, 1999.
———. "Writing the Roman Metropolis." In *Roman Urbanism: Beyond the Consumer City*, edited by Helen Parkins, 1–20. London: Routledge, 1997.

Bibliography

Lazenby, John F. *The First Punic War: A Military History.* Stanford: Stanford University Press, 1996.

———. *Hannibal's War: A Military History of the Second Punic War.* Warminster: Aris & Phillips, 1978.

Lévêque, Pierre. *Pyrrhos.* Bibliothèque des écoles françaises d'Athènes et de Rome, fasc. 185. Paris: E. de Boccard, 1957.

Lewis, Oscar. "Plow Culture and Hoe Culture—A Study in Contrasts." *Rural Sociology* 14 (1949): 116–27.

Libourel, J. "Galley Slaves in the Second Punic War." *CPh* 68 (1973): 116–19.

Lindegren, Jan. "The Swedish 'Military State,' 1560–1720." *Scandinavian Journal of History* 10 (1985): 305–27.

Lirb, Huib J. "Partners in Agriculture." In *De Agricultura: In Memoriam Pieter Willem de Neeve (1945–1990)*, edited by H. Sancisi-Weerdenberg, R. J. van der Spek, H. C. Teitler, and H. T. Wallinga, 263–95. Dutch Monographs on Ancient History and Archaeology, edited by F. J. A. M. Meijer and H. W. Pleket, vol. 10. Amsterdam: J. C. Gieben, 1993.

Littman, R. J. "The Plague at Syracuse, 396 BC." *Mnemosyne* 87 (1984): 110–16.

Littman, R. J., and M. L. Littman. "Galen and the Antonine Plague." *AJP* 94 (1973): 243–55.

———. "The Athenian Plague: Smallpox." *TAPA* 100 (1969): 261–75.

Liverani, Paolo. "L'Ager veienatnus in età repubblicana." *PBSR* 52 (n.s. 39) (1984): 36–48.

Livi Bacci, Massimo. *The Population of Europe: A History.* Translated by Cynthia De Nardi Ipsen and Carl Ipsen. Oxford: Blackwell, 1999.

———. *Population and Nutrition: An Essay on European Demographic History.* Translated by Tania Croft-Murray. Cambridge Studies in Population, Economy, and Society in Past Time 14. Cambridge: Cambridge University Press, 1991.

Lo Cascio, Elio. "Recruitment and the Size of the Roman Population from the Third to the First Century BCE." In *Debating Roman Demography*, edited by Walter Scheidel, 111–38. Mnemosyne, Bibliotheca Classica Batava, suppl. 211. Leiden: Brill, 2001.

———. "The Population of Roman Italy in Town and Country." In *Reconstructing Past Population Trends in Mediterranean Europe (3000 BC–AD 1800)*, edited by John Bintliff and Kostas Sbonias, 161–71. Oxford: Oxbow Books, 1999.

———. "Populazione e risorse agricole nell'Italia del II secolo a.C." In *Demografia, sistemi agrari, regimi alimentari nel mondo antico*, edited by D. Vera, 217–45. Atti del Convegno Internazionale di Studi (Parma, October 17–19, 1997). Pragmateiai 3. Bari: Edipuglia, 1999.

———. "The Size of the Roman Population: Beloch and the Meaning of the Augustan Census Figures." *JRS* 84 (1994): 23–40.

———. "Ancora sui Censi Minimi delle Classi Cinque 'Serviane.'" *Ath*, n.s., 76 (1988): 273–302.

Loomis, William T. "The Introduction of the Denarius." In *Transitions to Empire: Essays in Greco-Roman History, 360–146 B.C., in Honor of E. Badian*, edited by Robert W. Wallace and Edward M. Harris, 338–55. Norman: University of Oklahoma Press, 1996.

Luce, T. James. *Livy: The Composition of His History*. Princeton: Princeton University Press, 1977.

MacMullen, Ramsey. "How Many Romans Voted?" *Ath* 58 (1980): 454–57.

Majno, Guido. *The Healing Hand: Man and Wound in the Ancient World*. Cambridge, Mass.: Harvard University Press, 1975.

Major, Ralph H. *Fatal Partners, War and Disease*. Garden City, N.Y.: Doubleday, Doran, 1941.

Malthus, Thomas Robert. *An Essay on the Principle of Population*. 6th ed. Edited by E. A. Wrigley and David Souden. In *The Works of Thomas Robert Malthus*, vol. 2. London: William Pickering, 1986.

Marchetti, Patrick. *Histoire économique et monétaire de la deuxième guerre punique*. Mémoires de la Classe des beaux-arts: Collection in-oct. 2. sér., t. 14, fasc. 4 et dernier. Brussels: Académie royale de Belgique, 1978.

———. "La Marche du calendrier romain de 203 à 190 (Années Varr. 551–564)." *AntClass* 42 (1973): 473–96.

Martin, Dale B. "The Construction of the Ancient Family: Methodological Considerations." *JRS* 56 (1996): 40–60.

Mattern, Susan P. *Rome and the Enemy: Imperial Strategy in the Principate*. Berkeley: University of California Press, 1999.

McCall, Jeremiah B. *Cavalry of the Roman Republic: Cavalry Combat and Elite Reputations in the Middle and Late Republic*. London: Routledge, 2002.

McDonnell, Myles. "The Speech of Numidicus at Gellius, *N.A.* 1.6." *AJP* 108 (1987): 81–94.

McNeill, John R. *The Mountains of the Mediterranean World: An Environmental History*. Cambridge: Cambridge University Press, 1992.

Meier, Christian. *Caesar: A Biography*. Translated by David McLintock. New York: Basic Books, 1982.

Meyer, Eduard. *Kleine Schriften*. 2 vols. Halle, Saale: M. Niemeyer, 1924.

Michels, Agnes Kirsopp. *The Calendar of the Roman Republic*. Princeton: Princeton University Press, 1967.

Millar, Fergus. "Political Power in Mid-Republican Rome: Curia or Comitium?" *JRS* 79 (1989): 138–50.

———. "The Political Character of the Classical Roman Republic, 200–151 B.C." *JRS* 74 (1984): 1–19.

Millar, H. F. "The Practical and Economic Background to the Greek Economic Explosion." *G&R* 31 (1984): 153–60.

Momigliano, Arnaldo. *Alien Wisdom: The Limits of Hellenization.* Cambridge: Cambridge University Press, 1975.

Mommsen, Theodor. *Römisches Staatsrecht.* Handbuch der römischen Altertümer. 3d ed. Bd. 1–3. Reprint, Basel: Schwabe, 1952.

———. *The History of Rome.* Translated by William P. Dickson. 4 vols. New York: C. Scribner's Sons, 1888.

———. *Die römische Chronologie bis auf Caesar.* 2d ed. Berlin: Weidmann, 1859.

Moore, Albert B. *Conscription and Conflict in the Confederacy.* New York: Macmillan, 1924.

Mora, Fabio. *Fasti e schemi chronologici: la riorganizzazione annalistica del passato remomto romano.* Historia Einzelschriften, Heft 125. Stuttgart: F. Steiner, 1999.

Morens, D., and R. Littman. "Epidemiology of the Plague at Athens." *TAPA* 122 (1992): 271–304.

Morgan, M. Gwyn. "Calendars and Chronology in the First Punic War." *Chiron* 7 (1977): 89–117.

Morley, Neville. "The Transformation of Italy, 225–28 B.C." *JRS* 91 (2001): 50–62.

———. *Metropolis and Hinterland: The City of Rome and the Italian Economy, 200 B.C.–A.D. 200.* Cambridge: Cambridge University Press, 1996.

Mouritsen, Henrik. *Italian Unification: A Study in Ancient and Modern Historiography.* Bulletin of the Institute of Classical Studies, suppl. 70. London: Institute of Classical Studies, School of Advanced Study, University of London, 1998.

Münzer, Friedrich. *Römische Adelsparteien und Adelsfamilien.* Stuttgart: J. B. Metzler, 1963.

Murdock, Eugene C. *One Million Men: The Civil War Draft in the North.* Madison: State Historical Society of Wisconsin, 1971.

Nagle, D. Brendan. "The Etruscan Journey of Tiberius Gracchus." *Historia* 25 (1976): 487–89.

Nicolet, Claude. "Economy and Society, 133–43 B.C." In *The Cambridge Ancient History*, 2d ed., vol. 9: *The Last Age of the Roman Republic, 146–43 B.C.*, edited by J. A. Crook, Andrew Lintott, and Elizabeth Rawson, 599–643. Cambridge: Cambridge University Press, 1994.

---. *The World of the Citizen in Republican Rome.* Translated by P. S. Falla. Berkeley: University of California Press, 1980.

---. *Tributum: recherches sur la fiscalité directe sous la république romaine.* Antiquitas. Reihe 1, Abhandlungen zur alten Geschichte, Bd. 24. Bonn: Habelt, 1976.

---. "A Rome pendant la seconde guerre punique: techniques financières et manipulations monétaires." *Annales* 18 (1963): 417–36.

Nippel, Wilfried. *Public Order in Ancient Rome.* Cambridge: Cambridge University Press, 1995.

---. "Policing Rome." *JRS* 74 (1984): 20–29.

North, John. "Religion and Rusticity." In *Urban Society in Roman Italy*, edited by Tim Cornell and Kathryn Lomas, 135–50. New York: St. Martin's Press, 1995.

---. "Deconstructing Stone Theaters." In *Apodosis: Essays Presented to Dr. W. W. Cruickshank to Mark His Eightieth Birthday*, 75–83. London: St. Paul's School, 1992.

Norton, Mary Beth. *Liberty's Daughters: The Revolutionary Experience of American Women, 1750–1800.* Boston: Little, Brown, 1980.

Nuttonson, Michael Y. *Wheat-Climate Relationships and the Use of Phenology in Ascertaining the Thermal and Photo-Thermal Requirements of Wheat. . . .* Washington, D.C.: American Institute of Crop Ecology, 1955.

Oakley, Stephen P. *A Commentary on Livy, Books VI–X.* 2 vols. Oxford: Oxford University Press, 1997–.

Ober, Josiah. *Fortress Attica: Defense of the Athenian Land Frontier, 404–322 B.C.* Mnemosyne, Bibliotheca Classica Batava, suppl. 84. Leiden: E. J. Brill, 1985.

Ogilvie, Robert M. *A Commentary on Livy, Books 1–5.* Oxford: Clarendon Press, 1965.

Osborne, Robin. "The Economics and Politics of Slavery at Athens." In *The Greek World*, edited by Anton Powell, 27–43. London: Routledge, 1995.

---. *Classical Landscape with Figures: The Ancient Greek City and Its Countryside.* London: Philips, 1987.

Otto, John S. *Southern Agriculture during the Civil War Era, 1860–1880.* Contributions in American History, no. 153. Westport, Conn.: Greenwood Press, 1994.

Pachtère, Félix Georges de. *La Table hypothécaire de Veleia: étude sur la propriété foncière dans l'Apennin de Plaisance.* Bibliothèque de l'Ecole des Hautes Etudes. IVe section, Sciences historiques et philologiques, fasc. 228. Paris: E. Champion, 1920.

Panella, Clementina. "La distribuzione e i mercati." In *Società romana e produzione*

Bibliography

schiavistica, vol. 2: *Merci, mercate e scambi nel Mediterraneo*, edited by Andrea Giardina and Aldo Shiavone, 55–80. Rome-Bari: Editori Laterza, 1981.

Parke, Herbert W. *Greek Mercenary Soldiers, from the Earliest Times to the Battle of Ipsus*. Oxford: Clarendon Press, 1933.

Parker, Geoffrey. *The Military Revolution: Military Innovation and the Rise of the West, 1500–1800*. 2d ed. Cambridge: Cambridge University Press, 1996.

———. *The Army of Flanders and the Spanish Road, 1567–1659: The Logistics of Spanish Victory and Defeat in the Low Countries' Wars*. Cambridge: Cambridge University Press, 1972.

Parkin, Tim G. *Demography and Roman Society*. Baltimore: Johns Hopkins University Press, 1992.

Pasquinucci, Marinella. "La Transumanza nell'Italia romana." In *Strutture agrarie e allevamento transumante nell'Italia romana (III–I Sec. a.C.)*, by Emilio Gabba and Marinella Pasquinucci, 79–182. Biblioteca di studi antichi 18. Pisa: Giardini, 1979.

Paterson, Jeremy. "Hellenistic Economies: The Case of Rome." In *Hellenistic Economies*, edited by Zofia H. Archibald, John Davies, Vincent Gabrielsen, and G. J. Oliver, 367–78. London: Routledge, 2001.

Patterson, W. F. "The Archers of Islam." *Journal of the Economic and Social History of the Orient* 9 (1966): 69–87.

Pédech, Paul. *La Méthode historique de Polybe*. Paris: Société d'édition "Les Belles Lettres," 1964.

Percival, John. *The Wheat Plant, a Monograph*. London: Duckworth, 1921.

Peter, Hermann. *Historicorum Romanorum Reliquiae*. 2 vols. Stuttgart: Teubner, 1967.

Peterson, Rudolph F. *Wheat: Botany, Cultivation and Utilization*. London: L. Hill Books; New York: Interscience Publishers, 1965.

Petrochilos, Nicholas. *Roman Attitudes to the Greeks*. Athens: National and Capodistrian University of Athens, Faculty of Arts, 1974.

Phang, Sara Elise. *Marriage of Roman Soldiers (13 B.C.–A.D. 235): Law and Family in the Imperial Army*. Columbia Studies in the Classical Tradition, vol. 24. Leiden: Brill, 2001.

Phillips, Jane E. "Roman Mothers and the Lives of Their Adult Daughters." *Helios*, n.s., 6 (1978): 69–80.

Piazza, A., N. Cappello, E. Olivetti, and S. Rendine. "A Genetic History of Italy." *Annals of Human Genetics* 52 (1988): 203–13.

Pleiner, Radomir. *The Celtic Sword*. Oxford: Oxford University Press, 1993.

Poole, J. C. F., and A. J. Holladay. "Thucydides and the Plague at Athens." *CQ* 29 (1979): 282–300.

Potter, Timothy W. "Towns and Territories in Southern Etruria." In *City and Country in the Ancient World*, edited by John Rich and Andrew Wallace-Hadrill, 191–209. Leicester-Nottingham Studies in Ancient Society, vol. 2. London: Routledge, 1991.

———. *Roman Italy*. Berkeley: University of California Press, 1987.

———. *The Changing Landscape of South Etruria*. New York: St. Martin's Press, 1979.

Prinzing, Friedrich. *Epidemics Resulting from Wars*. Oxford: Clarendon Press; London: H. Milford, 1916.

Pritchett, W. Kendrick. *The Greek State at War, Part I*. Berkeley: University of California Press, 1971.

Proctor, Dennis. *Hannibal's March in History*. Oxford: Clarendon Press, 1971.

Purcell, Nicholas. "Wine and Wealth in Ancient Italy." *JRS* 75 (1985): 1–20.

Quilici, Lorenzo, and Stefania Quilici Gigli. *Ficulea*. Latium vetus 6. Rome: Consiglio nazionale delle ricerche, 1993.

———. *Fidenae*. Latium vetus 5. Rome: Consiglio nazionale delle ricerche, 1986.

———. *Crustumerium*. Latium Vetus 3. Rome: Consiglio nazionale delle ricerche, Centro di studio per l'archeologia etrusco-italica, 1980.

Raaflaub, Kurt. "Born to Be Wolves? Origins of Roman Imperialism." In *Transitions to Empire: Essays in Greco-Roman History, 360–146 B.C., in Honor of E. Badian*, edited by Robert W. Wallace and Edward M. Harris, 273–314. Norman: University of Oklahoma Press, 1996.

Rable, George C. "Despair, Hope, and Delusion: The Collapse of Confederate Morale Reexamined." In *The Collapse of the Confederacy*, edited by Mark Grimsley and Brooks D. Simpson, 129–67. Lincoln: University of Nebraska Press, 2001.

———. *Civil Wars: Women and the Crisis of Southern Nationalism*. Urbana: University of Illinois Press, 1989.

Ramsdell, Charles W. *Behind the Lines in the Southern Confederacy*. Edited by Wendell H. Stephenson. New York: Greenwood Press, 1969.

Rathbone, Dominic. "The Census Qualifications of the *Assidui* and the *Prima Classis*." In *De Agricultura: In Memoriam Pieter Willem de Neeve (1945–1990)*, edited by H. Sancisi-Weerdenberg, R. J. van der Spek, H. C. Teitler, and H. T. Wallinga, 121–52. Dutch Monographs on Ancient History and Archaeology, edited by F. J. A. M. Meijer and H. W. Pleket, vol. 10. Amsterdam: J. C. Gieben, 1993.

———. "The Italian Countryside and the Gracchan Crisis." *JACT Review*, ser. 2, 13 (1993): 18–20.

———. Review of *Römische Agrargeschichte*, by Dieter Flach. *Göttingische Gelehrte Anzeiger* 254 (1993): 26–38.

———. "The Slave Mode of Production in Italy: Review of *Società romana e produzione schiavistica*, edited by Andrea Giardina and Aldo Schiavone." *JRS* 73 (1983): 160–69.

———. "The Development of Agriculture in the 'Ager Cosanus' during the Roman Republic: Problems of Evidence and Interpretation." *JRS* 71 (1981): 10–23.

Rawson, Elizabeth. *Roman Culture and Society: Collected Papers of Elizabeth Rawson.* Oxford: Oxford University Press, 1991.

———. "Literary Sources for the Pre-Marian Army." *PBSR* 39 (1971): 13–31.

Reinhold, Meyer. *From Republic to Principate: An Historical Commentary on Cassius Dio's Roman History Books 49–52 (36–29 B.C.).* Historical Commentary on Cassius Dio's Roman History, vol. 6. American Philological Association Monograph Series, no. 34. Atlanta: Scholars Press, 1988.

Reynolds, Peter J. "Deadstock and Livestock." In *Farming Practice in British Prehistory*, edited by Roger Mercer, 97–112. Edinburgh: University Press, 1981.

———. *Iron Age Farm: The Butser Experiment.* London: British Museum Publications, 1979.

Rich, John. "The Supposed Roman Manpower Shortage of the Later Second Century B.C." *Historia* 32 (1983): 287–331.

Richardson, John S. *The Romans in Spain.* Oxford: Blackwell, 1996.

———. *Hispaniae: Spain and the Development of Roman Imperialism, 218–82 BC.* Cambridge: Cambridge University Press, 1986.

———. "The Ownership of Roman Land: Tiberius Gracchus and the Italians." *JRS* 70 (1980): 1–11.

———. "The Triumph, the Praetors and the Senate in the Early Second Century B.C." *JRS* 65 (1975): 50–63.

Rickman, Geoffrey. *The Corn Supply of Ancient Rome.* Oxford: Oxford University Press, 1980.

Roberts, Michael. *The Swedish Imperial Experience, 1560–1718.* Cambridge: Cambridge University Press, 1979.

Rosenstein, Nathan. "Marriage and Manpower in the Hannibalic War: *Assidui, Proletarii* and Livy 24.18.7–8." *Historia* 51 (2002): 163–91.

———. "Sorting Out the Lot in Republican Rome." *AJPh* 116 (1995): 43–75.

———. "Competition and Crisis in Mid-Republican Rome." *Phoenix* 47 (1993): 313–38.

———. *Imperatores Victi: Military Defeat and Aristocratic Competition in the Middle and Late Republic.* Berkeley: University of California Press, 1990.

Rosivach, Vincent J. "Manning the Athenian Fleet, 433–426 BC." *AJAH* 10 (1985): 41–66.
Roth, Jonathan. *The Logistics of the Roman Army at War (264 B.C.–A.D. 235)*. Columbia Studies in the Classical Tradition, vol. 23. Leiden: Brill, 1999.
———. "The Size and Organization of the Imperial Legion." *Historia* 43 (1994): 346–62.
Rotondi, Giovanni. *Leges Publicae Populi Romani: elenco cronologico con una introduzione sull'attività legislativa dei comizi romani*. Reprint, Hildesheim: G. Olms, 1962.
Russell, E. John, and J. A. Voelcker. *Fifty Years of Field Experiments at the the Woburn Experimental Station*. London: Longmans, Green, 1936.
Sabin, Philip. "The Face of Roman Battle." *JRS* 90 (2000): 1–17.
Sachs, Carolyn E. *Gendered Fields: Rural Women, Agriculture, and Environment*. Boulder, Colo.: Westview Press, 1996.
Salazar, Christine F. *The Treatment of War Wounds in Graeco-Roman Antiquity*. Studies in Ancient Medicine, vol. 21. Leiden: Brill, 2000.
Sallares, Robert. *Malaria and Rome: A History of Malaria in Ancient Italy*. Oxford: Oxford University Press, 2002.
———. *The Ecology of the Ancient Greek World*. London: Duckworth, 1991.
Saller, Richard P. *Patriarchy, Property and Death in the Roman Family*. Cambridge Studies in Populations, Economy and Society in Past Time 25. Cambridge: Cambridge University Press, 1994.
———. "Men's Age at Marriage and Its Consequences in the Roman Family." *CP* 82 (1987): 21–34.
———. "Slavery and the Roman Family." *Slavery and Abolition* 8 (1978): 64–87.
Saller, Richard P., and Brent D. Shaw. "Close-Kin Marriage in Roman Society?" *Man*, n.s., 19 (1984): 432–44.
Salmon, Edward T. *Roman Colonization under the Republic*. Ithaca, N.Y.: Cornell University Press, 1970.
———. *Samnium and the Samnites*. Cambridge: Cambridge University Press, 1967.
———. "The *Coloniae Maritimae*." *Ath*, n.s., 41 (1963): 3–38.
Sanders, John. *Basic Demographic Measures: A Practical Guide for Users*. Lanham, Md.: University Press of America, 1988.
Sandison, A. T., and Edmund Tapp. "Disease in Ancient Egypt." In *Mummies, Disease and Ancient Cultures*, edited by Adian Cockburn, Eve Cockburn, and Theodore A. Reyman, 38–58. 2d ed. New York: Cambridge University Press, 1998.
Scheidel, Walter. "Germs for Rome." In *Rome the Cosmopolis*, edited by Catherine

Edwards and Greg Woolf, 158–76. Cambridge: Cambridge University Press, 2003.

———. "Progress and Problems in Roman Demography." In *Debating Roman Demography*, edited by Walter Scheidel, 1–82. Mnemosyne, Bibliotheca Classica Batava, suppl. 211. Leiden: Brill, 2001.

———. "Roman Age Structure: Evidence and Models." *JRS* 91 (2001): 1–26.

———. "The Slave Population of Roman Italy: Speculation and Constraints." *Topoi* 91 (1999): 129–44.

———. "Quantifying the Sources of Slaves in the Early Roman Empire." *JRS* 87 (1997): 156–69.

———. "Finances, Figures and Fiction." *CQ* 46 (1996): 222–38.

———. *Measuring Sex, Age and Death in the Roman Empire: Explorations in Ancient Demography*. Journal of Roman Archaeology, suppl. ser., no. 21. Ann Arbor: Journal of Roman Archaeology, 1996.

———. "The Most Silent Women of Greece and Rome: Rural Labour and Women's Life in the Ancient World." *G&R* 42 (1995): 202–17 and 43 (1996): 1–10.

———. *Grundpacht und Lohnarbeit in der Landwirtschaft des römischen Italien*. Europäische Hochschulschriften. Reihe III, Geschichte und ihre Hilfswissenschaften, Bd. 624. Frankfurt am Main: Lang, 1994.

———. "*Coloni* und Pächter in den römischen literarischen Quellen vom 2. Jh. v. Chr. bis zur Severerzeit. Eine kritische Betrachtung (*Colonus*-Studien I)." *Ath* 80 (1992): 331–70.

———. "Feldarbeit von Frauen in der antiken Landwirtschaft." *Gymnasium* 97 (1990): 405–31.

Schulten, Adolf. *Geschichte von Numantia*. Reprint, New York: Arno Press, 1975.

———. *Numantia: die ergebnisse der ausgrabungen 1905–1912*. 4 vols. Munich: F. Bruckmann a.-g., 1914–31.

Scobie, A. "Slums, Sanitation, and Mortality in the Roman World." *Klio* 68 (1986): 399–433.

Scott, Tom, ed. *The Peasantries of Europe: From the Fourteenth to the Eighteenth Centuries*. London: Longman, 1998.

Scullard, Howard H. "The Carthaginians in Spain." In *The Cambridge Ancient History*, 2d ed., vol. 8: *Rome and the Mediterranean to 133 B.C.*, edited by A. E. Astin, F. W. Walbank, M. W. Frederiksen, and R. M. Ogilvie, 17–43. Cambridge: Cambridge University Press, 1989.

———. "Rome and Carthage." In *The Cambridge Ancient History*, 2d ed., vol. 7, pt. 2: *The Rise of Rome to 220 B.C.*, edited by F. W. Walbank, A. E. Astin, M. W.

Frederiksen, and R. M. Ogilvie, 486–572. Cambridge: Cambridge University Press, 1989.

———. *Scipio Africanus in the Second Punic War*. Cambridge: University Press, 1930.

Segalen, Martine. *Love and Power in the Peasant Family: Rural France in the Nineteenth Century*. Translated by Sarah Matthews. Oxford: Blackwell, 1983.

Shanin, Teodor. "The Nature and Logic of the Peasant Economy." *Journal of Peasant Studies* 1 (1973): 63–80.

———. *The Awkward Class: Political Sociology of Peasantry in a Developing Society: Russia, 1910–1925*. Oxford: Clarendon Press, 1972.

———, ed. *Peasants and Peasant Societies: Selected Readings*. Oxford: Blackwell, 1987.

Shatzman, Israel. *Senatorial Wealth and Roman Politics*. Collection Latomus, vol. 142. Brussels: Latomus, Revue d'Etudes Latines, 1975.

———. "The Roman General's Authority over Booty." *Historia* 21 (1972): 177–205.

Shaw, Brent D. "The Age of Roman Girls at Marriage: Some Reconsiderations." *JRS* 77 (1987): 30–46.

Sherwin-White, A. N. *The Roman Citizenship*. 2d ed. Oxford: Clarendon Press, 1973.

Shochat, Yanir. *Recruitment and the Programme of Tiberius Gracchus*. Collection Latomus, vol. 169. Brussels: Latomus Revue d'Etudes Latines, 1980.

Simon, Helmut. *Roms Kriege in Spanien, 154–133 v. Chr.* Frankfurter wissenschaftliche Abhandlungen. Kulturwissenschaftliche Reihe, Bd. 2. Frankfurt am Main: V. Klostermann, 1962.

Skutsch, Otto. *The Annals of Q. Ennius*. Oxford: Oxford University Press, 1985.

Smith, Richard E. *Service in the Post-Marian Roman Army*. Publications of the Faculty of Arts of the University of Manchester, no. 9. Manchester: Manchester University Press, 1958.

Soltau, Wilhelm. *Römische Chronologie*. Freiburg: J. C. B. Mohr, 1889.

Southern, Pat, and Karen Ramsey Dixon. *The Late Roman Army*. New Haven: Yale University Press, 1996.

Spence, I. G. "Perikles and the Defense of Attika during the Peloponnesian War." *JHS* 110 (1990): 99–109.

Spurr, M. S. "Agriculture and the *Georgics*." In *Virgil*, edited by Ian McAuslan and Peter Walcot, 69–93. Oxford: Oxford University Press on behalf of the Classical Association, 1990.

———. *Arable Cultivation in Roman Italy, c. 200 B.C.–c. A.D. 100*. Journal of

Roman Studies Monographs, no. 3. London: Society for the Promotion of Roman Studies, 1986.

———. "Slavery and the Economy in Roman Italy: Review of *Società romana e produzione schiavistica*, edited by Andrea Giardina and Aldo Schiavone." *CR* 35 (1985): 123–31.

Stanley, Autumn. *Mothers and Daughters of Invention: Notes for a Revised History of Technology*. Metuchen, N.J.: Scarecrow Press, 1993.

Steiner, G. "The Fortunate Farmer: Life on the Small Farm in Ancient Italy." *Classical Journal* 51 (1955): 57–67.

Steinwender, Th. "Altersklassen und reguläre Dienstzeit des Legionars." *Philologus* 48 (1889): 285–305.

Stockton, David. *The Gracchi*. Oxford: Oxford University Press, 1979.

Sumner, G. V. *Orators in Cicero's Brutus: Prosopography and Chronology*. Phoenix, suppl. 11. Toronto: University of Toronto Press, 1973.

———. "The Chronology of the Outbreak of the Second Punic War." *PACA* 9 (1966): 5–30.

Syme, Ronald. *Roman Papers*. Edited by E. Badian and Antony Birley. 7 vols. Oxford: Oxford University Press, 1979–91.

———. "Marriage Ages for Roman Senators." *Historia* 36 (1987): 318–32.

Taylor, Lily Ross. "Forerunners of the Gracchi." *JRS* 52 (1962): 19–27.

———. *Voting Districts of the Roman Republic: The Thirty-Five Urban and Rural Tribes*. American Academy in Rome. Papers and Monographs, vol. 20. Rome: American Academy, 1960.

Tchernia, André. *Le Vin de l'Italie romaine: essai d'histoire économique d'après les amphores*. Bibliothèque des écoles françaises d'Athènes et de Rome, fasc. 261. Rome: Ecole française de Rome, 1986.

———. "Italian Wine in Gaul at the End of the Republic." In *Trade in the Ancient Economy*, edited by Peter Garnsey, Keith Hopkins, and C. R. Whittaker, 87–104. London: Chatto & Windus, 1983.

Thiel, J. H. *A History of Roman Sea-Power before the Second Punic War*. Amsterdam: North-Holland, 1954.

Tibiletti, Gianfranco. "Ricerche di storia agraria Romana." *Ath*, n.s., 28 (1950): 183–266.

———. "Il Possesso dell'*Ager Publicus* e le norme *De Modo Agrorum* sino ai Gracchi." *Athenaeum* 26 (1948): 173–236 and 27 (1949): 3–42.

Toynbee, Arnold J. *Hannibal's Legacy: The Hannibalic War's Effects on Roman Life*. 2 vols. London: Oxford University Press, 1965.

Treggiari, Susan. *Roman Marriage: Iusti Coniuges from the Time of Cicero to the Time of Ulpian*. Oxford: Clarendon, 1991.

Bibliography

———. *Roman Freedmen during the Late Republic.* Oxford: Clarendon Press, 1969.
Tulard, Jean, ed. *Dictionnaire Napoléon.* Paris: Fayard, 1989.
Ulrich, Laurel T. *Good Wives: Image and Reality in the Lives of Women in Northern New England, 1650–1750.* New York: Knopf, 1982.
United Confederate Veterans. Arkansas Division. *Confederate Women of Arkansas in the Civil War, 1861–'65: Memorial Reminiscences.* Little Rock, Ark.: H. G. Pugh, 1907.
United States. Surgeon-General's Office. *The Medical and Surgical History of the War of Rebellion.* . . . 2 vols. Washington, D.C.: Government Printing Office, 1875–83.
Upton, Anthony F. *Charles XI and Swedish Absolutism.* Cambridge: Cambridge University Press, 1998.
Varese, Prospero. *Il calendario romano all'età della prima guerra punica.* Studi di Storia Antica, fasc. 3, edited by Giulio Beloch. Rome: Ermanno Loescher, 1902.
Vassberg, David E. *The Village and the Outside World in Golden Age Castile: Mobility and Migration in Everyday Rural Life.* Cambridge: Cambridge University Press, 1996.
Vaughn, Pamela. "The Identification and Retrieval of the Hoplite Battle-Dead." In *Hoplites: The Classical Greek Battle Experience,* edited by Victor Hanson, 38–62. London: Routledge, 1991.
Vergopoulos, Kostas. "Capitalism and Peasant Productivity." *Journal of Peasant Studies* 5 (1978): 446–65.
Vidal-Naquet, Pierre. "La Tradition de l'hoplite athénien." In *Problèmes de la guerre en Grèce ancienne,* edited by Jean-Pierre Vernant, 161–81. Civilisations et Sociétés, vol. 11. The Hague: Mouton, 1968.
Walbank, Frank W. *A Historical Commentary on Polybius.* 3 vols. Oxford: Clarendon Press, 1957–79.
Wallace, Robert W. "Diskussion, Sektion VI: Aussenbeziehungen und innere Entwicklung." In *Staat und Staatlichkeit in der frühen römischen Republik: Akten eines Symposiums, 12.–15. Juli, 1988, Freie Universität Berlin,* edited by Walter Eder, 556–57. Stuttgart: Franz Steiner Verlag, 1990.
Walsh, P. G., ed. *Livy Book XL.* Edited with an introduction, translation, and commentary by P. G. Walsh. Wiltshire: Aris & Phillips, 1996.
Warrior, Valerie M. "Intercalation and the Action of M'. Acilius Glabrio (cos. 191 B.C.)." *Studies in Latin Literature and Roman History* 6 (1992): 118–44.
———. "Notes on Intercalation." *Latomus* 50 (1991): 82–87.
Watt, Bernice K., and Annabel L. Merrill. *Composition of Foods.* United States

Department of Agriculture, Agricultural Handbook, no. 8, 1964. Reprint, New York: Dover Publications, 1975.

Weissenborn, Wilhelm, and Mauritius Müller, eds. *Titi Livi Ab Urbe Condita Libri*. 4 vols. Leipzig: B. G. Teubner, 1923.

Wells, Colin. "Celibate Soldiers: Augustus and the Army." *AJAH* 14 (1989 [1998]): 180–90.

Westrup, Carl W. *Introduction to Early Roman Law: Comparative Sociological Studies, the Patriarchal Joint Family*. Copenhagen: Levin & Munksgaard; London: H. Milford, Oxford University Press, 1934.

Wheeler, Everett L. "The Laxity of Syrian Legions." In *The Roman Army in the East*, edited by David L. Kennedy, 229–76. Journal of Roman Archaeology, suppl. ser., no. 18. Ann Arbor: Journal of Roman Archaeology, 1996.

White, K. D. "Food Requirements and Food Supplies in Classical Times in Relation to the Diet of the Various Classes." *Progress in Food and Nutrition Science* 2 (1976): 143–91.

———. *Roman Farming*. Ithaca, N.Y.: Cornell University Press, 1970.

———. *Agricultural Implements of the Roman World*. London: Cambridge University Press, 1967.

———. "The Productivity of Labour in Roman Agriculture." *Antiquity* 39 (1965): 102–7.

Wiley, Bell I. *The Road to Appomattox*. Memphis: Memphis State College Press, 1956.

Will, Elizabeth. "The Roman Amphoras." In *The Roman Port and Fishery of Cosa: A Center of Ancient Trade*, by Anna McCann, Joanne Bourgeois, Elaine K. Gazda, John Peter Oleson, and Elizabeth Lyding Will, 171–220. Princeton: Princeton University, 1987.

Williams, Craig A. *Roman Homosexuality: Ideologies of Masculinity in Classical Antiquity*. Oxford: Oxford University Press, 1999.

Williams, J. H. C. *Beyond the Rubicon: Romans and Gauls in Republican Italy*. Oxford: Oxford University Press, 2001.

Williams, John H. *The Rise and Fall of the Paraguayan Republic, 1800–1870*. Latin American Monographs, no. 48. Austin: Institute of Latin American Studies, University of Texas at Austin, 1979.

Wolf, Eric R. *Peasants*. Englewood Cliffs, N.J.: Prentice-Hall, 1966.

Wood, Ellen M. *Peasant-Citizen and Slave: The Foundations of Athenian Democracy*. London: Verso, 1988.

Wood, James W. *Dynamics of Human Reproduction: Biology, Biometry, Demography*. New York: Aldine de Gruyter, 1994.

Wrigley, Edward A. "Fertility Strategy for the Individual and the Group." In

People, Cities and Wealth: The Transformation of Traditional Society, by Edward A. Wrigley, 197–214. Oxford: Blackwell, 1987.

———. "Family Limitation in Pre-industrial England." In *People, Cities and Wealth: The Transformation of Traditional Society*, by Edward A. Wrigley, 242–69. Oxford: Blackwell, 1987.

Wrigley, Edward A., and R. S. Schofield. *Population History of England, 1541–1871: A Reconstruction*. Cambridge, Mass.: Harvard University Press, 1981.

Yadin, Yigael. *The Art of Warfare in Biblical Lands in the Light of Archaeological Discovery*. London: Weidenfeld and Nicolson, 1963.

Yeo, C. E. "The Development of the Roman Plantation and Marketing of Farm Products." *Finanzarchiv*, n.s., 13 (1951–52): 321–42.

Zelener, Y. "The Slavish Gene." A paper presented at the Second Finley Colloquium on Ancient Social and Economic History: Comparative Approaches to Ancient Slavery, Darwin College, Cambridge, July 16–17, 1999.

Ziechmann, Jürgen, ed. *Panorama der fridericianischen Zeit: Friedrich der Grosse und seine Epoche: ein Handbuch*. Forschungen und Studien zur Fridericianischen Zeit, Bd. 1. Bremen: Edition Ziechmann, 1985.

Ziegler, Philip. *The Black Death*. New York: John Day, 1969.

Zinsser, Hans. *Rats, Lice and History*. . . . Boston: Little, Brown for the Atlantic Monthly Press, 1935.

Ziolkowski, Adam. "*Urbs direpta*, or How the Romans Sacked Cities." In *War and Society in the Roman World*, edited by John Rich and Graham Shipley, 69–91. London: Routledge, 1993.

———. "The Plundering of Epirus in 167 B.C.: Economic Considerations." *PBSR* 54 (1986): 69–80.

Index

Abstinence, 144
Acerrae, 40
Achaean League, 136
Acilius Glabrio, M'. (cos. 191), 9
Aelii, 165
Aemilius Barbula, L. (cos. 281), 32
Aemilius Lepidus, M. (cos. 187), 122 (n. 17)
Aemilius Lepidus, M. (cos. 175), 123 (n. 34), 285 (n. 2)
Aemilius Paullus, L. (cos. 182), 10, 122 (n. 13), 123 (n. 23), 285 (n. 2)
Aerarium, 109, 166
Aesernia, 222 (n. 189)
Africa, 43, 203 (n. 20), 212 (n. 86), 220 (n. 168), 265 (n. 130); seasons in, 281 (n. 6)
Ager Capenatus, 195 (n. 12)
Ager Faliscus, 195 (n. 12)
Ager Gallicus, 59, 154, 166, 223 (n. 196)
Ager publicus, 92–93, 143, 158, 164–66, 169, 236 (n. 82), 237 (n. 86), 246 (n. 164), 271 (n. 10), 279 (n. 105); dependence of small-scale agriculture on, 77–79
Ager Romanus, 150
Ager Veietanus, 234 (n. 68)
Ages at first marriage. *See* Marriage ages
Agesilaus, 217 (n. 138)

Age structure: of *alae*, 142; of male Roman citizen population, 22, 86–87, 136, 242 (n. 133)
Agincourt, battle of, 126
Agrarian crisis, 26, 52, 58, 61, 108–9, 141, 155
Agriculture
—changes in, 4–5; and seasons, 65; and working year, 69
—large-scale, 5–7, 64, 108, 153–54, 163, 168, 223 (n. 197), 231 (n. 49), 246 (n. 164), 279 (n. 105); and competition with small-scale agriculture, 15–17, 155; and free labor, 15, 78, 236 (n. 85); geographic extent of, 15; growth of, 103, 155–56; labor on, 107; and livestock grazing, 8; and markets for products, 8, 15; and pace of expansion, 8, 17–18; productive potential of, 16; and slave labor, 7–12, 58, 97, 199 (n. 50); storage facilities of, 16; and villa system, 280 (n. 110). *See also* Slaves and slavery
—small-scale, 167, 231 (n. 49); alternate sources of labor for, 91–94, 101–2, 105; and competition with large farms, 5, 15–17, 197 (n. 30), 224 (n. 197); and conscription, 107; labor requirements of, 3–4, 19–20, 22, 50–51, 54, 61, 65, 107,

Index

149–50; models of, 65; and need for money, 246 (n. 163); and partition of farms, 161, 164; productive potential of, 16; productivity of, 144; and sizes of farms, 75; slaves in, 163, 220 (n. 163); sources of labor for, 113, 246 (n. 159), 270 (n. 6); storage facilities of, 16; surplus labor in, 167; and travel to and from fields, 74; underemployment in, 18–19, 65, 229 (n. 30); and war, 3–5, 18–20, 52, 55, 61, 89. *See also* Families; Smallholders
—and war: in classical and Hellenistic Greece, 27–29; conventional view of, 3–4, 53, 61, 108, 141, 148–49, 155, 167, 220 (n. 165), 224 (n. 197); relationship between, 18–20, 26–29, 31–52, 105
Agrigentum, siege of, 33, 264 (n. 122)
Akarnanians, 217 (n. 138)
Alcibiades, 55
Alexander the Great, 55, 128, 220 (n. 164)
Allied contingents in Roman army. See *Socii*
Amphoras, Dressel Type 1, 6–7
Andriscus, 136
Anicius Gallus, L. (pr. 168), 215 (n. 122)
Annales Maximi, 207 (n. 40)
Annalists, Roman (and annalistic tradition), 23, 31–32, 53–54, 113–16, 117, 138–39, 219 (n. 156), 269 (nn. 155, 165)
Antiochus the Great, 253 (n. 9), 280 (n. 113)
Anti-Roman historians, 114, 117
Antony, Marc (M. Antonius, triumvir), 253 (n. 7), 262 (n. 106), 267 (n. 142)
Apuani, 166
Apulia and Apulians, 29, 40, 42, 51, 195 (n. 12), 205 (n. 32), 264 (n. 118), 286 (n. 4)
Aquileia, 77, 278 (n. 88)
Aquilius Florus, C. (cos. 259), 33, 206 (n. 35)
Arausio, battle at, 85
Arboriculture, 144, 196 (n. 29). *See also* Fruit trees; Orchards
Archaeological evidence for agricultural and economic change, 6–7, 17
Ariminum, 46, 130, 211 (n. 74), 222 (n. 189)
Aristocratic competition, 3, 167
Armies
—British, 125, 129, 131
—Carthaginian, 50, 131; of Hannibal, 134
—hoplite, 28, 140, 202 (n. 7)
—Japanese, 131
—Macedonian, 128; phalanx of, 127
—of Pyrrhus, 50
—Roman: discharge of, 103, 214 (n. 116); during empire, 133; travel to theater of combat, 32; weapons training of, 28, 31, 203 (n. 20). *See also* Camps: of Roman army; Legions
—Spanish, 134–35
—Union, 125–26, 129, 130, 132, 135; and disease, 264 (n. 121). *See also* Camps: of Union army
Arpi, 40–41
Asculum, siege of, 262 (n. 107)
Asia Minor, 134

Assidui, 53–54, 85, 105, 219 (n. 155), 221 (n. 178), 237 (n. 89), 244 (n. 148); minimum census for, 14, 26–27, 56–58, 76, 156–57, 199 (n. 60), 233 (n. 64), 235 (n. 75), 276 (n. 76), 283 (n. 6), 285 (n. 18); numbers of, 14, 25, 240 (n. 121), 273 (n. 31); as proportion of citizen population, 283 (n. 3)
Athens, 55, 63
Atilius Caiatinus, A. (cos. 258), 206 (n. 36)
Atilius Regulus, M. (cos. 267), 4, 33, 21, 206 (n. 37), 207 (n. 41), 234 (n. 68), 242 (n. 137), 245 (n. 155), 249 (n. 193)
Atinius, C. (pr. 188), 122 (n. 18)
Attalus, King of Pergamum, 167; bequest of, 166
Attica, 202 (n. 8)
Augurs, 215 (n. 122)
Augustus, 94; legislation on marriage and adultery, 239 (n. 119); length of military service under, 84; prohibition on soldiers' marrying, 84, 239 (n. 116); reforms of military, 83
Aurunici, 204 (n. 22)
Auspicia, 205 (n. 32)
Auxilia, 115, 118, 136, 138–39; Gallic, 255 (n. 39); Greek, 255 (n. 39); Spanish, 111 (n. 4), 255 (n. 39), 287 (n. 5)
Auximum, 222 (n. 185)
Aztecs, 253 (n. 14)

Baebius Tamphilus, Cn. (cos. 182), 123 (n. 23)
Barley, 28, 216 (n. 128); cultivation of, 47–48; labor to grow, 229 (n. 38)
Beneventum, 40, 222 (n. 189)
Biferno valley, 232 (n. 52)
Birthrate, 25, 144, 151–52, 168; aristocratic fears concerning, 156–57
Births, 148–49; intervals between, 153. *See also* Women and wives
Black Death, 145, 271 (n. 14)
Body weight, 73, 226 (n. 16), 230 (n. 45)
Boer War, 24, 131, 265 (n. 128)
Boii, 43, 112, 124, 287 (n. 6)
Bononia, 77, 236 (n. 82), 278 (n. 88)
Booty, 55, 76, 81, 101, 163, 197 (nn. 32, 36), 217 (n. 136), 240 (n. 122)
Bovianum, 204 (n. 22)
Bows, 126–27, 260 (n. 84)
Boys: food requirements of, 225 (n. 12); labor of, 228 (n. 27). *See also* Smallholders: food requirements of
Breast-feeding, 97, 152, 274 (nn. 55, 56)
Brothers, 247 (n. 168), 279 (n. 107); survival of, 92, 142; younger, 91. *See also* Families: members of
Brundesium, 209 (n. 50), 222 (n. 189), 223 (n. 195)
Bruttium, 130, 274 (n. 47)
Buxentum, 195 (n. 12), 272 (n. 22)
Bygdea (parish in Sweden): losses of in war, 134–35; women of, 250 (n. 197)

Caecilius Metellus, L. (cos. 251), 33
Caecilius Metellus, Q. (cos. 206), 42

Index

Caecilius Metellus Macedonicus, Q. (cos. 143), 104, 277 (n. 81); urges citizens to marry, 156
Caecilius Metellus Numidicus, Q. (cos. 109), 277 (n. 81)
Caesar (C. Iulius Caesar, cos. 59 etc.), 128, 253 (nn. 7, 12)
Calabria, 32–33, 205 (n. 32)
Calendar, Roman, 34–35, 40–41, 174–80, 208 (n. 44), 213 (nn. 96, 102, 103), 217 (n. 145)
Cales, 36
Callinicus, battle of, 113–14, 117, 255 (n. 39)
Calpurinius Piso, C. (cos. 180), 122 (n. 20), 269 (n. 160)
Campania, 29, 31, 36, 38, 42, 50, 149, 195 (nn. 12, 14), 205 (n. 27), 210 (n. 55), 211 (n. 77), 217 (n. 138), 274 (n. 47)
Camps: of Roman army, 24, 132–33; sanitation in, 132–35, 140, 266 (n. 134), 267 (n. 141); of Union army, 132; water supplies of, 133; winter, 32–33, 36–38, 41–42, 44–46, 48, 103–4, 211 (nn. 74, 77), 264 (n. 122), 266 (n. 134), 267 (n. 141)
Cannae, battle of, 36, 40, 90, 112, 114, 116, 152, 215 (n. 122), 256 (nn. 45, 46), 260 (n. 101), 283 (n. 1), 285 (n. 13)
Canusium, 269 (n. 165)
Capital: economic, 8, 15; social, 159; symbolic, 15
Captives: Roman soldiers as, 113, 118, 255 (nn. 33, 37)
Capua and Capuans, 40–41, 50, 100, 134, 211 (nn. 81, 83), 217 (n. 138)

Carseoli, 222 (n. 189)
Carthage, 33, 43, 63, 130, 136; siege of, 130, 133, 264 (n. 122); slaves in, 11
Casilinum, 40
Cassius Longinus, C. (cos. 171), 222 (n. 185)
Castra claudiana, 36–38, 40
Casualties: in Roman battles, 23–24, 108–30, 269 (n. 161); Appian on, 110 (n. 2), 114; and burial of soldiers killed in battles, 112, 128; French, 125; numbers reported (of enemies), 112–13; numbers reported (of Romans), 111, 118, 130, 138, 254 (n. 24); numbers reported by Livy, 109, 114; numbers reported by Polybius, 114–15; rates of, 117, 125. *See also* Classical Greece: casualty figures for battles in
Cavalry, 50, 216 (n. 135), 243 (n. 146), 244 (n. 148); booty and donatives of, 76; combat between, 258 (n. 66)
Celiberia and Celtiberians, 122 (n. 18), 123 (n. 22)
Celts: Greek perceptions of, 256 (n. 42). *See also* Mercenaries: Celts as
Censors, 89, 103, 156–57, 197 (n. 36), 277 (n. 81), 278 (n. 86)
Census, 56–57, 148, 160, 272 (n. 39); numbers of citizens reported in, 12–13, 87, 146–47, 151–52, 154, 169, 244 (n. 152), 272 (nn. 31, 34), 279 (n. 105); under-registration, 157
Centenius Paenula, M., 216 (n. 133)

324

Centurions, 113, 117, 240 (n. 122)
Chickpeas: labor to grow, 229 (n. 38)
Childbirth. *See* Women and wives
Children, 66–68, 81–82, 88, 107, 148–49, 154, 156, 161, 165, 168, 245 (n. 160), 270 (nn. 5, 6), 278 (n. 94); and fieldwork, 98; food requirements of, 96; labor of, 96, 249 (n. 193); sale into slavery, 144
Cholera, 131
Civil War, American, 22–23, 98–100, 125, 129, 131–32, 135, 243 (n. 142)
Classical Greece, 140; casualty figures for battles in, 117
Claudius Centho, Ap. (pr. 175), 255 (n. 39)
Claudius Marcellus, M. (cos. 222 etc.), 33, 36–38, 40–42, 208 (n. 43), 209 (n. 53), 210 (nn. 61, 63), 213 (n. 107), 266 (n. 134), 269 (n. 165)
Claudius Marcellus, M. (cos. 196), 112, 116, 121 (n. 3), 124, 257 (n. 59), 285 (n. 2)
Claudius Marcellus, M. (cos. 183), 44, 123 (n. 23)
Claudius Nero, C. (cos. 207), 42
Claudius Pulcher, Ap. (cos. 212), 41
Claudius Pulcher, Ap. (cos. 185), 122 (n. 19)
Claudius Pulcher, Ap. (cos. 143), 267 (n. 146)
Claudius Pulcher, C. (cos. 177), 123 (n. 30), 285 (n. 2)
Claudius Sabinus Inregillensis, Ap. (cos. 495), 218 (n. 151)
Clients and *clientela*, 15, 103
Coinage, 36
Coloni. *See* Tenants
Colonies, 26–27, 43, 90, 153–54, 162–64, 166, 167–69, 221 (nn. 172, 179), 222 (nn. 185, 189), 223 (n. 195), 234 (n. 68), 236 (n. 82), 252 (n. 3), 270 (n. 8), 272 (n. 22), 274 (n. 53), 275 (n. 61), 278 (nn. 88, 102), 279 (nn. 105, 106), 283 (n. 5), 285 (n. 18); emigration to, 155; maritime, 57–58
Colonists, 235 (n. 78)
Colonization, 59–61, 77, 104, 145
Colstridium bacteria, 127
Columella (L. Iunius Columella), 20, 67–69, 73–74
Combat Mortality. *See* Mortality, military: in combat
Comentarii, 112
Comitia centuriata, 60
Commissariat, 49
Conception, rates of, 149–50
Confederate States of America, 58, 107; and experiences of civilians during Civil War, 99–100; mobilization rate of, 98, 244 (n. 150), 251 (n. 1)
Conscription, 6, 19, 52, 56–58, 91, 94, 100, 164, 219 (n. 156), 280 (n. 110); age of eligibility for, 21, 82, 183–84; attitudes of citizens towards, 14, 59–61, 82, 101, 106, 156–57, 218 (n. 150); and Augustus' military reforms, 83–84; crisis in, 106; and deaths of fathers, 89; in early modern Sweden, 76; and economic ruin of small farmers, 59–61; effects on conception, 149–50; increased demands for, 26; and married men, 82–88; maximum age for, 84–85; num-

Index

bers drafted, 60; resistance to, 14; of veterans, 56; voluntary compliance with, 82. *See also* Tenants and tenancy
Consumption demand, 66
Cornelius Blasio, Cn. (cos. 257), 206 (n. 36)
Cornelius Cethegus, M. (cos. 204), 42
Cornelius Lentulus Caudinus, L. (cos. 275), 207 (n. 41)
Cornelius Merula, L. (cos. 193), 112, 116
Cornelius Scipio, P. (cos. 218), 51, 217 (n. 144), 255 (n. 39)
Cornelius Scipio Aemilianus, P. (cos. 147), 133
Cornelius Scipio Africanus, P. (cos. 205), 82, 130, 203 (n. 20), 220 (n. 168), 223 (n. 192), 255 (n. 30), 256 (n. 44), 266 (n. 133), 288 (n. 11); trial of, 9
Cornelius Scipio Asiagenus, L. (cos. 190), 253 (n. 9); trial of, 9
Cornelius Scipio Asina, P. (cos. 260), 33
Cornelius Scipio Nasica, P., (cos. 191), 43–44, 122 (n. 11), 209 (n. 50), 213 (n. 103)
Cornelius Scipio Nasica Corculum, P. (cos. 162), 23, 114, 117
Cornelius Sulla, L. *See* Sulla
Coruncanius, Ti. (cos. 280), 32
Cosa, 195 (n. 14), 222 (n. 189), 234 (n. 68); mortality rate at, 252 (n. 3)
Cousins, 245 (n. 157). *See also* Families
Cows, 73
Cremona, 40, 222 (n. 189), 223 (n. 195)

Crimean War, 125, 259 (n. 73)
Crisis, agrarian or rural. *See* Agrarian crisis
Cultivation by hand, 232 (n. 53)
Curius Dentatus, M'. (cos. 290 etc.), 206 (n. 32), 207 (n. 41), 223 (n. 196), 234 (n. 68)
Cyclical mobility, 160. *See also* Smallholders
Cynoscephalae, battle of, 253 (n. 6), 255 (n. 39)

Daughters, 93–95, 246 (nn. 165, 166); as heirs, 247 (n. 168); marriages of, 102; mothers of, 95
De Agricultura, 8, 17
Debt, 4, 53–55, 108, 162
Debt bondage. See *Nexum* and *Nexi*
Decius Mus, P. (cos. 312 etc.), 204 (n. 23)
Defensive equipment of Roman soldiers, 127, 129. See also *Scutum*
Demographic regime, 153–54
Depopulation, 273 (n. 31)
De Requensens, Don Luis, 126, 129–30
Deserters, Roman, 113, 118
Devastation of farms and farmland, 27–29, 30, 49, 51, 107
Diet: of smallholders, 144, 154; of soldiers, 132
Digitius, Sex. (pr. 194), 122 (n. 10)
Discipline, military, 107, 266 (n. 133)
Disease, 24, 108, 125, 130–36, 140, 152, 259 (n. 73)
Dispatches of generals to senate, 111–13
Donatives, 55, 76, 81, 101, 163, 251 (n. 208)

326

Donkeys, 73
Dowries, 94–95, 102, 143, 160–61, 251 (n. 211)
Dyrrachium, siege of, 264 (n. 122), 267 (n. 141)
Dysentery, 132–35, 264 (n. 121)

Ecnomus, 33
Elections: consular, 211 (n. 77), 212 (n. 86); timing of, 36, 41–45, 207 (n. 42)
Emmer. *See* Wheat
Endogamy, 245 (n. 157)
Enslavements, 10–11
Epidemics, 10, 25, 42, 131, 145, 147, 149–50, 267 (n. 142), 272 (n. 20)
Equipment, military, 219 (n. 151). *See also* Defensive equipment of Roman soldiers; *Scutum*
Eryx, Mt., 33
Etruria, 29, 31–32, 42, 195 (n. 12); Tiberius Gracchus's journey through, 155
Eumenes, 255 (n. 39)
Exemptions from military service: in Rome, 91, 105, 285 (nn. 15, 18); in Sweden, 243 (n. 142); in U.S. Civil War, 243 (n. 142). See also *Vacationes*
Exposure of infants, 25, 94, 144, 271 (n. 16), 275 (n. 58). *See also* Infanticide

Fabius Buteo, Q. (pr. 181), 123 (n. 23)
Fabius Labeo, Q. (cos. 183), 123 (n. 23)
Fabius Maximus Cunctator, Q. (cos. 233 etc.), 36–37, 217 (n. 138)
Fabius Maximus Gurges, Q. (cos. 292), 31

Fabius Maximus Rullianus, Q. (cos. 322 etc.), 31, 42, 179, 212 (n. 96)
Fabius Pictor, Q., 114, 186
Fabius Pictor, N. (cos. 266), 32, 205 (n. 32)
Fallow, 37, 101, 231 (nn. 47, 49), 233 (n. 63), 251 (n. 209)
Families, 91, 219 (n. 155), 242 (n. 137), 270 (n. 6), 279 (n. 107); formation patterns of, 167; life cycles of, 88, 105, 141; limitation of, 147, 152; members of, 245 (n. 160); partitioning of, 161; sizes of, 89, 92; structure of, 142. *See also* Marriage ages; Smallholders
Famine, 149
Far. *See* Wheat: emmer
Farm equipment, 102, 166–67
Farmers, small. *See* Smallholders
Farming. *See* Agriculture
Farmland, availability of, 162
Farms. *See* Agriculture—large-scale; Agriculture—small-scale
Fasti triumphales, 19, 33–34, 42, 46, 139, 205 (n. 32), 207 (n. 40), 209 (n. 50), 213 (n. 103)
Fathers, 82, 88, 141–42, 156, 168, 233 (n. 63), 238 (n. 101), 271 (n. 18); deaths of, 83, 89
Fertility regimes, 18, 147
Fertilizer, 231 (n. 47). *See also* Manure
Firearms and artillery, 125–26, 129
Firmum, 222 (n. 189)
Flaminius, C. (cos. 187), 122 (n. 17)
Flaminius, C. (cos. 223), 51, 85, 154, 210 (n. 63), 223 (n. 196)
Food: for soldiers, 64, 109, 224 (n. 6); shortages, 274 (n. 47)

327

Index

Foraging, 30, 203 (n. 17), 216 (n. 135), 217 (n. 136)
Freedmen, 51, 85, 146–47, 240 (nn. 120, 121), 272 (n. 31), 280 (n. 2)
Fruit trees, 70, 75. *See also* Arboriculture; Orchards
Fulvius Centumalus Maximus, Cn. (cos. 211), 38–39
Fulvius Flaccus, Q. (cos. 237 etc.), 40
Fulvius Flaccus, Q. (cos. 180), 213 (n. 108)
Fulvius Flaccus, Q. (cos. 179), 123 (n. 22), 138–39, 262 (n. 107), 285 (n. 2)
Fulvius Nobilior, M. (cos. 189), 122 (n. 16), 269 (n. 160), 285 (n. 2)
Fulvius Nobilior, Q. (cos. 153), 132
Furloughs, 26, 35–52, 103–4, 208 (n. 43)

Gangrene, 125–27, 260 (n. 91)
Gardens, kitchen, 70, 75, 96, 230 (n. 39), 232 (n. 58), 233 (n. 61)
Garrisons, 32, 125, 134, 216 (n. 133), 253 (n. 6), 266 (nn. 135, 136)
Gaul and Gauls, 19, 43, 45–46, 51, 115–16, 122, 124, 138, 140, 211 (n. 77), 215 (n. 122), 217 (n. 144), 257 (n. 59), 268 (n. 151), 270 (n. 166); courage of, 256 (n. 44); image of, 256 (n. 43). *See also* Celts
Germany, 134
Girls: food requirements of, 225 (n. 12). *See also* Smallholders: food requirements of
Glory, 167
Graviscae, 77
Grazing, 144

Great Plains, battle of, 192
Greece and Greeks, 63, 103–4, 115, 124, 130, 138, 208 (n. 48), 213 (n. 102), 214 (n. 108), 215 (n. 122), 237 (n. 95); as foes, 116; and recovery of dead, 128. *See also* Classical Greece
Grenada, 131
Gustavus Adolphus, 134

Hadria, 222 (n. 189)
Hannibal, 36–38, 40–43, 50–51, 130, 150, 152, 203 (n. 20), 256 (n. 45), 264 (n. 122), 268 (n. 151), 269 (n. 165), 281 (n. 15); losses in battles, 116; marches on Rome, 38
Hannibalic War: 3, 5, 26–27, 34–36, 39, 43, 46, 50, 52, 54–55, 58–59, 86–87, 91, 108, 114, 130, 132, 135, 141, 143, 148, 151, 167; and demography, 154; doubtful battles in, 269 (n. 165); experiences of Roman civilians during, 99–100; financial problems of Rome during, 54; Rome's demands for manpower during, 87, 99
Hanno, 40
Harvest, 27–28, 49, 50–52, 69, 202 (n. 9), 204 (n. 21)
Hasdrubal, 42
Hastati, 39, 141, 243 (n. 146); ages of, 85; numbers of required, 85–86, 89–91
Heirs, 142–43, 271 (n. 12)
Helvius, M. (pr. 197), 122 (n. 5), 130
Herdonia, 41, 114
Heredium, 230 (n. 39)
Histria and Histrians, 45, 123 (n. 28), 213 (n. 107)

328

Hoeing, 74
Hoplites armies. *See* Armies—hoplite
Hostilius Mancinus, A. (cos. 170), 257 (n. 58), 258 (n. 65)
Hostilius Mancinus, C. (cos. 137), 253 (n. 9)
Households, 66, 165, 271 (n. 18); composition of, 92
Husbands, 92, 95, 142, 168, 238 (n. 101), 246 (nn. 159, 167). *See also* Families; Marriage: and ratio of potential brides to prospective husbands; Smallholders
Hygiene, 132, 135

Infanticide, 152, 247 (n. 169). *See also* Exposure of infants
Inflation, 99
Inheritances, 51, 94, 142–43, 155–56, 158, 160–61, 164, 168, 247 (n. 168)
Inkermann, battle of, 259 (n. 75)
Intercalation, 207 (nn. 41, 42), 208 (n. 48), 213 (nn. 102, 107), 214 (n. 117). *See also* Calendar, Roman
Italians. See *Socii*
Iulius Caesar, Sex. (cos. 91), 262 (n. 107)
Iulius Lubo, L. (cos. 267), 207 (n. 41)
Iuniores, 39, 243 (n. 147), 244 (n. 148), 283 (n. 12); defined, 56; numbers of, 56–57
Iunius Brutus, M. (cos. 178), 45
Iunius Pera, D. (cos. 266), 32, 205 (n. 32)
Iunius Pera, M. (cos. 230), 40

Javelins, 127
Jefferson, Thomas, 97
Junonia, 279 (n. 106)

Kin, 92, 103, 105; agnate and cognate, 142. *See also* Families: members of
Kin structure, 22, 92, 245 (n. 158), 246 (n. 159). *See also* Families: structure of

Labici, 234 (n. 68)
Labor: agricultural, 201 (n. 2); division of between sexes, 97–98; drudgery of, 92–93; hired, 18, 92, 105, 154, 165–66, 242 (n. 137), 249 (n. 193), 280 (n. 110); of neighbors, 91–93; surplus, 93, 105; wages for, 144, 165–66. *See also* Agriculture—large-scale; Agriculture—small-scale; Families; Slaves and slavery: labor of; Smallholders
Laelius, C. (cos. 140), 157
Land: availability of, 145; demand for, 153; productivity of, 66, 79, 154; redistribution of, 158; shortage of, 164; viritane distributions of, 153. *See also* Yields and yield ratios
Land commission, Gracchan, 166
Landlessness, 58–59, 106
Landless proletariat, 5
Landlords, 77, 107
Large estates. *See* Agriculture—large-scale
Latifundia. *See* Agriculture—large-scale
Latin allies, 151. See also *Socii*
Latium, 29, 132–33, 195 (n. 16), 198 (n. 50), 226 (n. 21)
Latrines, 132–33, 265 (n. 125), 267 (n. 142)
Leaves, military. *See* Furloughs
Legio Campana, 216 (n. 133)

Legio classica, 286 (n. 4)
Legionaries. See *Hastati*; *Principes*; *Triarii*; *Velites*
Legiones Cannenses, 266 (n. 135)
Legions: age structure of, 22; combat experience of, 119, 137; demobilization of, 45, 150; dismissal of, 43–45, 48–49; levying of, 32, 38–39, 45; size of, 252 (n. 5); structured by age groups, 85–86; travel of, 45–46; "victorious" defined, 257 (n. 59). *See also* Armies—Roman
Legumes, 70, 229 (n. 36), 230 (n. 38); labor to grow, 229 (n. 38)
Lex Claudia, 7, 224 (n. 4)
Lex de modo agrorum, 196 (n. 28)
Lex Iulia Campania, 233 (n. 63)
Lex Licinia, 77, 234 (n. 68)
Lex Poetilia, 77, 236 (n. 79)
Lex Sempronia agraria, 17, 155–56, 166, 275 (n. 65), 276 (n. 72)
Licinius Crassus, C. (cos. 168), 46, 215 (n. 122)
Licinius Crassus, M., (cos. 70), 279 (n. 107)
Licinius Crassus, P. (cos. 205), 42
Licinius Lucullus, L. (cos. 151), 262 (n. 106)
Licinius Lucullus, L. (cos. 74), 253 (n. 13)
Licinius Stolo, C. (tr. pl. 376 etc.), 234 (n. 68)
Ligures Baebiani, 215 (n. 213)
Liguria and Ligurians, 19, 43–44, 115–16, 123 (n. 28), 254 (n. 17), 268 (n. 151), 270 (n. 166)
Ligustinus, Sp., 240 (n. 122), 285 (n. 1)
Lilybaeum, 33

Livestock, 101, 144
Livius Drusus, M. (cos. 112), 279 (n. 106)
Livius Salinator, C. (cos. 188), 122 (n. 15)
Livius Salinator, M. (cos. 219), 42, 285 (n. 2)
Livy, 109, 117, 138–39
Logistics, 28–29, 30, 49, 107, 109, 112, 253 (n. 9)
Low Countries: Spanish wars in, 126, 134–35
Lucania, 29, 195 (n. 12), 201 (n. 5)
Luna, 278 (n. 88)
Lusitanians, 122 (n. 18), 123 (n. 27)
Lutatius Catulus, C. (cos. 242), 33
Lysimachia, 280 (n. 113)
Lysine, 229 (n. 34)

Macedon and Macedonians, 55, 60, 114. *See also* Armies—Macedonian
Magna Graecia, 63
Magnesia, 110 (n. 2)
Malaria, 198 (n. 50); and military mortality, 263 (n. 118)
Malthus, Thomas, 153
Mamertines, 60, 222 (n. 191)
Maniples, 29–31, 48–49; date of introduction, 203 (n. 19)
Manlius Acidinus Fulvianus, L. (cos. 179), 122 (n. 18), 124, 269 (n. 160)
Manlius Vulso, A. (cos. 178), 45, 124
Manlius Vulso, Cn. (cos. 189), 9, 112, 122 (n. 14), 285 (n. 2)
Manpower, military, 107, 153; demands for, 22; fears concerning, 157; and *Lex Sempronia agraria*, 156; Rome's potential, 21

Manumission, 146–47, 283 (n. 2)
Manure, 127, 228 (n. 26), 233 (n. 62), 260 (n. 93)
Marcius Censorinus, L. (cos. 149), 130, 133
Marcius Philippus, Q. (cos. 186), 44, 116–17
Marcus Aurelius, 131
Marius, C. (cos. 107 etc.), 128
Markets for agricultural products, 8
Marriage, 156–57; between close kin, 245 (n. 157); *cum manu*, 142, 247 (nn. 168, 175); of daughters, 93–95, 105; opportunities for, 143, 145; patterns, 106; rates of, 148–50, 152; and ratio of potential brides to prospective husbands, 95, 104, 143, 151, 247 (n. 169); and remarriage of widows, 25, 246 (n. 159); *sine manu*, 95, 142, 247 (nn. 168, 175); of soldiers, 82, 86, 238 (n. 101), 239 (n. 119); of veterans, 104. *See also* Husbands; Men: of marriageable age; Women and wives
Marriage ages: differences between men's and women's, 102; in Greece and Balkans, 83; men's and women's, 238 (n. 101), 239 (n. 115), 246 (n. 159); men's at first, 22, 82–84, 89, 106–7, 161; military service and, 83, 86, 88, 106; urban vs. rural, 238 (n. 105), 239 (n. 115); in Western Europe and Mediterranean, 82; women's at first, 25, 93–94, 152–54, 239 (n. 108), 246 (n. 167), 275 (n. 57). *See also* Men; Nuptuality; Women and wives

Massilia, siege of, 264 (n. 122)
Measles, 131
Medical treatment. *See* Wounds: treatment of
Medical writers, Greek and Latin, 126, 132
Men: between ages 30 and 46, 241–42 (nn. 130, 133); between ages 17 and 30, 243 (nn. 144, 148), 244 (n. 150); of marriageable age, 86, 88, 151, 247 (n. 169); and plowing, 230 (n. 44), 249 (n. 192); proportion over age 30, 241 (n. 128). *See also Iuniores*; Marriage ages: men's at first
—married, 142, 148, 150, 220 (n. 164); conscription of, 82–88; exemptions for, 243 (n. 139, 142). *See also* Conscription; Exemptions from military service; *Vacationes*
—unmarried, 88, 141–42, 145, 148, 150, 163; seek wives following military service, 102. *See also* Never-marrieds
Mercenaries, 49, 63–65, 237 (n. 97); Celts as, 256 (n. 42)
Messapians, 205 (n. 32)
Messene, 33
Metaurus, battle of, 42, 212 (n. 93), 285 (n. 2)
Military revolution of fifth-fourth century Greece, 63
Military service: length of, 26–53, 55–56, 90, 101, 103, 245 (n. 153); payment for, 29–30, 36, 49, 54, 64, 80–81, 92, 99, 101, 107, 109, 203 (n. 16), 219 (nn. 151, 153, 155), 220 (n. 161), 224 (n. 3), 251 (n. 208), 253 (n. 9); under Augus-

331

Index

tus, 84. *See also* Conscription; War: timing of
Military tribunes, 113, 117, 210 (n. 71), 214 (n. 108)
Millet, 47, 210 (n. 55)
Minucius Rufus, M. (cos. 221), 38
Minucius Rufus, Q. (cos. 197), 112, 254 (n. 22)
Minucius Rufus, Q. (cos. 193), 46
Minucius Thermus, Q. (cos. 193), 121 (n. 4), 122 (n. 12)
Mobility, economic and social, 160
Mobilization, military, 251 (n. 1); Roman, 10, 22, 61, 105, 107, 164, 167, 244 (n. 150), 252 (n. 1); American Confederacy, 98, 107. *See also* Hannibalic War
Mortality, military, 23–24, 95, 102, 108–40, 147, 162, 168, 246 (n. 159), 246 (n. 167), 272 (n. 29); British, 267 (n. 140); in combat, 109–25, 253 (n. 6), 272 (n. 29), 288 (n. 13); demographic effects of, 24–25; excess due to war, 136–37, 145–46; and population growth, 153–55; rates of, 151; Swedish, 135
Mothers, 95, 246 (nn. 165, 166). *See also* Daughters: mothers of; Women and wives
Mucius Scaevola, Q. (cos. 220), 130
Mumps, 131
Mutina, 77, 123 (n. 30), 236 (n. 82)
Mutinies, 286 (n. 10)

Naples, 40
Napoleonic Wars, 125
Near East, 63
Neighbors: labor of, 91–93

Neolocality, 165
Nephews, 92
Never-marrieds, 242 (136), 274 (n. 50). *See also* Men—unmarried
New Carthage, 203 (n. 20)
Nexum and *Nexi*, 64, 77, 107, 224 (n. 197), 236 (n. 79)
Nicias, 55
Nieces, 92
Nola, 36–37, 40, 209 (n. 53), 269 (n. 165)
North Africa, 29
Nuceria, 40
Numantia, 130, 133, 253 (n. 9), 264 (n. 122)
Nuptuality, 242 (n. 136)

Oats, 47
Olive, 70; oil, 78, 236 (n. 79); presses, 78
Orchards, 96, 101, 159, 232 (n. 58), 233 (n. 61). *See also* Arboriculture; Fruit trees
Orphans, 238 (n. 101)
Overpopulation, 18, 25, 156
Oxen, 70, 73, 78, 230 (nn. 42, 43), 231 (n. 51)

Paeligni, 111
Paestum, 222 (n. 189)
Panic, 47
Panormus, 33
Parents, 104, 142, 161, 163, 238 (n. 104), 245 (n. 160), 270 (n. 6), 278 (n. 94), 279 (n. 107); death or illness of, 101. *See also* Fathers; Mothers
Parthia, 126, 253 (n. 8), 262 (n. 106)
Partition of farms. *See* Agriculture—

332

small-scale: and partition of farms; Families: partitioning of
Pasture, 70, 73, 96, 232 (n. 58), 233 (n. 61)
Patavium, 123 (n. 34)
Patres familiae, 57, 238 (n. 101). *See also* Fathers
Patria potestas, 95
Patronage, 15, 102–3
Peasants: definition of, 193 (n. 3); French, 159; Russian, 159–60
Peritonitis, 126
Persius, 113–14
Petillius Spurinus, Q. (cos. 176), 123 (n. 29)
Phalanx. *See* Armies—hoplite; Armies—Macedonian: phalanx of
Philip V, 265 (n. 128)
Pillaging, 50, 202 (n. 8), 216 (n. 135), 217 (n. 136)
Pinarius Rusca, M. (pr. 181), 123 (n. 23)
Pisa, 44–45, 123 (n. 23), 214 (n. 108)
Placentia, 40, 214 (n. 108), 222 (n. 189), 223 (n. 195)
Plague, 131, 135; at Athens, 263 (n. 115)
Plancius, Cn. (tr. pl. 56), 279 (n. 107)
Plantation agriculture. *See* Agriculture—large-scale
Plebeians, 234 (n. 68)
Plebs, 77–78, 155
Plowing, 27, 35, 46–48, 50, 70, 73, 211 (n. 74), 228 (n. 23), 232 (n. 53), 248 (nn. 180, 182). *See also* Men: and plowing; Women and wives: and plowing
Plows, sole-ard, 97
Plundering. *See* Pillaging
Poland, 134

Police, 82
Polybius: and reports of battles, 138–39; text of, 286 (n. 12)
Pomerium, 205 (n. 32)
Pompeius, Q. (cos. 141), 130
Pompey (Pompeius Magnus, Cn. cos. 70 etc.), 204 (n. 26), 209 (n. 50)
Pontifices, 208 (n. 48)
Popillius Laenas, M. (cos. 173), 45
Population, Roman: controls on growth of, 152–54; decline in, 145, 149, 156–58, 169; growth of, 146, 154–55, 158, 161, 164, 169, 273 (n. 31); numbers of male citizens in, 89, 91, 242 (n. 135); proportion of slaves among, 10; rates of increase in, 146, 148; structure of women's kin in, 92. *See also* Age structure: of male Roman citizen population
Population of Italy, 271 (n. 19); decrease of, 12–13, 149; genetic make up of modern, 12; growth of, 8, 17–18, 25, 198 (n. 43); urban and rural, 10, 13
Porcius Cato the Elder, M. (cos. 195), 7–8, 17, 122 (n. 8), 128, 138–39, 195 (n. 19), 203 (n. 20), 224 (n. 4), 228 (n. 25), 251 (n. 212), 276 (n. 76), 279 (n. 107), 285 (n. 2)
Porcius Laeca, P. (pr. 195), 122 (n. 7)
Posidonius, 114, 117
Postumius Albinus, A. (cos. 180), 44, 213 (n. 108)
Postumius Albinus, L. (cos. 234), 211 (n. 77)
Postumius Albinus, L. (cos. 173), 123 (n. 25), 138–39

333

Index

Po Valley, 272 (n. 21). See also *Ager Gallicus*
Praetor's court, 102
Praetutii, 234 (n. 68)
Prefects of allies, 113, 117
Preventive checks, 168. See also Population, Roman: controls on growth of
Principes, 39, 141, 240 (n. 123), 243 (n. 146); ages of, 85; numbers required of, 85–86, 89–91
Prisoners of war. See Captives: Roman soldiers as
Proletariate, landless rural, 5, 102, 105
Proletarii, 151, 156, 185–86, 221 (n. 172), 245 (n. 152), 272 (n. 31), 277 (n. 76), 283 (nn. 5, 6), 284 (n. 12); numbers of, 57–58; as rowers, 90, 185
Provisioning. See Logistics
Publicani, 78
Punic War, First, 35
Punic War, Second. See Hannibalic War
Purpurio Furius, L. (cos. 196), 121 (n. 3)
Pydna, battle of, 111, 114, 124, 127, 128, 215 (n. 122)
Pyrrhus, 32, 50, 265 (n. 128), 266 (n. 134)

Quaestors, 109
Quinctius Cincinnatus, L. (cos. 460), 234 (n. 68)
Quinctius Crispinus, L. (pr. 186), 122 (n. 20)
Quinctius Crispinus, L. (pr. 168), 269 (n. 160)

Quinctius Flamininus, T. (cos. 198), 103, 124, 253 (n. 6), 255 (n. 39), 258 (n. 66), 266 (n. 134), 285 (n. 2)

Rameses V, 263 (n. 115)
Rationes, 109
Records, food and pay, 130
Regulus. See Atilius Regulus, M.
Remarriage, 143
Rhegium, 211 (n. 74), 216 (n. 133)
Risk, 80–81, 167; management of, 201 (n. 3)
Rome, city of, 149–50, 155, 214 (n. 108); emigration to, 144–45, 147, 162; food supply of, 200; population of, 10, 13–14, 144–45
Romulus, 234 (n. 68)
Rounded approximations, 117
Rowers, 185, 196 (n. 20), 197 (n. 37), 221 (n. 172), 224 (nn. 1, 2), 283 (nn. 5, 6). See also *Proletarii*
Rubella, 131
Russo-Japanese War, 131
Rutilius Rufus, P. (cos. 105), 85, 240 (n. 121)

Sabinum and Sabines, 206 (n. 32), 234 (n. 68)
Salassi, 267 (n. 146)
Sallantini, 205 (n. 32)
Same, 122 (n. 16)
Samnite Wars, 31, 33, 60, 166, 203 (n. 15), 206 (n. 32)
Sanitation in Roman camps. See Camps: of Roman army; Camps: sanitation in
Sardinia and Sardinians, 36, 123 (n. 31), 125, 130, 149, 263 (n. 118), 266 (n. 136), 267 (n. 143)

Index

Saserna, 235 (n. 71)
Sassini, 205 (n. 32)
Satricum, 234 (n. 68)
Scutum, 127, 260 (nn. 89, 90). *See also* Defensive equipment of Roman soldiers
Seed-to-crop ratio. *See* Yields and yield ratios
Sempronius Gracchus, C. (tr. pl. 123), 155, 279 (n. 106)
Sempronius Gracchus, Ti. (cos. 215), 40, 215 (n. 122)
Sempronius Gracchus, Ti. (cos. 177), 10, 123 (nn. 24, 31), 285 (n. 2)
Sempronius Gracchus, Ti. (tr. pl. 133), 5, 17–18, 25, 52–53, 76, 106, 108–9, 137, 141, 155–57, 164, 166, 169, 253 (n. 9), 275 (n. 65), 277 (n. 78), 279 (n. 106), 280 (n. 111); supporters of, 155, 165
Sempronius Longus, P. (pr. 184), 130
Sempronius Longus, Ti. (cos. 218), 46, 205 (n. 32), 211 (n. 74)
Sempronius Longus, Ti. (cos. 194), 122 (n. 9), 124
Sempronius Tuditanus, C. (pr. 197), 112, 121 (n. 2), 257 (n. 58), 258 (n. 65)
Senate and Senators, 5, 7, 31–46 *passim*, 48, 51, 59, 76–77, 82, 112, 145, 150–51, 153–54, 156–57, 160, 163–64, 166–67, 206 (n. 32), 208 (n. 43), 212 (n. 86), 213 (n. 107), 214 (n. 116), 215 (n. 122), 221 (nn. 172, 179), 222 (n. 191), 224 (n. 4), 240 (n. 122), 242 (n. 137), 245 (n. 155), 258 (n. 65), 263 (n. 118), 266 (n. 135), 275 (n. 65), 277 (n. 76), 283 (nn. 1, 5, 6)

Sentium, battle of, 42
Sepsis, 126
Septicemia, 125, 127
Sertorius, Q. (pr. 83), 204 (n. 26)
Servilius Caepio, Cn. (cos. 203), 43
Servilius Pulex Geminus, M. (cos. 202), 128
Sesame, 229 (n. 27)
Sextius Sextinus Lateranus, L. (tr. pl. 376 etc.), 234 (n. 68)
Sharecroppers and sharecropping, 74–78, 93, 107, 235 (nn. 71, 75, 78), 246 (n. 166)
Sherman, William Tecumseh, 100
Shields. See *Scutum*
Sicilian expedition, 55
Sicily, 29, 33, 36, 43, 46, 50–51, 63, 125, 149, 156, 266 (n. 136), 267 (n. 143); slaves in, 11
Sieges, 63, 132–33, 267 (n. 141)
Sino-Japanese War, 131
Sinuessa, 215 (n. 122)
Sipontium, 272 (n. 22)
Sisters, 92, 142. *See also* Families: members of
Skirmishes: losses in, 124. *See also* Casualties: in Roman battles
Slave labor. *See* Slaves and slavery: labor of
Slave mode of production. *See* Agriculture—large scale
Slaves and slavery, 4–6, 58, 76, 105, 107, 159, 164, 195 (n. 16), 220 (n. 163), 221 (n. 183), 224 (n. 197), 236 (n. 79), 278 (n. 103); and children's labor, 237 (n. 99); enslavement of, 81; female, and fieldwork, 97–98, 228 (n. 27), 248 (n. 185); and free labor, 165–66;

335

Index

labor of, 77, 169, 244 (n. 150); numbers of, 7–12, 147, 198 (nn. 39, 44, 45, 48, 49); ownership of, 58, 163, 242 (n. 137), 245 (n. 155); rate of increase in, 11–12; replace conscripted sons, 91; revolts of, 156; as rowers, 163, 283 (n. 6); as soldiers, 182, 197 (n. 37), 215 (n. 122), 242 (n. 131). *See also* Agriculture—large-scale; Rowers

Smallholders, 66; assets of, 162, 165; and burdens of military service, 89–91; cyclical mobility of, 25, 160–62, 168; and death of son, 142, 165; defined, 193 (n. 3); diet of, 69–70, 229 (n. 38); displacement of, 4–5; effect of conscription on, 69–70, 72, 74–75, 79, 81; family life cycles of, 20–22; family size of, 159; food requirements of, 20–21, 65–67, 69, 73–74, 94, 95–96, 158, 226 (n. 21), 234 (n. 67); and labor of sons, 158, 161–64; labor potential of, 21, 66, 68–69, 73, 79, 81, 89, 93, 95–96, 159, 162, 234 (n. 67); labor requirements of, 67, 69–73, 75, 78–79, 81, 93, 95–96, 229 (n. 38), 234 (n. 67); labor surplus of, 15, 69–70, 229 (n. 30); and markets, 15; and military mortality, 198; in Northern U.S., 144; poverty and crisis among, 6, 14, 17–19, 25; and risk, 80–81; self-exploitation of, 15–16; and sharecropping, 74–78; and slaves, 163–64; social position of, 158–59; social reproduction of, 145; and taxes, 78–79; and weights of members, 66–67, 73. *See also* Agrarian crisis; Agriculture—small-scale; Risk

Smallpox, 131–32, 135, 263 (n. 115)

Socii, 44–45, 49–51, 63, 79–80, 82, 111, 117, 121 (n. 2), 123 (n. 23), 125, 128, 130, 139, 149, 151, 158, 165, 218 (nn. 147, 149), 219 (n. 154), 252 (n. 5), 255 (n. 39), 258 (nn. 61, 65), 267 (n. 143), 270 (nn. 4, 8), 275 (n. 61), 276 (n. 72), 287 (n. 5)

Soldiers, Roman. *See* Captives: Roman soldiers as; *Hastati*; *Principes*; *Triarii*; *Velites*

Sons, 148, 271 (n. 18). *See also* Families; Smallholders

Sowing, 27, 30, 35–37, 46–50, 53, 68, 211 (n. 74), 217 (n. 138), 226 (n. 20), 233 (n. 61), 274 (n. 47)

Spain and Spaniards, 100, 104, 111, 113, 115, 117, 122 (n. 18), 123 (n. 33), 128, 130, 132, 134, 136, 138, 168, 208 (n. 48), 223 (n. 193), 256 (n. 57), 262 (n. 106), 269 (n. 165), 286 (nn. 8, 11, 12)

Spanish wars: Roman defeats in, 138; Roman losses in, 116, 156

Spears, 127, 129

Spelt, 47

Spoletium, 222 (n. 189), 223 (n. 195)

Statelliates, 45

Stertinius, L. (proc. 199), 121 (n. 1)

Stipendium. See Military service: payment for

Stockrearing: labor requirements of, 196 (n. 29); large-scale, 197 (n. 30). *See also* Agriculture—large scale

336

Storage of food, 80
Struggle of the Orders, 64, 79
Succession to estates, 270 (n. 8). *See also* Inheritances
Suessula, 36–37, 40
Sulla (L. Cornelius Sulla, cos. 81), 169, 219 (n. 152)
Sulpicius Galba Maximus, P. (cos. 211), 38–39
Supplementa, 111, 113
Surpluses, agricultural, 237 (n. 94)
Sweden: conscription in early modern, 76; mobilization rate in, 252 (n. 1)
Swedish soldiers: deaths of. *See* Mortality, military: Swedish
Swords, 129; Gallic, 126; "Spanish," 127
Syracuse, 63, 100, 130–31, 149, 203 (n. 20), 208 (n. 43), 211 (n. 83), 266 (n. 133); siege of, 264 (n. 122)

Tarentum, 32, 100, 217 (n. 139), 286 (n. 4)
Tarracina, 234 (n. 68)
Taxes and taxation, 27, 63, 219 (n. 154); on *ager publicus*, 237 (n. 86); *vectigal*, 78
Telemon River, battle of, 256 (n. 44)
Tenants and tenancy, 75, 107, 181–82, 228 (n. 25), 235 (nn. 71, 75, 78), 283 (n. 6); conscription of, 76–77
Terentius Varro, A. (pr. 184), 269 (n. 160)
Terentius Varro, C. (cos. 216), 112, 286 (n. 4)
Tetanus, 125, 127
Thermopylae, 255 (n. 39)
Thetes, 63

Thucydides, 118
Tigranes, 260 (n. 99)
Tirones, 56
Titinius Curvus, M. (pr. 178), 123 (n. 33)
Training, military, 203 (n. 20)
Transportation costs, 15
Trasimene, battle of, 51, 85, 114, 116, 256 (n. 46)
Travel. *See* Agriculture—small-scale: and travel to and from fields; Armies—Roman: travel to theater of combat; Legions: travel of
Trebia River, battle of, 40
Triarii, 141, 238 (n. 101), 240 (n. 123), 270 (n. 3), 283 (n. 3); ages of, 85–86; conscription of, 87–88; numbers of required, 85–87
Tribunals, local, 102
Tribunes, military. *See* Military tribunes
Tributum, 53–54, 219 (nn. 152, 153, 155), 237 (n. 89)
Triumphs, 23, 34, 42, 139, 204 (n. 24), 206 (n. 32), 207 (nn. 40, 41, 43, 44), 209 (n. 50), 212 (n. 96), 213 (n. 103), 269 (n. 160)
Truces, 128
Typhoid fever, 132–35, 264 (n. 121), 265 (n. 128)
Typhus, 131–32, 135, 263 (n. 117); types of, 262 (n. 114)

Umbria, 29, 32, 205 (n. 32)
United States: southern states of, 198 (n. 44), 237 (n. 99). *See also* Confederate States of America
Usurpation of small farms, 102–3
Utica, 266 (n. 136)

Index

Vacationes, 57, 89, 99, 221 (nn. 173, 179). *See also* Furloughs
Valerius Antias, 138–39, 255 (n. 30), 269 (n. 163)
Valerius Flaccus, L. (cos. 195), 122 (n. 6), 195 (n. 19)
Valerius Laevinus, M. (cos. 210), 130, 212 (n. 86)
Valerius Laevinus, C. (cos. 176), 123 (n. 29)
Valerius Laevinus, P. (cos. 280), 32
Valerius Messalla, M. (cos. 188), 122 (n. 15)
Varro (M. Terentius Varro), 20
Vectigal. See Taxes and taxation
Veii, 54
Velia, 215 (n. 123)
Velites, 39, 49–50, 141, 240 (n. 123), 243 (n. 146); ages of, 85; numbers of required, 85–86, 89–91
Venusia, 32, 112, 222 (n. 189)
Veterans, 56, 103, 105
Villa system. *See* Agriculture— large-scale
Villius Tappulus, P. (cos. 199), 138–39
Vines and viticulture, 6, 70, 74–75, 101, 144, 159, 236 (n. 79); labor to cultivate, 96
Vinyards, 196 (n. 29), 230 (n. 39), 232 (n. 58), 233 (n. 61)
Viritane distributions of land, 58–59, 77, 222 (n. 189), 223 (n. 196). *See also* Colonies
Volones. See Slaves and slavery: as soldiers
Volsci, 31

Wage labor. *See* Labor: hired
Wages. *See* Labor: wages for

War: location of, 3–4, 18–19, 29–30, 35, 64; timing of, 19, 26, 28, 65, 77. *See also* Agriculture— and war: conventional view of; Agriculture—small-scale: and war; Conscription; Military service: length of; Mobilization, military; Mortality, military; Smallholders: and military mortality
Water supplies. *See* Camps: of Roman army
Wealth, 5, 8, 9, 167
Weeds, 74
Wet nurses, 274 (n. 56)
Wheat, 28, 69, 93, 210 (n. 55), 228 (n. 25); bread, 216 (n. 128); caloric content of, 67; continuous cropping of, 231 (n. 47); cultivation of, 20, 36–37, 47–48; einkorn, 216 (n. 128); emmer, 47–48, 216 (n. 128), 226 (nn. 21, 22); labor required to grow, 68, 73, 228 (n. 26), 229 (n. 37); naked, 226 (nn. 21, 22); as rations, 109; weight of, 225 (n. 19); yields of, 67–68, 233 (nn. 61, 63). *See also* Sowing; Yields and yield ratios
Widows, 89, 92–100, 238 (n. 101). *See also* Marriage: and remarriage of widows
Wills, 142, 247 (n. 168), 270 (n. 7). *See also* Heirs; Inheritances
Winter encampments and winter quarters. *See* Camps: of Roman army
Women and wives, 66–68, 81–82, 92, 107, 149, 156, 165, 220 (n. 164), 238 (n. 104), 246 (n. 159), 249 (n. 193), 279 (n. 107); of American

Confederacy, 98–100; and birth intervals of children, 25; deaths of due to childbirth, 247 (n. 169); and dishonor, 97, 249 (n. 190); and fieldwork, 97–100, 105, 250 (n. 197); and husbands' absences, 21, 248 (n. 180), 249 (n. 193); kin of, 92, 245 (n. 158); labor potential of, 68, 96–97, 228 (n. 27); and plowing, 97, 248 (nn. 180, 185); and pregnancy and childbearing, 97; and ties to natal families, 95. *See also* Breast-feeding; Families; Marriage; Marriage ages; Smallholders; Widows

World War I, 127

Wounds, 24, 108, 125, 131, 140, 259 (nn. 73, 75, 76, 80), 260 (nn. 86, 90), 270 (n. 3); gunshot, 126; treatment of, 128, 130, 140, 261 (nn. 95, 96)

Yields and yield ratios, 73–74, 93, 158, 226 (nn. 20, 21), 246 (n. 166), 251 (n. 209). *See also* Sowing; Wheat: yields of

Zama, battle of, 114, 191, 256 (n. 44), 287 (n. 6)

www.ingramcontent.com/pod-product-compliance
Lightning Source LLC
Chambersburg PA
CBHW021353290426
44108CB00010B/216